SILK

An Exposé of Commercial Fishing

LIMITED EDITION

Limited to One Thousand Numbered
and Registered Copies of which
this is Number_____*277*_____

Sprenloeghke

Kay Bese

i

SILK

An Exposé of Commercial Fishing

by

Soren Roegdke and Kay Busse

W.K.B. Enterprises, Inc.
Portland, Oregon

Publisher's Note

Except where actual persons and events are
named, the material in this story is entirely fictional.

......for the fishermen

SILK

An Exposé of Commercial Fishing

Copyright © 1995 by Soren Roegdke and Kay Busse

Printed in the United States of America

Library Of Congress Catalog Card Number: 94-61842

ISBN:0-9629242-3-7

First Edition 1995

W.K.B. Enterprises, Inc.
790 S.E. Webber #206
Portland, Oregon 97202

CONTENTS

PROLOGUE

CAPTAIN ALEXI SEMYONOV

Captain Alexi Semyonov stood in the wheelhouse of the Soviet refrigerator/factory mothership called the MURMANSK. He was a tall man, about six feet, and looked like a mass of bones beneath the navy turtle-necked sweater and Levi's. He cursed in Russian and looked at the dusk-dimmed lights of the nearest catcher boat. As the Captain General he was not only responsibile for the needs of hundreds of workers on the MURMANSK, he was also responsible for every fisherman on the Soviet trawlers. The weight of this responsibility lay like lead upon his shoulders.

He rubbed his dry eyes and wondered how long it had been since he had slept without nightmares or eaten without pain in his stomach. In his younger days he could have worked around the clock without eating, and he could have slept comfortably on a hard bunk with his knees hunched against his chest. But now he was past fifty, and he felt old and tired. The long months at sea, the tedium of watching over a fleet every day, and the endless reports of quotas and figures had worn him down.

Captain Semyonov had been a deep sea fisherman in the Soviet fishing fleet for seventeen years, and this was to be his last trip as Captain General. So far the catch had not been good. He knew if he did not fill his production target of 4,700 tons of frozen fish, he and his crew would not earn enough roubles to pay for their rent, and the KGB would call for an investigation of the entire fleet. For this reason he had kept the twelve Soviet trawlers streaming back and forth in various directions paying no attention to the American fishermen who were fishing nearby on the grounds they called the "rock pile". But tonight they would circle around inside the American trollers and hope that the catch would be better.

He felt the acid in his stomach rise as he swallowed the last dregs of the black coffee. It was cold. He cursed and spit the grounds out on the floor. He was sick and he knew it. Only the vodka in his cabin would help his condition.

1

It was near dusk before the ship's bell sounded and Vladmir Usyenko, the stocky First Mate, stepped into the pilot house with another pot of black coffee. He stood in the doorway for a moment, adjusting his eyes to the change of light, then he smiled with his wide mouth and scattered teeth.

"Coffee, captain?" he said in a gutteral voice.

"No, thank you. I must sleep," Captain Semyonov said. "It is imperative, Vladmir, that we meet our quota. My orders are to stay on the coordinates that I have set out until 2300 hours, and then take the ship shoreward twelve kilometers. Hold her steady on a south to north course."

"Yes, captain," said Vladmir.

"Keep a close eye for any American fishing boats. Sometimes when they are drifting, they may be in the dead zone of our radar and cannot be seen until the last possible instant."

"Do not worry, captain," Vladmir smiled again. "I will give a full strength signal by radio should a dangerous circumstance arise, and if they cross our path, I will go up on the bridge and shout, 'why the hell don't you give way!'"

"Do not make light of this order, Vladmir," said Captain Semyonov sternly. "What we are attempting to do can be very dangerous, but if we are to make our quota, I can see no alternative. You are in charge."

Yes, captain," said Vladmir.

Captain Semyonov tried to hide his disgust as he left the wheelhouse and unlocked the door to his cabin. Vladmir was dependable and would follow his orders. He was the most efficient first mate he had ever had, and the oldest, but in truth, he was an insensitive bore. 'They are all impossible,' he thought. 'Why must I continue to be plagued with eight year old minds?'

He threw his hat on the bunk and reached into the small cabinet beside the desk for the bottle of vodka and a large glass. No, he decided. He would not drink vodka again tonight. Instead, he would open his last bottle of rare, 5 star cognac, the last of the twelve crates that he had drunk so prudently since the day he received his prestigious medal, The Order of Lenin.

He felt the blood pounding in his temples and his throat tightened with panic. There was no peace. There was only control. The damnable illusive sleep was evading him again tonight. He pushed

2

his wirey gray hair under the cossack hat and went up on the observation deck to smoke.

Dense fog had drifted in. The atmosphere was heavy as silent ghosts slithered across the deck from the northwest fog, the cursed everlasting fog. At least he could depend upon Vladmir to hold her steady at twelve knots, and perhaps the catcher boats on their new coordinates would supply them with abundant fish this time.

The Marlboro tasted good. He watched it sail through the air as he thumped it toward the dark shadow in the water.

The captain stared. There should be no dark shadow in the water! 'It cannot be possible!' he thought, 'but it appears to be a small troller.' Another dark shadow formed to the left.

"My God, we're in the middle of the American fishing fleet," he whispered. "They have drifted southeast with the wind."

He began to run toward the pilot house calling Vladmir's name.

"Stop engines!" he yelled. "Stop engines!"

As his foot hit the first step of the steel ladder leading up to the pilot house, the MURMANSK shuddered and rocked.

"What have you hit?" he roared as he burst into the control room.

"Probably just a large log, captain," said Vladmir calmly. "You see the radar. There are only boats far to the left. There is nothing straight ahead."

"You damned fool," shouted the captain. "We are directly in the path of the American boats. We're supposed to be twenty kilometers shoreward."

"But we are on the coordinates you set out, captain," said Vladmir. "It is that the American boats must have moved."

"Tack ten kilometers starboard," the captain shouted. "then hold her steady due west until we are well beyond the thirty-five mile zone,"

"At once, captain," said Vladmir.

"And radio the other ships. We must move outside the American fleet at once. We are violating the safety of the American boats. I only hope we have not struck one."

"Yes, captain," said Vladmir.

"We will say nothing about this incident to anyone, Vladmir, do you understand?" said Semyonov. "Nothing. But I fear that we have destroyed innocent lives."

3

Semyonov rubbed his long boney hand over his bushy eyebrows and across the thick mustache as if to erase the image of what he knew was true. The heinous crime he had committed against the ocean's resources haunted him. Had he understood the consequences of the Russian fishing industry at nineteen, he would never have left Orlov. Now, this was the final violation. 'I will not take human lives to fulfill their quotas,' he thought.

PART I

JOHN SILK

CHAPTER

1

JOHN SILK

Fifty fathoms down in the murky waters off the Pacific Coast of Oregon lay a long shadow, darker than the water around it. Except for a slight wafting of his tail the big Chinook salmon lay motionless in the strong current. His eyes were fixed on a twelve inch herring sliding through the water on a steady course. At this moment his belly contained three of the small silvery fish, but there was still room for one more.

The Chinook, known as the king salmon by fishermen, was beautiful, but he was a killer in his own right. He preyed on fish smaller than himself. During his five years in the ocean he had devoured thousands of fish, but he had never seen a herring like this one. His senses told him that something was wrong with the fish. It was alone, not swimming in a school with other herring. It was moving in a straight line, unwavering, darting neither to the right nor to the left. He had no way of knowing that this particular fish was thrust through the body with a metal baiter held fast in its jaws by an escutcheon pin which was then wrapped tight around the gills with a rubber band. At the herring's head the baiter was attached to a colorless leader. Beneath his tail a large hook lay hidden. The tail fin itself had been pinched off to keep the herring moving in a straight line.

The Chinook eyed the small creature for a moment more, then his instinct took over. A thrust of his powerful body brought the salmon parallel with his victim; a second thrust shot him in front of it. As he streaked past, he slapped it on the nose with his tail, then he circled once more. The herring moved unerringly forward.

The Chinook stretched and shook himself vigorously. Two more quick swipes with his tail, and he swept down on the small fish with the voracity of a shark. In one savage bite he snapped off its head. Another quick gulp and the small fish disappeared.

Suddenly, the big Chinook was jerked straight forward by a

force far greater than his own. For a moment he thrashed furiously in an attempt to free himself from a sharp hook imbedded in his left gill, but the weight of a sixty pound lead cannonball dragged by the boat above kept the line taut. After a while he ceased fighting. He let himself be dragged steadily through the water on the same course as the herring. Now he was bait for a sea lion or dinner for a man.

On the ocean's surface a solitary fisherman stood in the stern pit of an old ocean troller named the SEAGULL. He had seen the spring at the end of the float line jerk when the fish hit, and he knew the salmon was large. Only a big king hit like that. He shoved the brass handle of the gurdy forward and watched the hydraulic engine drag the thrashing salmon to the surface and into the wake of the boat. The salmon leaped free of the water shaking frantically at the hook.

Pitting his strength against the failing salmon, the fisherman steadily pulled in the nylon line hand over hand until the fish was close. He wrapped the line twice around his left hand and reached for the oak-handled gaff leaning against the side of the pit.

For a moment he stared into the orange eyes of the salmon as it looked up the line at him, and he felt the desperation of the big king and pity that a creature so majestic had to die. Then, raising the gaff like a club, he struck the fish squarely between the eyes.

The force of the blow stunned the big Chinook momentarily, and the light went out of his glowing eyes. Then the fisherman hooked the point of the gaff through an upturned gill; and, with perfect coordination born of long practice, he heaved the heavy fish over his left shoulder onto the deck. The Chinook flopped twice, slipping toward the low rail of the SEAGULL. The fisherman struck it again, solidly above the eyes.

When the salmon lay still, he removed the hook. He thought it curious the way a salmon changed color when it was out of the water. Now in the sunlight the fish shown a beautiful efflorescent. This was a big splitter, over twenty pounds, a good Chinook to end the day.

Within twenty minutes the fisherman had pulled in the last float line, gaffed off a hake and a shaker coho too small to keep, stowed the styrofoam floats on deck, set the hooks in their racks, and covered the fishing gear on the stern of the boat with burlap sacks.

Last, he cleaned the seven bloody salmon still lying on the back deck and arranged them neatly in the ice hold with a solid layer of ice between each one. He replaced the hatch cover and stepped into the wheelhouse of the old troller.

As he shoved the throttle of the SEAGULL forward, she gained speed and began to slice through the heavy swells. He felt the same sense of exhilaration he had felt ten years before, when on his first fishing trip, the boat had met the large waves of the open sea.

His name was John Silk, but he had been called Silk by his friends in Port Bane since the night he had arrived from Boston ten years before. He was 6' 3" and stringy, with a lean boney face, deep-set gray eyes, and a whimsical smile. Upon first meeting Silk, he smacked of a gentleman in every respect, but closer study testified to more violent behavior. It was obvious, when he was viewed squarely in the face, that his straight nose had been knocked off kilter several times and was only straight now because the last breakage had healed in a line parallel with his forehead. His full red beard and pinpoint freckles hid several scars from his more violent encounters, and his forehead was etched with thin white lines beneath the sun-burnt skin. He was quiet spoken, polite, enthusiastic, powerful, loyal, proud, noble, and brilliant. But trouble had seemed to be his middle name since he could remember.

Silk's problems had started long before he was born when his great, great grandmother, Isabel, bred two gray-eyed, copper-haired babies. The strain had been strong and had resulted in at least one red-haired kid per generation; Silk's misfortune was to have been that kid for his generation. His aristocratic mother had tried to overlook the fact that her son had strange colored hair; his physician father had ignored it; but the kids at school had found it an everlasting joke; that is, until they experienced the wrath of the kid with the glowing locks. Carrot top, copper penny, rooster comb, juice head, or red had been names that brought instant reprisal by the skinny, long-legged boy who was sensitive about his hair. As a result Silk had found himself labeled a troublemaker by his teachers and a frequent visitor to the principal's office.

In high school Silk's reputation as a potential juvenile delinquent had preceeded him, and instead of becoming friends with the upper class in school, which was the desire of his mother and father, he had become the class among the lower echelon of students. In his sophomore year he had fought the leader of a street gang called the Insiders, and he had won. This exploit made him a boy to be reckoned with and had enlarged his circle of the wrong kind of friends. He had become a hero in the streets and a terror in the classroom, and it had been only a keen intelligence that had kept him out of the reform schools. He had discovered early in life that good grades kept his parents off his back and his teachers at a distance.

The result of these early experiences had made an anomaly out of Silk. On the one hand he had missed all the excitement of being a normal high school student, or a band member, or a good athlete, or the student body president; but on the other hand he had developed his leadership ability, his fighting skills, and his flair for new experiences. His only accomplishment in high school had been high grades, high enough to get him into Harvard. That, of course, and his father's money.

Harvard hadn't done Silk a bit of good as far as trouble was concerned. He had graduated with a degree in business administration from this prestigeous institution and had immediately accepted a good job as manager of the Baird Motor Company. At twenty-six he had a new car every six months, a three-bedroom house in Boston suburbia, a promotion coming up, and Elizabeth Baird, his beautiful fiancee, apple of her father's eye, who spent his money faster than he could make it.

Had Silk been an interested participant, it might have made all the difference. Instead, his stomach had gone to hell on him, and he had developed a bleeding ulcer. The doctor had recommended complete rest and a change of jobs. When Baird had refused to give him time off, and told him to take rolaids and forget his ulcers, the scene had ended with Silk's explicit "fuck you."

It was his temper as usual; Silk had never learned to control his temper. He had walked out the door of the Baird Motor Company, past the new and used car section on the highway until he had come to the Solar Truck Stop. As luck would have it, there, in the parking lot, had stood a blue and chrome, eighteen-wheel Kenworth, with its diesel motor running and its driver sitting at the wheel. For some strange reason that he had never been able to explain to himself, Silk had thumbed a ride and had climbed into the high front seat beside the driver. Once he started, he hadn't stopped until he had

8

ridden completely across the continent to Portland, Oregon. From Portland he had hitched-hiked to Port Bane where he started his new life as a commercial fisherman on the ocean troller called CRACKANOON.

Jake "Cap" Carver owned and skippered the CRACKANOON at this time, and he had hired Silk as a deck hand until his son Trevor graduated from Oregon State University two years later and had taken Silk's place. By that time Silk had learned everything there was to learn about commercial trolling. He had been taught by the master.

After leaving the CRACKANOON, Silk had leased his own trollers, and had become a highliner a time or two. Trollers were smaller boats that fished with lines controlled by hydrolic reels on either side of the stern as opposed to larger boats called trawlers or draggers which fished with nets rolled on a drum. This year, however, he had leased the SEAGULL, an old brute of a boat built in 1928. Silk had sworn that every mechanical part on the old wooden troller had worn out. During the season he had replaced two high-pressure fuel lines, the hydraulic hoses, and the water pump. He had overhauled the diesel engine and replaced one trolling pole. Now the battery was shot. He had been out five days on this trip, and had had to run the diesel engine full throttle for fifteen minutes every night to charge the battery up enough to have power for the cabin lights. A dead battery on the ocean meant real trouble, and real trouble meant the coast guard.

However, even with the expensive repairs, the real source of the problem had been the fish. Try as he might, he had not caught enough salmon to pay for repairs or to make ends meet. On this trip he had fewer than thirty salmon on board to show for his fourteen hour days. Not enough cash to buy a new battery let alone enough to put some money in the old sock for the lean months ahead.

Tonight Silk felt bone tired, and the unhealed cut on his left hand burned from the salt water and from the fishing lines biting into it all day. He switched on the C.B. radio and turned the channel to 37. Without thinking, he reached for the roll of black electrician tape on the wall shelf behind his head, broke off a piece and double-wrapped it around the deep slice in his index finger. As he listened to the crackling static over the steady roar of the diesel engine, he heard the gravelly voice of old Jake "Cap" Carver come over the speaker.

"Hello, SEAGULL. This is the CRACKANOON calling the

SEAGULL. Do you read me, Silk? O.K."

Silk smiled. He could hear Cap grumbling to his son Trevor in the background. The old man had forgotten to release the microphone button.

Cap was a towering brute of a man in his late fifties with thick, wirey gray hair and beard. He had a booming voice that could be heard across any room as he laughed uproariously at the fishing tales he told. Cap was magic; wherever he went an audience gathered to bask in a new story or to relive one of the old tales. It was fascinating to watch him clench his long-stemmed pipe between two gold teeth and puff the words out through his lips and up past his bushy eyebrows.

The CRACKANOON, a sixty year old troller made of Alaskan yellow cedar, was the most prosperous troller in the fleet and the idol of every skipper and deck hand alike. But there were only two fishermen, besides his son, Trevor, that Cap had ever allowed on board the CRACKANOON: Joel "Cowboy" Scott, who grew up with Trevor, and John Silk himself.

"Gollydamnamighty! Where is that boy?" Cap roared. "I've been tryin' to raise him for thirty minutes. Heyoh! Silk! Damn it! This is Cap. O.K."

Silk opened his own mike and spoke with a clipped Boston accent.

"Good evening Cap. You don't have to shout; I'm receiving you loud and clear. Beautiful evening, isn't it? Did you see the sunset? O.K."

"Don't beautiful evenin' me!" Cap's voice had calmed to its usual drawl. "I've been tryin' to raise you ever since we shut down tonight. O.K."

"Where are you located? O.K." Silk asked.

"We're just off the rock pile with some of the fleet. I can make out the lights on three of them big Russian catcher boats, but they're due west of us. At least they're out where they belong. Where are you at? O.K."

"I'm located about 60 Mics north of you, Cap. I have a weak battery, and I have to continually charge her up. In a minute now I'll spin her around and point her south. This spot looks as good as any to shut her down for the evening. Are you up from California for the winter? O.K."

10

"Yep, I suspect me and Trevor'll pull in for a long rest come mornin'. We just about got her plugged, and Trevor's kinda hankerin' to see his woman. How'd you do today? O.K."

"Not so good, Cap. I don't know where they are. I can't seem to find them. I've been following the radio calls like a damn fool, and I've gone through every patch of birds I could find. I got less than a dozen today, a couple of splitters, but the rest were small. Did you find them? O.K."

"We run into a main line about 3:00 this afternoon, and it was hot and heavy for awhile. Trevor was punchin' bait, and I was haulin' them lines in. I decided last night to go solid herrin'. To hell with them spoons and hoochies; we had to feed them babies real meat. We went through a couple of cases of herrin' but she paid off. Yep. She ain't been a bad season. If I'm lucky, I'll be able to turn the CRACKANOON over to Trevor next season. Hell, I've been old enough to retire for years. Yes, sir. Old Trevor's got to be a good one. Who you fishin' with, Silk? I heard you was goin' it alone. O.K."

"You heard right, Cap. I haven't made enough money this season to hire a deck hand, but if I'm lucky and catch a few more fish tomorrow, I'll have enough money to buy a new battery. O.K."

"Didn't I tell you not to lease that old wreck? Hell, there's worms in her wood, and ever' piece of equipment on her's worn out. She's just a hole in the ocean for you to pour your money into. How long you been out? O.K."

"Four days. I'll probably head into port tomorrow afternoon. I want to fish the morning. And you were right, as usual. I should never have leased the SEAGULL. But the lease terms were low, and I thought with a good season I could make her pay. She's sure not the CRACKANOON though. O.K."

"You just keep fishin', Silk. Take your time, and you'll get your own boat. Your times a comin'. If you're plannin' to stay out longer, keep an eye on this weather. I can't say as I ever remember seein' her 'zactly like this. She's calm all day, then at night she starts a blowin', and the fog rolls in. I 'spect it means we got a storm a brewin', but I can't say for sure. Keep your drift lights on, too. If a tug comes through, or one of them Russian ships strays inside, at least they'll see you. O.K."

"I don't have drift lights, Cap. I've been using the running lights and the anchor light. O.K."

11

"Geehosiphat! You do have a tub there! Well, keep the channels open on 9 and 16. If you don't hear me first, give me a holler in the mornin', at the crack a dawn, now, not the crack a noon. O.K."

"Will do, Cap. O.K."

Silk turned the C.B. to emergency channel 9, and switched the VHF to its emergency channel 16. If he were to be warned about an approaching vessel or bad weather, the message would come over one of these two channels. He released the automatic pilot and cranked the wheel to the starboard. As soon as the SEAGULL had made a wide circle and her bow was pointed due south, he shut off the engine and went out on deck. In the distance toward the southwest, he could see a faint gleam of light against the solid horizon. He knew they were the same lights Cap was talking about, lights from catcher boats which surrounded one of the large foreign motherships.

At night the Port Bane fishermen slept while they let their small salmon trollers drift with the wind and the ocean currents. At a distance of over thirty miles from land, they were beyond the continental shelf, and the ocean bottom lay far below the length of the anchor chain. Left by themselves this practice held little danger for the fishing fleet unless one boat shut down too close to another boat. Then there might be a collision if the skippers weren't on their toes. Usually such a mishap resulted in, perhaps, the snapping of a trolling pole and the loss of a stabelizer and some lines. Large freighters or tugs and barges constituted a real hazard, however. If one of these huge vessels came through the fleet during the night, a collision might result, and if a fishing boat were struck, it would be sunk.

To the fishermen in the Port Bane fleet other fishing boats or American ships were not their concern. American ships had the same rights as they did. But the foreign fishing fleets, that was a real problem. The Russian catcher boats fished the coastal waters off the west coast day and night. Usually this fleet stayed outside the American boats, but sometimes during the night they would come inside in an effort to increase their quota. Most of the time there was no danger because the small trollers could be tracked on the radar, and on a clear night their drift lights or mast lights could be spotted at a considerable distance. This practice by the Russians made the Port Bane fishermen madder than hell. These foreigners had no business being in American waters inside the two-hundred mile limit. But curse as they might, the Port Bane fishermen could do nothing about the problem. After all, the foreign fishing ships were protected by the U.S. Government, and they were in joint venture with American fishing companies.

Silk looked up at the sky. The tiny stars were just beginning to dot the heavens, but further west, and north, where the big dipper should appear, there was nothing except solid darkness. He felt the northwest wind begin to stir. By midnight the thick gray bank would be on him. If the damn fog ever left them, he could sleep on one of the bunks in the fo'c'sle, but tonight he'd better sleep on the short bunk in the cabin, just in case.

He crawled into the sleeping bag fully clothed except for his fishing boots and the ragged sweat shirt which he spread over his pea jacket for a pillow. He listened to the intermittent slapping of waves against the hull of the old SEAGULL, and to the creak and moan of her aged planks as the black ocean swells surged beneath her. What old Cap had said ten years ago was true, he decided. "If you want to understand a boat, just listen to the music she makes at night when she's a driftin'. You can tell whether she's happy or hurtin'. She'll let you know if there's anything botherin' her."

Tomorrow should be quite a homecoming. Now they would all meet together again. Cowboy coming down from Alaska, and Cap and Trevor up from California. The Albatross Tavern would be buzzing with pitchers of beer and more of Cap's yarns that stretched over forty years of fishing the Pacific Ocean from Cape Fairweather Alaska to San Francisco.

Silk closed his eyes as the first rolling banks of gray fog shrouded the SEAGULL. Maybe the catch would be better in the morning. What was it Cap said it took to be a highliner?

"Fishermen can't tell shit about what they see on the bottom through their gadgets. You got to get beneath the surface and think like a fish. A fish's got his own world where he lives, and until you can git down there in the dark and become a salmon, you can't get the bastard on board your boat."

13

CHAPTER

2

CAP AND TREVOR

Oblivious to the fifteen knot ocean breeze that pushed the thick fog bank inland where it would rise at the dawn before the green forested mountains of the coastal range, oblivious to the clear sky above the cotton mass where the sliver moon and diminutive stars shown down, old Jake Carver lay snoring peacefully on the narrow bunk inside the wheelhouse of the CRACKANOON. Like John Silk, before he shut down for the evening he had taken every precaution for any situation that might arise while he drifted. The VCR and two CB radios were tuned to emergency stations, and his boots and coat were hanging ready on the hook just inside the cabin door. Cap lay flat on his back, his big hands folded together across his gray beard.

Below in the fo'c'sle, Trevor lay comfortable in his bunk. His heavy breathing matched the rhythm of the flowing waves. Neither man was aware that the MURMANSK had completed her great circle, and that she was now cruising due north on her captain's precise coordinates. Neither man had felt the continuous force of the wind and current that had pushed the CRACKANOON twelve miles toward the ocean beaches and the sucking lips of the jaws. High on the crosstreed mast above the troller, two red drift lights tried vainly to make themselves visible in the thickening mist.

Without reason the old man's eyes suddenly blinked open, and for a moment he stared motionless into almost total darkness. A quick glance at the luminous dial on the clock radio told him it was nigh on to 3:00 A.M., too early to put out the lines. Strange. He never awoke without a reason. Something was wrong, certainly, but for the life of him he couldn't explain what that something was. Trevor was alright. He could hear his steady breathing as clearly as if he were lying beside him. He held his own breath and listened. There was no sloshing of internal water in the bilge; no grinding or grating of planks separating. He knew his boat perfectly. There was not a loose screw

14

or broken nail on the solidly constructed troller.

Over the years of fishing on the ocean, of daily facing the dangers of her tempestuous nature, Cap had honed his senses to a fine edge. It wasn't the explainable circumstances that got you, but the ones that couldn't be explained.

Then he heard it, dimly at first, then louder. It sounded like a great threshing machine he had heard one time on a trip to the midwest. Something ominous and inhuman was coming toward them. What the hell could it be? It certainly wasn't any sea creature, and they were too far inside the shipping lane to be in the path of a tug or a freighter.

He threw back the heavy quilt and rolled off the bunk. He could barely see the bow of the boat through the thick soup. Nothing else was visible. But the sound was growing steadily louder. Whatever it was, he'd better move, because from the threshing, the thing was getting closer.

He had just opened his mouth to wake Trevor when the excited voice came over the CB.

"God damn me," it said. "One of them Russian ships just passed me about a hundred yards off my starboard bow. What in the hell are they doing this far inside? Do you read me anyone? I know there's boats closer in than me. They've come inside! Looks like the one that passed me is coming straight through us. If you're listening, you'd better get your ass out of the way, either inside or outside, don't make no difference. Just get the hell out of here. O.K."

Cap clicked on the loran and read the numbers. Christ, they had drifted in twelve miles. He switched on the cabin lights and pressed the starter switch.

"Trevor," he shouted. "Get your ass out of bed and help me up here. We got to move. Them fuckin' Russians are coming right at us."

Those were the last words Cap ever spoke. The engine caught, and the forward gear snapped the troller into motion. Cap's sigh of relief caught in his throat as a huge black shape suddenly loomed before him. High above the small troller it came through the fog. The giant MURMANSK, steaming due north at a steady twelve knots, was headed straight into the starboard bow of the CRACKANOON.

Hearing his father's voice, Trevor threw the blankets back and started to sit up when he felt a brief flash of pain as both his legs

15

disappeared, or was it the whole lower half of his body. His eyes shut, then his heart stopped beating.

Cap stared in disbelief as the wheelhouse windows shattered and the bow of the boat disappeared. He turned to escape, but before he could reach the cabin doorway the force of the impact rolled the boat over, and he found himself lying on the wheelhouse ceiling. His whole world had turned upside down.

'Jesus! It's cold', he thought as the icy water began to seep up over his body and cover his face. 'Why's it have to be so cold?'

He gasped and choked and struggled to keep his head above the water. The back of his neck hurt. God, it hurt when he tried to move his head. It burned like a hot poker was set to it. But he had to try! He had to try to get to some air, quick! And then he felt his neck crack and pop and separate from his spine and his head fell lifelessly back into the water. Now there was no pain anywhere. No way to lift his arms or his legs or to move his heavy body. There was no way to hold on to this life any longer.

'Oh, God! Oh, God in Heaven! So this is it, God! Well, it don't matter. Trevor's dead. Shit, there's worse ways to die than this.'

And then the proud CRACKANOON lifted her stern high into the air, as if to salute the world goodbye, and she plunged herself straight down into that eternal otherworld. "She's in good company," ol' Cap would have said.

CHAPTER

3

THE SEARCH

Silk, sleeping uncomfortably on the SEAGULL, heard the unknown fisherman's warning about the Russian ships. He knew that he was north of the main fleet, but one could never be too sure when there was a large vessel coming through. He got to his feet, checked his wrist watch, and turned on the loran. By the coordinates he had drifted twelve miles southeast. Which way to go? If the Russians had come inside, the best place to be was back where he had started drifting.

By the time he had moved the old troller and turned on one burner of the propane stove, it was almost dawn. He might as well have some coffee and a couple of rolls; it was only an hour until it would be time to set his lines.

As the light gradually permeated the dark gray canope, Silk cranked up the troller, set the automatic pilot, and went to the stern to set out his lines. That was when the wake-up calls of the Port Bane fishing fleet began to come over the outside speaker. He listened as he worked.

"Hey, Neil," said one voice. "Time to roll out and get that gear in the water. Ed's already got four lines in before his first cup of coffee. O.K."

"Morning," said another voice. "We're right with you. Did you hear that those Russian ships came right through the middle of us during the night? Hell, we had to move three times, and I'm too tired to fish. O.K."

"Yeah. We moved too." It sounded like Doug's voice. "I'm tired of this shit. When I get back in port, I'm going to file another complaint with the fish and game."

"Maybe it'd be a good idea for us to go into port and start hunting instead of fishin'," said Neil. "Hunting season's open, you know, and that venison's mighty good for a change. Besides, I think it's a whole lot safer in them woods with them fuckin' hunters then

17

out here playin' chicken with them Russian ships. O.K."

"Yeah, well you know, if we're lucky, we may be able to afford a huntin' license when we get home," said Doug. "Talk at you later about that. Gotta get our lines in. O.K."

As Silk listened to the radio calls, he was meticulously setting out his own fishing gear. Old Cap should be calling him any minute now. For once he was ahead of the old man. If he worked like hell, he could already have his lines in before Cap got on the horn.

With the tag lines, cannonballs, flashers, hoochies, spoons, clothespins and floats on both outer lines in place, Silk took one look at the springs to see if a salmon had hit, then stepped into the cabin for a second cup of coffee. By now the fog had lifted a hundred feet above the SEAGULL, and he could see a short distance over the dull horizon. The day would be gray; the wind had picked up, causing tiny white caps to form on the waves, and the barometer had dropped a point. Cap was probably right. There was a storm brewing out there somewhere. He switched the radio to the coast guard weather channel, and waited another ten minutes for Cap's call. Cap must be into them and busy, or they had their signals mixed up. Either way, Silk decided to wait no longer. He took the mike from her catch and switched the channel to call the CRACKANOON.

"This is the SEAGULL calling the CRACKANOON. Good morning Cap. You're sleeping late this morning. It's the crack a dawn, and I've already set my lines. Do you read me, Cap? This is Silk. O.K."

Silence. 'Probably on the back deck,' Silk thought. He waited for five minutes then tried to raise the CRACKANOON again. Still there was no answer. This was unusual. Cap would never have gone into port last night without calling him first. Silk got on the wire at five minute intervals, but thirty minutes went by and there was still no response from Cap. Maybe the damned radio was out of whack. What should he do?

Silk decided to call some of the other trollers fishing the rock pile with the CRACKANOON. He knew Doug Burbank on the JUANITA was there.

"Hello, Doug. Do you read me? This is Silk on the SEAGULL. O.K."

"Loud and clear, Silk. How're you goin'. Got her plugged yet? O.K."

18

"Not yet, Doug. Say, I was just wondering whether you've heard from Cap and Trevor. I can't raise them this morning. Could be the radio's on the blink. Would you give them a shout on channel 37? O.K."

"Will do, Silk. I saw the CRACKANOON yesterday and talked with Cap. In fact, they were in sight when we shut down last evening. As usual they were hauling em'. Old Cap sure knows how to get those salmon. Yeah, Silk, sure. I'll give 'em a holler and get right back to you."

"Thanks, Doug. I'll stay close by. O.K."

"Silk, this is Doug. Do you have me there? O.K."

"Yeah, Doug, loud and clear. O.K."

"I can't raise 'em either, Silk. I don't know where they went. Want me to keep tryin'? O.K."

"Yes, Doug. And do me a favor. Call all the boats in the vicinity and have them try to trace Cap. He wouldn't have gone in without calling somebody. O.K."

"Will do, Silk. I'm on her. O.K."

As the voices in the fleet continued to call the CRACKANOON, Silk began to worry. He called the marine channel to see if they had any information, then he checked with the two port docks. Each assured him that the CRACKANOON had neither contacted them nor pulled in during the night.

By this time the barometer had dropped another point, and the wind was shipping the waves to a moderate height giving them definite form. Doug's call came back.

"Hello, Silk. Doug again. Do you read me? O.K."

"Yea, Doug. O.K."

"No one in the fleet can raise them, and no one's spotted the CRACKANOON this morning. What the hell's goin' on? What do you think we should do? O.K."

"I called the docks and they're not in port. We'd better alert the coast guard. I'll make the call, then I'm coming down. I'm about 60 Mics north of you. Alert the rest of the fleet and tell them the problem. O.K."

Silk switched to channel 16.

"Hello. Port Bane Coast Guard? This is John Silk on the SEAGULL calling the Port Bane Coast Guard. Are you receiving me? Over."

19

"This is the Port Bane Coast Guard. Over."

"This is the SEAGULL about twenty miles out. I'd like to report a missing vessel. Over."

"This is the Port Bane Coast Guard back to the SEAGULL," said the impassive voice. "What vessel are you reporting? Over."

"The CRACKANOON shut down last night off the rock pile. I have been calling her all morning and there is no response. No one in the fleet has seen or heard from the troller this morning. Have there been any maydays or emergency calls during the night? Over."

"This is the Port Bane Coast Guard back to the SEAGULL. We have received no maydays or emergency calls. What is the type and color of the vessel, and how many people are on board? Over."

"The vessel is a forty foot ocean troller, white with black trim. There are two people on board, Skipper Jake Carver and his son, Trevor. Over."

"This is the Port Bane Coast Guard back to the SEAGULL. What was their shut down number? Over."

"They shut down just off the end of the rock pile, so the number is probably...just a minute here while I check my chart...probably about 12704. Over."

"This is the Port Bane Coast Guard back to the SEAGULL. Roger, skipper. We have that information without a distress call. Ask any vessels in the vicinity to keep an eye out for debris or oil slicks. We'll put out a distress call for the CRACKANOON on all channels. If we have no reception, we'll dispense a vessel to the immediate area. This is Port Bane Coast Guard on channel 16, over and out."

Silk changed the channel back to 37 just as a familiar voice came over the speaker. The call was from his old fishing partner, Joel "Cowboy" Scott, who had heard his conversation with the coast guard.

"Hey there, Silk. What's goin' on down there? Did I hear you say Old Cap's missin'? O.K."

"Yeah, Cowboy. Glad to hear your voice. You heard right. Nobody in the fleet can raise the CRACKANOON, and no one has seen her since last night. I'm pulling gear and heading down that direction. I'm worried as hell. O.K."

"We're off Cape Foulmouth right now so it'll take us a spell to get down there. We'll crank up the old OTTER full bore and meet you at the rock pile. O.K."

It was good to hear his old friend's voice. For some reason

Silk felt an intangible strength in the knowledge that Cowboy was coming to join in the search.

Cowboy Joel Scott looked like a character out of the old west. He sported a thick Wild Bill Hickock mustache and had a long skinny body that looked like a pair of six-guns should be hanging from his hips. But his mind was futuristic. Cowboy was a man with burning ambition. He wanted to experiment with new ideas and explore distant places. He never intended to get tied down like the other fishermen in Port Bane.

By the time he had met Silk, he had saved enough money to buy the OTTER, a sixty-two foot trawler, and had set up a joint venture program with the foreign ships off the coast of Alaska, and Alaska had become his fishing grounds for the season. But every fall when the Chinook fishing season ended, Cowboy pulled into Port Bane to relax and visit his old buddies again. Silk, at this moment, could think of no better timing than now.

As the aged SEAGULL sloshed her way south to the rock pile, the wind velocity had increased to over thirty knots. She wallowed in the northwest wind, now shipping in gusts, riding in and out of the troughs. By the time the SEAGULL reached the co-ordinates and came into view of the Port Bane trollers, knee-deep water was washing over her back deck.

There were fishing boats moving, it seemed, in all directions, and in the distance a sleek coast guard cutter threw white spray high against the dark horizon as she cut neatly through the granite gray scud. She closed in on one slow moving troller, then another. Overhead a coast guard helicopter flew in ever-widening circles.

At noon, the OTTER arrived and Cowboy joined in the search. By four o'clock a twenty mile area had been combed foot by foot. The wind was now up to forty knots, and showed signs of turning into a full nor'wester.

The waves became greater in length. Edges of the crests were breaking into spindrifts, and great streaks of foam were sliding south into the troughs. The spray over the bow had rendered Silk's visability to almost zero, and that was when the Port Bane Coast Guard issued the order.

"This is the Port Bane Coast Guard to all vessels involved in the search and rescue mission. There is a strong gale warning, and all small and moderate boats are ordered to cease searching and proceed to Port Bane. Port Bane Coast Guard. Over and Out."

Silk felt sick. There was nothing he could do. There was nothing anybody could do. He wanted to smash his fists through the glass. Instead, he took down the mike and said,

"This is Silk on the SEAGULL. Did everyone receive the message? O.K.?"

The fishermen's voices began to come over the CB one at a time.

"We received," came Doug's voice.

"Loud and clear." Neil, Al, Spence and Duane reported.

"O.K." said Cowboy's enraged voice. "The bastards."

With great reluctance the fishing boats from Port Bane headed into port without the CRACKANOON, without the Cap, without Trevor. It was almost dark. That night the sea launched her fury against the land with a screeching gale. The waves stretched forward in great overhanging crests. The whole ocean was white with foam.

CHAPTER

4

PORT BANE

Halfway up the rugged coast of Oregon a clear river pours her water into the Pacific Ocean. High above the ocean table at the mouth of the river craggy mountains poke their crests into the sky; steep sandstone cliffs form their base against the booming breakers which shoot spumes of spray up the eroded cliffs; and in the distance, shaggy-green basaltic rocks rear their ugly heads out of the surf. Narrow beaches dot this area, but because of the tumultuous tides, they are not accessible from the ocean. Inland, a short distance from the mouth of this river, lies the fishing community of Port Bane.

Port Bane was developed much like any other seaport on the west coast. The pioneers walked across the continent to the western states; the Spaniards got the hell knocked out of them by the United States; and gold was discovered in California. After the gold petered out, there were many starving miners wandering about looking for work.

Since there were few jobs to be had, these prospectors had to have some other way to make money. Eventually they noticed the abundance of tall trees in the area, and someone got the idea that he could make money out of wood from those trees if he could cut them down and saw them into boards to build houses and ships. This was how the logging industry on the west coast was born. Being a hearty breed and unafraid of hard work, these original loggers soon made short work of the California trees after which they moved on up the coast to Oregon where there were great forests of giant Sitka spruce and tall firs.

Over the years these logging entrepreneurs succeeded in their purpose, which was to destroy almost all the big trees in Oregon. Big lumber companies like Windover and Central Pacific moved into the area from back east. They built a railroad, and their sawmills burned day and night. Millions of board feet of lumber were shipped over-

23

seas to places like Korea, China, Japan, and Cambodia. Soon all the giant trees in the area were cut down, and some of the lumber companies went broke. The general managers of these large concerns felt terrible when their companies either had to move back east or go into the fish raising business.

The fishermen, of course, were certainly not going to be outdone by the loggers. They were determined to build a fishing port so that their boats could reap the ripe harvest of fish from the ocean. About the turn of the century the U.S. Government appropriated funds for two rock jetties to be built from the river's mouth about a mile or two out into the ocean. A deep channel was then dredged between the two jetties and on up the bay to where fishing docks were constructed for the Port Bane fishing fleet and ocean-going steamers. Now Port Bane was a real honest to God seaport with a channel across the bar. Much money changed hands over the years, and some of the successful fishermen became rich and built big houses. Then the United States Government, the United States Coast Guard, and the United States Fish and Game Department stepped into the picture and passed numerous fishing regulations.

Fishing seemed to go to hell after that, and most of the fishermen went broke and lost their boats. The fish processing plants and the canneries shut their doors, and the fifteen hundred boats in the Port Bane fishing fleet quickly dwindled to three hundred.

During this declining period of time even one or two bars closed their doors although the sale of liquor still flourished. Most of the broke fishermen were too poor to move from Port Bane, so they moved instead to the many taverns that lined the bay front. There they spent their days and nights telling fish stories about the time when fishing was great.

In the middle of this general economical development an event took place that changed the face of the community even more than the phenomenal effects of the fishermen and the loggers. Sometime after the automobile had been invented, engineers, congressmen, businessmen, and real estate people got together and figured out a way to make a highway all the way up the Pacific Coast from Los Angeles to Port Bane. This highway was named 101 and was symbolic of man's ability to conquer his environment. After a while the same brilliant engineers figured out how to build a bridge across Bane Bay just in front of the jetties which lined the Port Bane River bar. The span was called a milestone of progress. Now the highway could continue north through the farming community of Tillamook, through the artist community of Cannon Beach, through Astoria, across the mighty Columbia River then on toward Seattle. Port Bane was now vulnerable to people by land as well as by sea. It was to this seaport that the fishermen were returning after the coast guard's announcement to give up the search for the CRACKANOON.

John Silk stared woodenly through the spray-spattered window as the white-caps broke across the bow of the old SEAGULL. He seemed oblivious to the boat's difficulties as she strained to cut across the swelling crests. He had forgotten about the weak battery and the crooked propeller shaft; his mind was on Cap.

It was not until a change in the rhythm of the boat that his attention was diverted. At first he thought something new had broken down, then he realized he was approaching the outpouring bar. The long troughs had narrowed to sharp twelve-foot chops coming from different directions. He eased up on the throttle and squinted his eyes in an effort to see into the rapidly approaching darkness. The stern light on a troller up ahead disappeared into the black trough of the high rock jetties that lined the jaws. To the left, making a wide circle, were the lights of a large trawler, an American catcher boat.

"Hey, Silk, you read me?" came Cowboy's slow drawl over the speaker.

"I read you, Cowboy," Silk said. "What are you doing sitting out here. I thought you'd have docked an hour ago. O.K."

"Been waitin' for you, buddy," Cowboy answered. "I seen thirty trollers go on in. I was beginnin' to think I'd have to come out, throw you a line, and tow you in. Doug said you've been having a little battery trouble and what all. O.K."

"Yeah, but I think she'll make it in, Cowboy. O.K."

"How about me leadin' the way and breakin' some of them waves? She'll be pretty rough goin' across that bar tonight. O.K."

"Thanks, Cowboy," said Silk. "I'll take you up on that offer. How do you want to work this operation? O.K."

"I'll just complete my circle and you fall in my wake. Soon's we get through them jetties and under the bridge, she'll calm down a little, and I'll go on in. While you're puttin' up your poles and your floppers, I'll tie-up at Port Dock 3 and have a cold beer waitin' for you at the Albatross. O.K.? O.K."

"Sounds good, Cowboy. Will do." Silk said. "I'll wait for your entry. O.K."

Silk's SEAGULL fell into the wake of Cowboy's OTTER. They passed through the jaws of the jetty and under the Bane Bay Bridge. When they reached the calmer expanse of Bane Bay, Silk put the engine on idle, released the forward gear, and stepped out on deck to pull up her poles into their upright position. For a moment he watched Cowboy's catcher boat cruise on down the bay toward the main dock. Through the mist he could see the shimmer of lights across the docks and the entire bay. On the hillside high above the bay were the dimmer lights of Victorian houses nestled in descending rows. Below the houses stretched the bay front area where bright flood-lights lit the canneries and all night shrimp plants. On a clear night islands of sea lions could be seen floating lazily, barking sporatically, while they waited for their next meal of fish heads and fish guts. Across the street from the plants were the lively taverns and the fresh fish restaurants. To the east lay the port docks, two of them, their ramps leading down to the fingers—square spaces under spotlights— where each boat rocked gently in her own slip. One seldom saw this structure of the docks. He only saw the big drum trawlers, the troll-ers with twin poles, and the flat-bottomed dories docked there. This was the fishing port. The whole spectrum was as beautiful as any fishing port scene on any eastern coast calendar.

Silk blinked the image out of his eyes, pulled in the last metal flopper and heaved it on board. He curled the stern rope, flipped over the side the old tires that served as float bumpers, and went inside the wheelhouse to prepare the SEAGULL for docking.

The other trollers had already tied-up by the time Silk shut down the diesel engine. He knew that most of the skippers would be at the Albatross tavern waiting for word of Cap and Trevor; Cowboy would be looking for him to show up. But for some reason Silk could not bring himself to walk up to the tavern to mingle with the other fishermen. Instead, he sat like a zombie in his captain's chair, staring at the ugly chain on his automatic pilot while he listened to the coast guard operator's impersonal voice droning on in monotones to other stations. There was no mention of the CRACKANOON. In fact, the voice seemed more interested in the weather conditions than a miss-ing troller.

Halfway through a pot of black coffee a line from one of Cap's teachings came to him.

"When you have a problem that you can't figure out, work. Work'll help better'n anythin'."

Well, Cap was the problem this time, but the aphorism was still true. Maybe if he worked like Cap said, he wouldn't think. Silk busied himself scrubbing the boat down in the rain and dark. He checked the fuel, the oil, and the water level of the engine. He washed his dirty dishes and rearranged all his cooking utensils in their proper racks. Finally, with the practical jobs done, he walked up to the fishermen's bath house and stood under the hot shower until the water ran cold. Nothing seemed to help. He dressed in clean clothes and returned to the boat to finish his coffee. He had never felt so forlorn in his life. Unexpectedly, reality struck Silk. Cap was not coming back! The concept was staggering. He couldn't encompass it. His mind went blank, and he bolted out of the chair, out the cabin door, and on to the dock. He didn't bother to put on his rain gear, lock the cabin door, or turn off the lights on the boat. He didn't give a damn if the SEAGULL sank with everything on her. The only thing left to do now was to go to the Albatross and get stinking drunk.

Like most fishing communities on the west coast, Port Bane had its favorite hangout where the fishermen came together to swap stories and discuss in detail the possibilities of catching more fish. The Albatross Tavern was a home away from home for the fishermen who lived on the bay front or on the boats that lined Port Dock 3 or Port Dock 6.

The bar and grill was opened every morning at exactly 5:00 A.M. For about three dollars they could get a breakfast of home-made biscuits, gravy, eggs, sausage, ham or bacon, hash browns, toast, and strong coffee. If they didn't wish to eat because of a hangover or some personal problem, they could begin their day with a pitcher or two of good beer: COORS or HAMMS or BUDWEISER or OLD ENGLISH 500. If they so desired, they could drink their beer from the pitcher and neglect the glass. Many of the old timers felt this was a trademark of the Albatross and one of the good reasons for living on the bay front.

While they drank their beer, they could read the morning paper, or think, or gaze at the many pictures and slogans that lined the bar walls; pictures of fishing trollers and trawlers, of giant salmon and tuna, sharks and whales, and even an Alaskan crab or two. Their favorite slogan was hung on a wooden plaque directly over the center of the bar. It read, "Dudes in fancy clothes should be prepared to defend themselves."

If there were storm warnings and the sea was too rough to fish, the Albatross Tavern became more and more lively as the day progressed. The bar was lined with fishermen sitting and standing. The wooden benches at every long table were crammed with men meshing shoulder against shoulder in order to make room for one more, and the aisles were filled with pacing men who were either looking for a space to sit down or seeking a pathway to and from the men's room as the flow of beer increased.

The tavern was located almost in the center of the bay front, directly across the street from the Pacific Marine Works and the Crab Packaging Plant. This establishment was owned and operated by Hattie McDougal and her husband Pat, and it had been for twenty years.

Hattie herself was a handsome woman, tall, blonde and slinky, with protruding hipbones. She walked with a slow easy grace and talked with a southern accent. But Hattie was the "wrath of God in a storm," if she was crossed. For indelicate, filthy language, or throwing a beer pitcher, some fishermen had been eighty-sixed for as long as six months. She eventually relented and let them back inside because they spent money, but the thought of no meaningful conversation, and no beer among friends made the average fisherman an easy going, mild mannered type of guy. As long as no flatlanders or tourists came in, the Albatross was as peaceful as any tavern on the coast. One thing was certain, Hattie had built a home for fishermen in Port Bane, and they liked her.

Pat, her husband, was a soft spoken man who worked part-time as a security guard at the docks. He came in early to keep the books and left early, and no one had heard him say more than three words at a time in the last ten years.

When Silk opened the door of the Albatross Tavern, he didn't recognize the place. There was standing room only. In the ten years he had been in Port Bane he had never seen it packed like this; everyone who was anyone in the bay front community was here. Men were lined up three deep and shoulder to shoulder at the bar.

28

Glancing around the room, Silk could see that every table was filled except one. In the place where old Cap used to sit, known as "Cap's Table", there was an empty bench. A CB radio tuned to the coast guard station sat on top of the table. Some damned skipper had already disconnected it from his boat and carried it up to the Albatross. Tonight there was no loud western music over the speakers, and the conversation, instead of being raucous, like when a live band was playing, was reduced to a murmur or a deep rumble.

The mood of the place hit Silk. He felt an inherent sense of unrest, of resentment and violence. For a moment he didn't know whether to yell like a fisherman and start a fight, or to be silent and head for the bar. The decision was made for him when he saw a wildly gesticulating man at the far end of the bar. It was his old friend Cowboy, the same lanky figure, the same drooping mustache, and the same crooked smile. Silk hadn't seen him in a year.

"Damnamighty, Silk!"said his friend. "I was about to give you up. Thought you'd decided not to show up tonight. I got you a pitcher, but it was gettin' stale, so I drunk it myself. Just a minute, I'll get you another one."

"It's good to see you, Cowboy," said Silk extending his hand. Then he wrapped the smelly man in his arms. "Thanks for leading me in. I needed the help and didn't realize it. I had my mind on Cap, I guess. I just can't seem to accept the fact that the Cap is missing. If we don't hear any news tonight, I'm going to continue searching tomorrow, or as soon as the wind dies down. I don't think the coast guard gives a damn about fishermen disappearing."

"I know how you feel, Silk," said Cowboy. "Cap and Trevor was like my own kin. We both know how Cap is about fishin', and it's for damn sure he knows more about that ol' mother ocean out there than both of us put together. I keep thinkin' he mighta hit a hot spot and headed for them fish without tellin' nobody. Hell, we just might hear from them yet."

"You're an eternal optimist, Cowboy," said Silk picking up the pitcher of beer that the barmaid had just delivered. "You know as well as I do that the word is already out all up and down the coast. Cap would have called in unless all three of his radios went out at once, and that's not very likely now, is it? I know that those Russian ships came through the fleet last night. Sure as hell one of them hit the CRACKANOON. I can feel it. The Cap is dead."

29

"O.K., Silk," said Cowboy. "Say I agree with you. I just can't stand givin' up so soon. You say you're goin' out looking for them? Then I'm goin' with you. Meantime, let's do what Cap would want us to do. He'd want us to drink, dance, and throw a hell of a party."

"That's what I came here to do," said Silk. "Get drunk. Where are we going to drink Cowboy? I can't see standing here all night."

The two men gazed around the crowded room. Neither could bring himself to sit at Cap's old table. The table had been purposefully left vacant in honor of the missing man. To the left in front of the men's restroom were three tables pulled together by the skippers in the fishing fleet. To Silk, they looked like a package of frozen herring that he periodically used for bait. In Cap's terms, "solid herrin'". On the night he disappeared he had said that he and Trevor had gone solid herring.

"We'll go back there," said Cowboy, gesturing at the three tables.

"Looks to me like those tables are taken, Cowboy!" said Silk.

"You know fishermen," said Cowboy. "There's always room for one more. They'll make room for us."

Silk stared at the three tables of fishermen. Doug from the JUANITA was there, Al from the DAWN, Neil from the VENTURA. God, the whole fleet was here tonight.

On a tall stool along the back wall sat Doc Ox, a hulking brute of a man who was the self-imposed surgeon-general of the fishermen. No one had ever questioned whether he had a medical degree or not since he worked for free when it was necessary.

Next to him sat Claude, the port's deep sea diver. He was big and he was mean. He seemed to be holding forth, but Silk couldn't hear what he was saying.

Over in the corner were Graveyard, Bent, and Rat. They weren't the current fishermen, but they had been skippers on their own boats at one time. They were known as the Locals along the bay front. There were a dozen of these guys waiting to get on a troller as a deckhand. Silk had never understood how they managed to stay alive around the Port. He guessed it was the age-old adage of love for the sea. "Once tasting the sea you could never leave her, either for love or for money."

Silk shook his head in amazement as Cowboy threaded his way through the crowded room to the three tables where the fishermen were sitting. High above his head he held a pitcher of beer in each hand. As he bounced off burly fishermen, he didn't spill a drop. Silk knew what Cowboy had in mind; he had seen his mustachioed friend worm his way into more than one table of unsuspecting females. Why not the fishermen?

"Scoot over, men," Cowboy ordered three feet from the table. "We just got here, and we're thirsty too. Anyone interested in a free pitcher of beer?"

Immediately, there was space. The movement on the bench started like a wave. The fishermen on that side of the table moved to their right, and the diminutive Rat, who was perched precariously on the very end of the bench, slid unceremoniously to the floor.

"Glad to see you, Silk," said Claude, the big man sitting next to him. "I just got here myself, and I'm pissed off. I was just thinking that I, for one, would like to blow them fucking Russians right out of the water."

"I know how you feel, Claude," said Silk. "Right now, I'm not too friendly myself, but we have to get a handle on our emotions. Instead of running off half-cocked with explosives, let's try to use our heads."

"O.K. Silk," said Claude. "You just tell me how I can help."

"It looks to me like we have very few options open to us, but we must use every one," Silk said. "First of all, Cowboy and I have agreed to continue the search. But we need someone in port who will carry out other important activities while we're gone."

"What do you have in mind, Silk?" asked Claude.

"What we need is someone to register protests against the Russian ships coming inside our boats at night," said Silk. "We need to report them to the Marine Science Center, the Port Bane Coast Guard, the Fish and Wildlife Department, and the Oregon State Legislature. Everyone here has to fish, Claude, except you and some of the Locals."

"We may need to fish," chimed in Doug. "but we want to do our part, too. Claude can check with the Marine Science Center since his place is right next door to them. Al, Neil and I will do the rest before we go out."

"What about me?" asked Spence. "I want to do something about this shit too."

"Fine, Spence," Silk said.

31

"I'm so fuckin' mad, I'd like to punch some bastard in the nose." Spence stared around the room for a target.

"Quiet down, Spence! There are ladies present," Silk said calmly. He had seen several of the local women come in the front door.

Although the bay front women wear boots and Levi's and carry knives on their belts, they were treated with respect. Phyllis was a fishermen's wife. Darren and Beth worked as deck hands on the boats; Babs, Sue Ellen, Patty, and Lev lived on the bay front and played on the ladie's pool team. Nobody could beat them. Jodie, the slender, pretty one, owned a house out on Bay Front Road. Whenever a fisherman got injured and couldn't afford to go to a hospital, he was taken to Jodie's house. Doc Ox, the big man sitting on the stool, called Jodie's house his hospital. The tiny one with the dark complexion was named Lillian. She claimed to be a white witch, and if there was anything in the occult she didn't know, no one was yet to hear of it. These were Port Bane women, and the men here wouldn't put up with anyone insulting them.

"Sorry," said Spence. "I was so dogged mad I didn't see 'em come in."

"Man," said Graveyard who was now feeling the effects of the beer. "I sure wish Cap was here so he could tell us the story about that English fella, Captain Cook."

"Yeah," said Rat. "How he come a sailin' up the coast in his two fancy sail boats, the RESOLUTION and the DISCOVERY."

"He was lookin' for the Northwest Passage that Ol' Francis Drake never did find," said Bent. "Fact is, Cap said as far's he knows, nobody's found it yet."

Silk smiled in spite of himself. Cap was half philosopher, half historian, and half liar, but once you sat down at his table you couldn't leave.

"I tried to run competition with Cap one time and fell flat on my face," Silk said.

"What happened?" asked Claude.

"When I first started working on the CRACKANOON," Silk said, "I knew Cap was stretching the story a bit, so I decided to teach him a lesson. After he finished the story one night, I told about Port Bane as it is today; about how the real estate people began to zone the land on both sides of Highway 101; how they built new houses, new businesses, condominiums, shopping centers, grocery stores, hotels, motels, R.V. Parks, five and dime stores, A.M. and P.M. Markets. Then the charterboat companies, which ran competition with the fishermen. Finally, I mentioned the commercial sea garden, Marine Science Center and called Port Bane a veritable middle class tourist trap like a hundred others along the coastal highway."

"And?" queried Claude.

"No one was listening," said Silk. "They got up and left the table. They didn't give a hoot about my story. They wanted to listen to the Cap. Cap was a unique person." Silk cursed himself for using the past tense.

The crowd was silent as they listened to another weather report from the Port Bane weather station.

"Finish your drinks," Hattie said. "It doesn't look like we're going to hear anything tonight, and we have to clean up and get ready for breakfast in the morning."

Ben, the local musician pulled a harmonica out of his pocket and began to play a haunting melody. He was several bars into the tune before Silk recognized it as Cap's favorite new song, A PIRATE LOOKS AT 40.

"We're closing now," Hattie announced.

The crowd murmured. That was it. The men chugged their beer and began to wander towards the front door.

Silk and Cowboy walked toward the port dock in silence. Silk was thinking about the inclement weather conditions and the struggle Cap and Trevor must be having if they were still alive. The usually garrulous Texan walked with his hands in the pockets of his raincoat, his face down against the driving rain. He seemed to be lost in thoughts of his own.

"We've got a long search ahead of us, Cowboy," Silk said.

"Yeah," said Cowboy. "Let's get us a bottle of that there scotch and make us a evenin' of it. By mornin' I'm bettin' this wind'll die down."

33

CHAPTER

5

BAZAAR

The feisty northwest gale that had blown the Port Bane fishing fleet into port and had ended the search and rescue mission for the CRACKANOON was replaced by a cautious breeze creeping in from the southwest. The moment the wind died down enough for the old SEAGULL to navigate the treacherous river bar, Silk and Cowboy began their own personal search for the CRACKANOON. They started at the line where Cap had shut down on the night of the tragedy, and in ever extending circles covered the whole area foot by foot as far south as Charleston and as far north as Astoria. An oil slick or any debris would indicate what had happened to the troller. Several times they saw coast guard cutters and coast guard helicopters involved in the same task, and they kept in constant communication with the other fishing boats. At the end of three days the coast guard called off its search and listed the CRACKANOON and her crew as lost at sea. At the end of five days, out of food and fuel, Silk and Cowboy surrendered to the inevitable and came back into port. Cowboy went directly to the main dock to check the OTTER and his crew while Silk walked to the Albatross to find out what had happened on the bay front while they had been gone.

When he entered the tavern, Hattie was standing behind the bar.

"Have you heard any news?" he asked Hattie.

"Not a word," she said handing him a pitcher of draft. "Did you and Cowboy find anything?"

"No," said Silk. "We checked out every possibility until we finally ran out of fuel and food. They just disappeared. That's all. Cap and Treavor are gone."

"Doug's registered protests with the fish and wildlife both times he's been in," Hattie said. "Al's checked with the local authorities, Neil the science center, and I've been in contact with the coast guard

34

every day."

"It's all over, Hattie," said Silk taking a deep swallow from his pitcher. "I just never figured Cap would end up drowned." Hattie watched him shake his red head from side to side and stare blindly into his pitcher of beer.

"You've done everything you could do, Silk," she said sympathetically. "We just have to accept fate. Life goes on. Now we have to make it as easy as possible for the loved ones. We have to think of Ann and Trevor's children and give them our support. They're going to be suffering financially."

"What do you want me to do,Hattie?" asked Silk.

"Just what you always do when one of our fishermen disappears," said Hattie, "direct a fund-raising benefit for the family."

The fishermen in Port Bane had a deep affinity for each other, and even more, they had an almost religious respect for the woman who had married one. They figured that a man who fished took his chances with the dangers of the ocean, but a woman who had cared enough about this kind of man to have married him and to have had children by him was special and would damn well be taken care of. The other fishermen would see to that.

If a fisherman were lost at sea, someone in the fleet took up a collection from the boat owners, the skippers, and the deck hands on the boats. These contributions were given to the widow immediately. Then the bay front community held a benefit for the deceased. Fishermen donated fish, crab, shrimp, and oysters which were sold to the tourists. Local craftsmen and artists were asked to donate paintings, leatherwork, and jewelry to help the widow.

For the past five years Silk had organized and directed every fund-raising activity in Port Bane for the families of fishermen who had died. Hattie had no doubt as to who would take charge of this one, since it was Cap and Trevor who were missing.

Although setting up tents, building booths, and preparing food was not his idea of a man's work, this time Silk had no choice.He was in for it.

"Is there anyone around I can use as a crew?" Silk asked.

"Just the usual," said Hattie. "The Locals, Doc Ox, Jodie, Lillian, Claude, and anyone else who is not fishing. I've already made the necessary contacts."

The next few days were busy for Silk. With the Cap on his mind, he went about preparing for the social bazaar. He contacted the fishermen on each boat to see what percentage of their catch they would donate; he organized the women into committees responsible for publicity; he contacted the local artisans; he made arrangements for the food and beverage preparation; and finally, he organized his work crews, mainly the Locals, Doc Ox, Claude, and himself.

By the time the fishing fleet returned with its catch, the whole dock across the street from the Albatross Tavern looked like a small carnival was about to take place. Wooden booths had been built on the west side of the dock. At the back of the dock over the water was a raised bandstand. In front was a dancing space, and in front of that was the portable restaurant. Blue, green, and silver streamers hung from the corner of every booth, a final touch added at the last moment.

The plan was to cook the food in the Albatross Tavern and carry it across the street to the old crab dock. Under the large canvass each table was arranged at a convenient angle so that the diners had a clear view of the band and the booths. Jodie and Sue Ellen volunteered to sit behind the long plywood table next to the street and sell tickets for food and booze. Lillian was bartender and would pour the drinks from the kegs of beer and gallons of wine and soft drinks that were provided by the Albatross.

By the time Cowboy returned to port with two thousand pounds of shrimp, Silk had a lot to tell him.

"We're going to have the best fund-raiser we've ever had," said Silk.

On the first Saturday after the close of the fishing season, at 5:00 A.M., the people who were going to help at Cap's bazaar were waiting outside the Albatross door. At 9:30 A.M., or there abouts, the tourists from inland had begun to arrive. By that time there were great vats of homemade clam chowder bubbling on the grill. The ovens were filled with garlic bread and baking Chinook salmon. Fresh oysters and chilled dungeness crabs were stacked in platters along the bar. Across the street on the dock every booth was ready with its salesman. Bearded fishermen were standing at their posts waiting for the first tourists to arrive. Old Ben and the blues band were already playing.

36

Their favorite song, the FISHIN' BLUES, was written by old one-eyed Ben himself. The band, already stoned on good bud, put their hearts into the soulful tune, and the guests began to pick up on the nostalgic mood of the fishermen as they listened to the soaring harmonica, the deep base guitar, and Ben's quavering voice.

I been out on that ocean
Fishin' most of my life.
Death ain't no stranger
To a fisherman's wife.
I got them cold, black fisherman's blues.
We hunt that mean old salmon,
But at the end of the line,
We end up with a dollar
To buy a jug of wine.
We got them dark, down fisherman's blues.
My mother is a widow,
And she lives all alone.
The troller of her husband
Went down like a stone.
Oh, yea! Went down like a stone.
She's got them low, down fisherman's blues.
Her husband is a dead man;
Her baby followed him.
I am the older son,
And I don't give a damn.
I got them cold, black fisherman's blues.
God is the ocean.
He's cold as the sea.
He drowns all the fishermen,
And He's got his eye on me.
Oh yea! I got them low, down fisherman's blues.

This lament for fishermen lasted for thirty minutes, then Ben and his band began to play some of the popular tunes that they knew the tourists would like. The word circulated up and down the bay front that there was a party going on, and soon the old dock was filled with neck to neck onlookers spending money. The crew in the Albatross was busy now, hauling steaming platters of food and beverages from the Albatross across the street. Soon they had to have an extra man run interference for them in order to shoulder their way through the growing crowd.

"By God," grunted Doc Ox as he hefted a large vat of chowder from off the burners. "This is the biggest crowd I ever did see. They're eating this fish faster than we can cook it, and they're buying those trinkets like they're some kind of treasure. Look at the crowd in front of them booths. They're going to buy everything we got for sale."

"We're doing well," said Silk. "The ladies are already starting to turn in the money to Hattie."

As the time passed, Ben and his blues band began to pick up the tempo of their playing, especially after every fifteen minute break when they went beneath the dock to smoke a killer bud and drink a couple of free beers. No one had ever heard Ben sing better. Then, in the middle of the fifth or sixth set, while he was singing a tune called OH TAKE ME AWAY BABE, something unexpected happened. Without warning Ben's eye flew wide open and he began to focus on some unknown object that only he could see. He himself seemed surprised, and his voice stopped in the middle of the word "Oh". Everyone could tell by the way his fingers were positioned on the guitar that he thought he was still playing. The killer bud mixed with the beer had finally got to old Ben, and he had passed out standing straight up. The band tried to carry on for a few more measures before they realized that the tune just didn't sound right without Ben's voice. They stopped playing and carried his stiff body off the band platform and around behind the first aid stand where they laid him gently on the ground.

When the music stopped, the spirit of the occasion started to die. People could hear themselves talk, and not being pumped up by the continuous sound of the loud amplifiers, they soon fell back into their humdrum existence. Realizing the problem, Claude came from behind the first aid station carrying Ben's guitar. He climbed onto the stage, settled the band, and launched into a hard rock tune. Someone else carried a large set of bongo drums onto the stage. Lillian, not to be outdone, produced a tambourine from somewhere, and began to whirl like some errant gypsy woman in front of the stage, whacking each skinny thigh in tune with every beat. Soon the couples from the crowd picked up the excitement, and they began to dance.

"Claude to the rescue," said Silk watching the gyrating crowd. "He's so much better than Ben as a musician. I wonder if there's anything he can't do."

"Here come the cops," said Cowboy.

Two police cars answering a call about a disturbance on the bay front had pulled up directly in front of the dock. When they saw the band and the dancing tourists, they nodded to each other and drove back to the station house. They returned a short while later and erected wooden barricades at either end of the block, then they began to patrol the street.

By this time most of the beer and wine from the Albatross had been drunk, so the two policemen volunteered to drive up to the Pacific North Distributorship for a new supply. On the way up the hill they passed what looked like unsavory members of a motorcycle gang from Los Angeles descending on the bay front. The bikers were passing through town on their way to Seattle for a yearly convention. At a service station on the strip where they had stopped for gas, they heard about the party on the bay front. They thought they might just as well have a few drinks and look around.

The first motorcycle crashed through the police barricade at the bottom of the hill and shoved its way through the dancers until it ran smack into Graveyard who was carrying a large vat of clam chowder across the street. The bike tipped over. Graveyard, the chowder, and the burley biker went down together. Since Graveyard landed on top, he figured he might as well get in the first lick.

"You flatlander. You bush-rider," he bellowed. "You caused me to spill my clam chowder." He hit the biker in the mouth and the long-haired guy spit out a tooth.

Graveyard, still on his knees, turned his head at the onlookers and smiled. He was still congratulating himself on the success of his first punch when a heavy motorcycle boot crashed into the side of his head. He had forgotten the first rule of a fisherman. "Always watch your back."

Silk, seeing the altercation, ran into the street to break up the fight. Graveyard had the first biker on top of him now, and the second one was trying to get a clear kick at his head. Silk had almost reached the struggling men when a fist from an unseen assailant struck him square on the jaw, and down he went.

The biker who hit Silk did not realize that he had made a serious mistake until he heard the angry voices of the fishermen who came pouring out of the Albatross. They were cursing him and saying ugly things that they were going to do to him.

At one time or another over the years Silk had come to the rescue of nearly every one of them. To see him struck from behind when he was only trying to break up a fight made them madder than hell.

Now the battle really began. Led by Cowboy, Bent, Rat, Ox, Claude and Neil, the angry fishermen showed the bikers from California that they were the real street-wise brawlers. Rat flung himself onto the back of the biker who had hit Silk and sank his protruding front teeth into the man's neck. Ox lifted one biker clear above his head with Bent still twisting on the man's leg, and with a roar threw them both on top of a fallen motorcycle.

Gradually the fight moved toward one end of the street. The motorcycle gang was pushed as far as the overturned barricade where the ones who were still standing broke and ran. In triumph the fishermen collected the overturned motorcycles and pushed them to the end of the street where they stacked them into a rather untidy pile.

While most of the fishermen were dealing with the motorcycles, Graveyard and the leader of the gang, whom he had first fought with, were in the men's restroom of the Albatross trying to wash the clam chowder off each other.

"As soon as we get cleaned up," Graveyard said to the beat up biker, "we'll go out and finish up the fight. Then you can come to Cap's bazaar if you'll buy somethin' and don't cause no more trouble. If I had any money, I'd buy you a pitcher of beer. You're a pretty good scrapper."

"Well, thank you, Graveyard," said the biker. "That's real neighborly of you, but I think we may have already been whipped. What say I buy you a beer?"

When the policemen came back down the hill with the backseat and trunk of the police car filled with kegs, they were amazed at the stack of motorcycles and the gang sitting quietly around the pile. The band was playing heavy rock now, and the crowd was showing it's appreciation by gathering around the bandstand and clapping their hands.

"I think our department's all wrong about these people," said one officer. "It's as peaceful as when we left."

"I don't know," said the other. "Those riders that came down the hill look like someone beat the hell out of them, and I've never

40

seen motorcycles stacked like that before."

"It doesn't matter," laughed the first policeman. "Everything seems alright now." By dusk almost all the tourists had gone home, and the band quit playing. Old Ben walked out from behind the first aid station and asked for a beer. He didn't know what had happened and didn't care. In his own mind he had been playing all afternoon, and now he was tired.

The fishermen and artisans dismantled the booths and stacked the tables before they retired to the Albatross to celebrate their success.

Hattie already had the figures ready and announced that they had made over five thousand dollars.

CHAPTER

6

MEL PLANK

After the charity bazaar for Cap and Trevor, the good weather held for three weeks in Port Bane. The days were warm and balmy, and the nights cool and clear. Taking advantage of the good working time, the fishermen labored day and night on their boats, preparing for the fall crab season, installing new equipment, or repairing the damages they had incurred during the past salmon season. Then the rains came. At first stringy gray whisps of vapor floated across the clear blue sky. Then heavy dark cumulus clouds filled with moisture piled up in great layers above the port. The northwest winds blew in, and the cold rain fell in torrents. It was like God didn't give a damn. The inhabitants on the bay front cheered when the tourists went home. Until the good weather came in the spring, they would be rid of funny people and the streams of traffic caused by their cars. Besides, this was the beginning of their favorite season, the mushroom season. There were the common basidiomycetes, the chanterelles, and the chicken in the woods, all of which could be sold for a large profit to fancy restaurants. And if one were lucky enough to find the tiny, hallucinogenic liberty caps, he could stay stoned all winter long. Most of the Locals on the bay front were in the midst of happy anticipation for an abundant harvest this season.

John Silk, however, sat at the bar in the Albatross Tavern contemplating the winter months ahead. At this time of the morning there was no one else in the place except Hattie who was pretending to do something useful with the cash register. She knew better than to say anything to Silk. She could see he was in a rotten mood. Out of the corner of her eye she watched the tall fisherman stare out the window at the rain, now streaking down in blinding sheets whipped by the violent gusts of whirling wind.

Silk watched this latest downpour, his third cup of black coffee growing colder by the minute. Like all the other fishermen in

Port Bane he had been busy on his boat—washing, scrubbing, arranging, and painting—getting ready for the winter months ahead.

The silence in the large rectangular room was broken by the loud ringing of the telephone behind the bar. Hattie answered the third ring then carried the phone the length of the bar to Silk.

"It's for you," she said and went back to the cash register.

"Hello," Silk said, then frowned as he recognized the falsetto voice of Mel Plank, owner of the SEAGULL.

"I've been trying to reach you for a week," Plank said, "and I finally find you in a tavern like I figured."

"What can I do for you?" Silk asked.

"Eh, I received your money order for my thirty percent for the last trip of the season, and that makes a sum of $4,730 for the six months you've been the skipper of the SEAGULL. That's somewhat less than the $10,000 you promised me when you leased the boat. That's not much of a return on my investment, now, is it? For a boat that size I expected the season's return to be in excess of $10,000."

"That $10,000 figure was based on my worst previous fishing season," said Silk. "This season was not that good. I promised you thirty percent of the gross catch, and you got your thirty percent."

"Like hell I did." The voice had increased in speed and intensity. "You've held out on some of your sales is my guess. You've been selling to a dock whore down there. I know they're there. And..."

Silk looked up the bar at Hattie who had her ears tuned to the conversation. He nodded his head and smiled and waited until he thought Plank had finished his tirade.

He had been halfway expecting this call since the last time he had talked to Plank. He was about to get kicked off the SEAGULL.

"Mr. Plank," Silk spoke softly into the receiver. The voice droned on. "Mr. Plank," Silk began again in the same quiet even voice. "There are some circumstances about this season that I don't think you understand. I'd like to explain."

"Mr. Silk," Plank's exasperated voice started up again. "I don't want to hear any of your explanations. I don't have time. We're talking about money here. To get down to brass tacks, when you leased the SEAGULL, she was in top shape, and she'd better have been properly maintained. I purchased that boat for a business investment, and I haven't received enough return to merit the sales price. I know there are more skippers in Port Bane than there are fishing boats, and

I might as well say it. Someone should teach you how to catch fish."

Silk removed the phone from his ear. He had heard enough from Mel Plank. He glanced at his reflection in the mirror behind the bar and realized that Cowboy was right. His face did turn stone white when he "got riled". Right now, in contrast with the red beard and hair, not even the freckles seemed to give his flesh any color. The sonofabitch had called him a liar.

Silk placed the phone back to his ear just in time to hear the end of Plank's speech.

"And I've made up my mind, Mr. Silk. Your lease on the SEAGULL is up. Next season she will have a new captain. What do you think of that, Mr. Silk?"

"Your decision is just fine with me, Mr. Plank," said Silk in a monotone. "To tell you the truth I wasn't going to fish the SEA-GULL next season anyway unless we signed another lease that required you to pay for the repairs, or else give me a greater percentage."

"To hell with you," Plank screamed over the phone. "I'll be down to inspect the SEAGULL this weekend, and she'd better be in good condition. If I have to talk to my lawyer, you'll find yourself in court, and I'll make sure you never fish again."

"Give it your best shot," said Silk. "I'll have my gear off the boat by Saturday. You'll find my key lying on the table inside the cabin."

Silk hung up the phone and sat in silence staring at the pictures of fishermen and boats on the wall above the bar. If Melvin Plank had been any kind of a decent man, Silk would have given him some good advice; not to hire a skipper who wasn't experienced, and not to fish the SEAGULL next season without having first made major repairs, namely a complete engine overhaul, replacement of the rotten ribs on the starboard side, and restoration of any weakened planks. But Mel Plank wasn't the listening type. The man was finished in Port Bane and didn't know it.

Hattie came back and picked up the phone. She had been watching Silk's face too, and she could tell by the frozen smile on his lips that the call hadn't been a good one. She had no intention of saying a word, but her curiousity got the best of her.

"What was that all about?" she asked.

Silk grinned showing clenched white teeth through the red

44

stubble of his beard.

"That was Mr. Melvin Plank, owner of the SEAGULL. He just fired me."

"I'm sorry Silk," she said. "Where are you going?"

"Well, I don't have a boat to work on anymore, so I guess I'll walk down to Bane Beach and take a look at the ocean."

"Don't you want to finish your coffee?" asked Hattie.

"Just throw it out," said Silk over his shoulder.

By the time Silk had reached the beach, he was soaking wet from the two mile walk in the rain, but his anger was beginning to subside. He had smiled at Hattie in the Albatross, but to tell the truth he was furious. For a moment he wished he had the striped-suited Melvin Plank here on the beach. He'd smash his nose all over his face. But that wouldn't do, probably break the skinny bastard's neck, then he would end up in jail with some lawyer threatening to ruin him for life. There was no way of dealing with a Melvin Plank. It was best to forget the whole thing and think of his next move.

For a moment Silk stood motionless. The water ran in streams off his blue knit fisherman's cap, down into his straggly copper hair, and down on his neck, and down his back beneath the ragged sweatshirt. He was miserable.

He squinted his gray eyes against the blowing rain and gazed up the five mile stretch of beach to where the lighthouse was barely visible. Today, he was the solitary human sojourner on the beach. Above him the white gulls soared silently on the wind's currents. He watched them a moment, then he thrust his scarred hands deep into the pockets of the sweatshirt and started walking north.

So Plank didn't think he knew his business. Hell, that flat-lander had no idea of a fisherman's problems. Not only had he had to deal with the erratic weather conditions and constant breakdown problems, but also he'd had to contend with changes in the Oregon fishing borders and a shortened salmon season.

For the past several years the U. S. Department of Interior, through the Oregon Fish and Wildlife, had pulled every political trick possible to legislate the commercial trollers off the Pacific Ocean. They took two weeks off the season here and a month away there. They took away April 15th to May. They took the month of October; they cut back to Labor Day; then this year they had shortened the

45

coho salmon season. God only knew what those political bastards behind their big desks in Washington D.C. were planning for the next season, and there wasn't one of them that knew the difference between a Chinook and a coho.

And they had changed the fishing borders. They had literally moved the Washington border from Neah Bay down to Ledbetter Point, and all of a sudden the Oregon border was no longer the Columbia River where it belonged, but Cape Falcon. Silk couldn't understand the limited minds that made these decisions, but he did understand these kinds of determinations were putting the fishermen out of business.

But these problems were no concern of his now. He'd lost his boat, and there was almost no chance that he'd be fishing next season. Now, his fishing problems were all in the past, and it was time to look ahead. Ahead was the lighthouse; ahead was the rain, driving now, from the strong wind; ahead was the dark cloud cover, thick and rolling and heavy. Ahead was most likely six months of constant unemployment. He was more broke than when he had come to Port Bane ten years ago. Hell, he was in the same financial shape as the Locals. He wondered who would buy him his beer.

He had really done it to himself this time. He was in deep shit. Everything seemed to be coming down at once and landing on him. The loss of Cap and Trevor, a bad fishing season, rotten Mel Plank and his bastard of a boat. What next? Looked like he was on a treadmill to oblivion. This time it was a question of whether he could pull himself out of it enough to survive. He had to use his head. He had to make a plan then follow that plan step by step.

When he looked at it, these were all minor problems that time would solve. But what about Cap and Trevor? Time had stopped for them. One minute they were on the radio, the next minute they were gone, disappeared without a board or an oil slick to tell him what had happened. Their tragedy was the curse! Since that morning when the storm had driven him from the search, nothing had gone right. And in the back of his mind was a haunting whisper, like Cap was telling him from the depths of the sea that unless he uncovered the circumstances that had killed them, he would continue to have troubles. Silk didn't believe in ghosts, but there was an intangible someone or something pushing on him.

Silk's logic told him that one of those giant Russian ships had

46

run down the CRACKANOON during the night when they had come through the fleet, but he couldn't prove it. Subconsciously, this knowledge had been eating him up. Suddenly, he realized he had been ignoring the obvious. His first obligation was to Cap, the second to himself. Unless he found a way to avenge Cap's death, he could never fish again.

Excited now, Silk knew what moves he had to make.

Over the years of fishing Silk had collected some eight thousand pages of research on the fishing industry. Those papers were stored in a box in the fo'c'sle on the SEAGULL. It was the only valuable asset he had. The research had exposed the conspiracy of the government and big business against the commercial fishermen, but he had never done anything with it. Now was the time to act. Here was information concerning foreign fishing in joint venture with American companies within the two hundred mile limit. Silk's mind was leapfrogging.

So what if he found that one of those big mothers had run down the CRACKANOON? What was he going to do about it? What was he going to do about his eight thousand pages of collected data proving conspiracy of key figures in the Department of Commerce, the banks and lending institutions, the American Fisheries Society, and the giant corporations? These sour apples were in cahoots to take control over all the protein in the world! He even had a hardcover book written by the Department of Fisheries that no longer existed in the Library of Congress. Someone had pulled it. That book showed the long range plans of Washington bureaucrats for fish farming sites along the Pacific Coast as far back as 1955.

But hiding that information from the public wasn't as bad as milking the ocean's resources dry for their profit. Profit bought power and power brought control. So, who had all the control? Simple, the banks and insurance companies that controlled all the money in the world. The problem didn't stop with the USA. Even the bank of Canada had a little suck with the house and senate if there were something going on in the fishing industry which concerned a Canadian bill.

The same control applied to every bank in the world. Whether it was the bank of Japan, the bank of England, or the bank of Iran. But talking about control on an international level didn't get the job done. Underneath the bureaucrats' soft bellies was where the damage was being done.

Silk knew human nature. There was just one instinct that every human was born with, and that was survival. Staying alive! Mankind would do anything to stay alive; steal, lie, cheat and kill. Some poor slobs would even eat their own grandmothers.

The power dogs understood this principle. They were the controllers! They knew that to stay alive a man had to have food. To get that food, man would do anything. And in the future the controllers were plotting to own the whole food supply thereby controlling the entire human race. This plot was a lot more powerful than the threat of any nuclear war.

Right now the controllers were after the Pacific Ocean's resources, and they wanted them exclusively. Commercial fishermen were the target. A fisherman could hardly go to a bank anymore and get a loan on a troller. And insurance? Hell, the companies had priced insurance for salmon trollers clear out of the market. Insurance costs were more now than a damn fishing boat was worth.

Three years before he had tried to get insurance on a thirty-four foot fishing troller from Stevens and Forsyth Insurance Company in Port Bane. Stevens had told him that there wasn't one boat in ten that had insurance.

"Hell," Stevens had laughed. "I had to give up the insurance on my own fishing boat, and I own the insurance company."

But loan and insurance problems were minor compared to the real sickness of the fishing industry. He knew the whole sordid mess, and he had the facts written down, but he couldn't fight the government and the big institutions by himself. To get the attention of the big boys a man had to have an organization behind him. He had to have money and power. That meant politics and paid lobbyists. One thing for sure, he was certainly no politician. He detested the breed.

But the task could be done with the right organization. And here was his second idea, right before his eyes. The answer was the Port Bane Fishermen's Association. The PBFA hadn't done anything in years, but that didn't mean it couldn't do something with the proper leadership. Their last big stir had been in 1982 when the Fish and Wildlife Department had suddenly closed the commercial salmon season early, and the fishermen in the PBFA were furious.

They went to the governor of Oregon. Five or six hundred fishermen had gathered together in the big auditorium in Portland.

The meeting had been inspiring. There were plainclothesmen and uniformed policemen all over the place to watch over the governor and to protect him from the fishermen. All the fishermen had wanted to know was why they had been shut down when they were doing two hundred to four hundred fish a day. The governor had told them there were no fish in the ocean. Either the governor was a liar, or he didn't have the correct information, but the state had shut the fishermen down anyway. Powerless, the PBFA had given up the fight and hadn't made a move since.

Now, the California Fishermen's Association had it's shit together, the way that Port Bane should. There was a saying down in California "We put our legislature on a high pedestal in Sacramento. There they're easier targets." When those people in California said they were going to strike, they meant it. Either you went on strike or your boat was sunk, or it was burnt and you had to haul it out of the water. That was the way the California fishermen operated. They had a fishery and Oregon didn't. And they allowed no joint venture with foreign fishing countries. California fishermen stood by their principles.

Silk smiled to himself and struck a match to a new cigarette. He inhaled deeply. For the first time in months, he felt like he was thinking straight.

With the poor fishing season this year and the tragedy of Cap and Trevor fresh in their minds, the fishermen in the PBFA could be reasoned with. The organization could send letters to the governor and state legislature demanding that foreign fishing practices inside the 200 mile limit of Oregon be investigated. They could send letters protesting the continual cut backs of the commercial salmon seasons. Someone was bound to listen. There was no limit to what might be accomplished when a group of individuals joined together and became a force.

Maybe they'd have a fishing industry yet. And maybe, just maybe, if these ideas worked, down the road he could lease another fishing troller.

When Silk entered the Albatross Tavern, Cowboy ordered two bottles of Rainier dark. While he waited for the barmaid to bring the beer, he turned to watch Silk head for the back table in the far corner. To him Silk was a closer friend than anybody he had ever known in his life, but there were still times when he didn't quite understand the man. Times like right now when he was sitting silently alone at a back table in the corner. Times like when he deserved to beat the shit out of a bastard like Plank, yet he had showed little or no emotion over losing the SEAGULL.

As he walked back toward the table, he glanced at Silk's face. The man had a lot on his mind.

"Here," he handed Silk an uncapped beer. "This'll take the edge off."

"Thanks," Silk said taking a long swig. "You know, Cowboy, while I was walking I had time to think, and I figured out I had my priorities twisted around. I had a dozen questions about what was going to happen to me, about what I was going to do. Then right in the middle of the personal questions, I realized that I was putting myself first and ignoring the real problems. I think I know now what we've got to do. Not only for Cap and Trevor but for all the rest of us."

"Go on," said Cowboy taking out a pouch of Bugler from his shirt pocket.

"We're pretty certain that the CRACKANOON was run down by one of those Russian ships that came through the fleet that night. We just can't prove it," Silk said. "These so called accidents are happening far too often and everyone is letting it slip by. The secret lies out there somewhere within that foreign fleet. I just wish I could get on one of those monsters out there and find out what's really going on."

Cowboy stuck the hand-rolled cigarette between his lips and cupped his hands around the flaming match. He puffed a couple of times and squinted through the smoke at his friend.

"Well, why don't you?" he said.

"Why don't I what?" asked Silk puzzled.

"Get on one of them Russian ships," said Cowboy.

"And just how do you figure I'm going to do that?"

"Simple," said Cowboy. "Get to be a foreign observer."

"How can I find out how to become an observer?" asked Silk.

"Ask me!" said Cowboy smiling broadly. "What do you think I been doin' with joint venture all these years? I been on those ships, and I've met a few of them observers. I know what they got to do to get on one of them bastards."

"Cowboy, you never cease to amaze me," said Silk. "Let's

have it."

"Well," said Cowboy. "The first thing you got to do is be a senior in some college and study marine biology. Then you fill out a application, and if they take you, you're in. They send you to school for three weeks in Seattle and put you on one of them ships. You observe the catch for three or four months and make fifteen hundred bucks a month to boot. When your term's up, they take you off, and you wait for another ship. Hell, I've been with guys who been doin' that for years. Beats the hell out of workin' for a livin', they say."

"I don't know, Cowboy," said Silk frowning. "I doubt if they'd let me in. I've been out of college for a long time, and a degree in business administration isn't exactly related to marine science. I doubt if they'd want a Port Bane fisherman."

Cowboy got up and went to the bar for two more beers. This time he brought back two pitchers.

"Dern it, Silk," he said. "If you want to be a observer, I think I can fix it. You don't have to have no credentials to get in that school if you do it right. You don't get nowhere with them government jobs if you depend on your credentials. You just got to have some suck with the right people."

"And I suppose you know those people?" asked Silk.

"Right," said Cowboy warming to his subject. "In my business we got observers too, 'cept we call them company representatives."

"Alright, I'll be a representative for your company," said Silk.

"Well," said Cowboy. "She ain't as easy as all that. Our reps have to know Russian real good. They have to know fish, and they have to know how to keep them books right and change rubles into dollars. To tell you the truth, without Fran I wouldn't have the least notion how much money I ought to get from one of our cod-ends."

"I don't speak Russian and I don't know accounting procedures for fish sales," said Silk. "So that disqualifies me from becoming a representative. But who is this Fran?"

"That's what I was leadin' up to," said Cowboy. "Fran was the rep for STAR INDUSTRIES, that's the American-Russian company I'm fishing for. She'd been workin' with that Russian factory ship out there for a long time until they gave her a big promotion. Right now she's in Seattle. She knows all about the NMFS school. Fact is, she's the assistant to the main guy who runs it. All I got to do

51

is put in a phone call to the lady and tell her I have a friend who's qualified and who's interested in bein' a observer. She'll do the rest. I tell you Silk, if they have an openin' in that program, she can get you in. I know Fran real good."

"Don't tell me you're getting serious about some lady after all these years, Cowboy," said Silk amazed at his friend.

"Ah, heck, Silk, I wouldn't go and do that. I'd just cramp some nice woman's style. Besides that, I'd likely always be gone. No woman likes bein' left alone." Cowboy studied his cigarette ash. "Want me to call Fran and see if she can fix things for you?"

"Bloody right," said Silk. "If there's any kind of chance, I want to be in that program."

For a while the two men drank their beer in silence, then Silk said,

"Being an observer on a foreign ship will be a giant step forward, Cowboy, but it's just the first step in solving our fishing problems. If we don't move now, there won't be a single fishing troller on the west coast in five years. Last night I came up with an idea that might work to the port's advantage if it's handled right."

"Shoot," said Cowboy.

"The problem is this," said Silk. "I can't fight the whole fishing mess myself, especially if I'm an observer on a Russian ship, but the job needs to be done now, and we have the organization, the PBFA. They haven't done a thing in years, but if they could be motivated, get all the fishermen in the fleet together, they might become an effective force.

Right now they're pissed because of last season's cutbacks. They've had a bad season this year, and the CRACKANOON is still fresh in their minds. If the right person talked to them, they might be enticed to send a representative to the Oregon State Legislature, and every one of them could write a letter. That's my idea. Have the PBFA write letters defining the fishing problems from our point of view. State our grievances; get them out in the open. They could send those letters to every department in the state. What do you think, Cowboy?"

"I think you went a little crazy," said Cowboy. "I know the fishermen in this port have a lot of bad problems that ought to be ironed out. That's why I went into joint venture in the first place. I'll tell you what, I'll support you. And if you pull this thing off, I'll sell

52

the OTTER and buy a troller. That kind of fishin's a sight more fun even though it ain't near as much money. It's a mighty big order. Takin' on all them big organizations. Like I said it's crazy, and I doubt if we get to base one. But someone's got to do the job. Tell you what. You get the ball rollin', and I'll follow up on it if Fran gets you in that observer school in Seattle. Hell, I'll be in and out of Port Bane all winter anyhow. I might just as well do somethin' useful. It'll keep me out of the bars."

"Done," said Silk, and the two men shook hands.

"We can move your gear on the OTTER this afternoon, then first thing in the mornin' I'll call up Fran."

At that moment Silk heard a soft trilling voice speak his name. He looked around to find Lillian and Sue Ellen standing behind him.

"Hello, Silk," Lillian repeated shyly. "We've been looking for you."

"Hello ladies," Silk said. "What can I do for you?"

"We were wondering if we might use your van this Saturday," she said. "The rains have come, and the liberty caps are at their prime right now."

"Sure, Lillian," he said.

"Don't you worry one second about your van," said Lillian. "I will take charge."

CHAPTER

7

NATIONAL MARINE FISHERIES SOCIETY

Cowboy had been right on all counts about Fran Oliver except two, at least two. He hadn't mentioned that Fran was from Sussex, England or that she was tall and beautiful. All he had said was that she was a friend who had worked with him as a representative on the OTTER, and that she had the necessary connections in Seattle to get him into the NMFS observer program.

When Fran met him at the bus depot in Seattle when he arrived for his interview with Dr. Blackburn, Silk had been surprised at the obvious charm of the woman. She wasn't at all like the women in Port Bane. She wore a hip-hugging red wool dress with red pumps, and her sun-bleached hair was long and straight. Her face was quite pretty, Silk thought, very tan and slender as she stared directly up at him through tinted designer glasses.

"John Silk?" she asked in a clipped British accent.

"Yes, I am John Silk," he muttered.

"Fran Oliver," she said. "I've come to fetch you. Sorry to be late. Got caught in the bloody traffic. You weren't in the queues at the gate so I just looked about for a tall man with flaming locks. Cowboy's description of you, you know. Come along. We are rather late for your appointment with Dr. Blackburn." And she turned on her heels and headed for the exit in long sexy strides.

The ride to the NOAH complex in north Seattle where the NMFS school was located had been rapid and hair-raising. Fran had concentrated on driving the department sedan, weaving in and out of the four lane freeway traffic like a race-car driver. Silk found himself holding on tightly to the armrest and taking deep breaths every time they closed in on a car in front. Fran would come almost to the rear bumper, then whip quickly past on either side.

"Where did you learn to drive like that?" he ventured at last.

She looked directly at him again through laughing blue eyes.

"London," she said. "Have you been to London, Mr. Silk?"

"No," said Silk, "and call me Silk, not mister. I guess I'm just used to the slower pace in Port Bane."

Fran smiled with her pretty teeth. Here was one British subject who had pretty teeth.

"You have a jolly good chauffeur here," Fran said. "No smash-ups to date for this lass. Although, I must admit, I do get the right and the left lanes rather confused at times."

At the NOAH complex Fran whipped the car into a parking place for the physically handicapped, but before he could warn her about a ticket, she slammed the door and was striding toward a long, rectangular building. He followed her through the double doors, down a wide hall, and past a pasty-faced secretary into an inner office.

The man behind the desk stood up when they entered. He was much shorter than Silk and about as wide as he was tall. He had a half circle of gray hair around a bald pate so that he looked like someone had tried to fit a halo about his head.

When he reached across the desk to shake hands, his hand felt like a soft salmon tail.

"Dr. Blackburn," said Fran. "John Silk here, from Port Bane. Sorry to be late. Beastly traffic this time of day. I'll dash along now so you gentlemen can get on with your business."

"Be back in thirty minutes, Fran," said Dr. Blackburn. "This session won't take long, and we still have to get Mr. Silk settled."

At that moment Silk knew he had already been accepted into the observer school. He also knew that as per his request, he would be assigned to a Russian ship off the Oregon Coast. The old story was the same; just like any modern American bureaucracy. It wasn't what you knew but whom you knew that opened doors.

Fran opened many doors for Silk after that. Mostly his own. She arranged for him to live near the NOAH complex in an old motel which she explained as "charmingly quaint and rather convenient." She introduced him to the gourmet restaurants in Seattle. "Fantastic view," she said. "You simply must taste Wang's duck." or "You will adore the ballet or this play or that movie." or "Let me help you with your shopping."

It was soon obvious to Silk that she intended to stop by every night, and stopping by every night was not exactly what Silk had in mind.

From out of nowhere Cowboy's words came to him, "Watch that Fran; she'll try to get next to you."

Silk never let himself be penned again. In truth Fran reminded him of Elizabeth Baird back in Boston. She even wore the same pungent perfume called Shocking de Schiaparelli; he should know; he had bought Elizabeth gallons of the stuff. Elizabeth was so beautiful, and he had loved her dearly; that is, until she proceded to bring him to his knees when she suddenly married his best friend Chad, a Harvard professor. That fiasco had happened the week before Silk left Boston. Looking back on it, Silk felt like he had been damn lucky things had turned out the way they had. Still, Elizabeth had taught him a valuable lesson about beautiful women. He had learned to stay the hell away from them!

Once Silk could concentrate solely on his classes, he approached the thirteen day program much like a sleuth in a detective novel. He kept his mouth shut about his fishing experience, and he never mentioned his real purpose to anyone, not even to Fran. He was taking the job for one reason only; to solve the mystery of whether or not Cap and Trevor, and numerous other fishermen who had vanished at sea, were victims of the Russian fishing fleet.

He had picked up the printed sheets handed out to the class and tried to make heads or tails of what he was supposed to be doing for the next few days.

1. View introductory slide show—an explanation of the fishery concerned, illustrated by slides.

2. Study manual which includes the forms and sampling procedures to be completed at sea.

3. Use color slides and identification guides, study the species and specifications.

4. Learn the methods of measuring length frequencies, removing otoliths, and determining the sex of fish.

5. Perform sample exercise in filling out data forms.

6. Review the reports written by previous observers.

7. Fill out and turn in Emergency Data forms.

8. Assemble and pack equipment which include a complete set of data forms, plastic on-deck sampling forms, and a set of otolity collection vials. These were all packed in new suit cases.

This stuff was definitely all Greek to him, and for a moment he was filled with self-doubt. There was no way that he or any of the other five observers could digest and assemble in thirteen days the hordes of material that had been given to them. In the back of his mind was the nagging question of whether or not this school was such a brilliant idea. The school soon degenerated into a series of boring presentations accompanied by great amounts of reading and homework.

With his ten years fishing experience and his accumulation of material concerning the fishing industry, Silk didn't take long to find serious defects in the N.M.F.S.observer program.

His first lecture was by a middle-aged professor named Pinkerton from the oceanic department at the University of Washington. His subject was the Magnuson Fisheries Conservation and Management Act of 1976.

The Magnuson Act was a plan to conserve and manage the fish resources off the coast of the United States. By this act, U.S. jurisdiction over fishery resources was extended to within our two-hundred mile zone, and a management policy was instituted by our government. This management policy was designed to (1) stop the decline in numbers of overfished stocks and to provide potential for new U.S. fisheries, (2) protect the declining halibut and crab resources, and (3) allow no foreign fisheries on surplus stocks within the new U.S. two-hundred mile zone.

Silk knew that the Magnuson Act had become a farce, a lie behind which the U.S. Government could hide and the big fishing companies could make money.

All an interested foreign country had to do to fish within the designated two-hundred mile boundry was to obtain a fishing permit from the Secretary of Commerce, to agree to accept and finance U.S. observers on their ships, and to set quotas through the U.S. State Department for each species they expected to catch.

Naturally the foreign countries agreed to finance observers and they agreed to set quotas. They had fish to catch and money to make. As a result of this Magnusun bill, the foreign vessels looked like honey bees swarming around the queen mother right off the North Pacific Coast.

And if foreign fishing companies went into a joint-venture with American trawlers, they didn't have to sit out beyond the two-hundred mile zone. They could steam right up to within three miles of the coastline. These joint-venture companies were legally raping the ocean with their great nets and factory ships. The Japanese, South Koreans, Chinese, Taiwanese, Polish, and the Russians were all out there, right at our front door, hundreds of them.

If the U.S. government did not want to destroy American fishing, then why did it allow these countries to fish out there in the first place? The Magnuson Act was destroying American fishing, not protecting it. That was the truth.

The observers, who were selected to attend the thirteen day training course, were supposed to be mature, responsible biologists. But the five candidates that sat in the classroom were young men from various universities across the country. Most of them didn't know a salmon from a cod. Where were the professionals and why weren't they here? How could these young people master in thirteen days what it took a marine science student years to learn. They were only supposed to obtain daily composition of the catch; conduct maturity studies; determine the incidence of Pacific halibut, salmon, and crab in catches; and collect other data concerning fishing operations and marine resources as required? Bullshit!

These data were compiled at the Northwest and Alaskan Fisheries Center in Seattle for the purpose of estimating foreign catches of groundfish that foreign fisheries were allowed to keep, and for measuring the incidental catch of species that foreign vessels were not allowed to keep. Marine mammals, salmon, king crab, tanner crab, and halibut were among these species which were considered illegal by the U.S. enforcement agencies.

What happened if the foreign vessels didn't save these species? In fact, once they were caught in a net, how could they be saved? Were the halibut, salmon, and sea lions thrown back in the sea, dead? They probably just froze the hell out of them or ground them up into meal and called it an incidental catch. There was no way a twenty-mile drag net would be selective of a certain specie of fish. The thing was obviously sucking up everything alive from the bottom of the ocean and killing it. The observer was supposed to recognize and report any transgressions! Preposterous.

Silk wondered if he would ever tell anyone what he really thought about the credibility of the NMFS observer school...except perhaps, maybe, Cowboy. He knew the school was a farce. No one could be well enough prepared in thirteen days to observe thousands of tons of fish a day on a nine hundred foot foreign vessel. It was also clear to Silk why, at the end of the thirteenth day session, they were cautioned not to divulge to the public or to any newsperson for television or newspapers anything about their experiences as observers on foreign fishing vessels. Dr. Blackburn's words went like this, "It is not that we have anything to hide, but in the interest of relations with foreign nations...well, we must keep a good rapport. They, being the foreigners, don't want their activities broadcast to every Tom, Dick, and Harry. Their fishing activities on our coasts are not known by the general public. We are not a policing agency but a scientific organization. Your jobs are to be good ambassadors. If you have anything to say, say it in your reports. Your observations will all be analyzed by the proper people here at NMFS. They will see to it that the information provided by you is given to the right sources."

Silk knew that last statement had been a warning not to talk, well camouflaged, but nevertheless a warning. Well, NMFS didn't have to worry about him. He'd never divulge anything about what he found out until he could expose the whole scheme. That's what he was here for, to observe.

And now, standing on the trawl deck of the two-hundred foot Russian catcher boat, John Silk saw the bulk of the refrigerator/factory ship loom out of the fog ahead like some prehistoric sea monster. With her great gurney and two towering bow cranes she mounted the gray water, searching in her perversion, for every living creature in the sea.

In his classes at the National Marine Fisheries Service Observer School, or NMFS as it's commonly called, Silk had read about these enormous ships manufactured in Japan for the Russians, but he was still surprised at the immensity of the fishing operation. The words from one of the many texts came back to him.

"These floating cities are as efficient as the best shore-based industries," the author had said. "Consider a Soviet factory ship. This high-rise titan takes aboard the catch of twenty to forty trawlers as many as four at a time. In one day this single ship can process and store 400,000 pounds of hake, produce 300,000 pounds of fish meal, fillet and freeze 100,000 pounds of fish, and press 10,000 pounds of fish oil. As soon as a hold-full of fish has been processed, the products are transferred to refrigerator transports that can carry as much as 14.5 million pounds of frozen fish back to the Soviet Union."

Silk could see all kinds of ships swarming around the enormous mothership: medium-sized and larger stern trawlers, small otter trawlers, and purse seiners, transferring their catches to the factory ship for processing. Close at hand was a support fleet of oil tankers, repair and salvage vessels, fresh water tankers, tug boats, and base ships moving in and among the vessels of the Soviet fleet. Then there were the reconnaissance vessels which reported to the fleet command ship the type of catch, catch size, weather, and water temperature. What amazed Silk the most was that this foreign operation was taking place in American waters.

Silk pulled the collar of the navy pea jacket up around his neck, turned his back to the cutting wind, and lit a cigarette. He took a deep drag then offered the cigarette to Fran Oliver who was standing next to him.

"That MURMANSK is a mighty big ship," he said.

"Rather! About nine-hundred feet!" Fran said taking the proffered cigarette. "Breathtaking wouldn't you say?"

"I feel like I'm going to a foreign country," said Silk.

"That you are, love," Fran said.

"Are you going to board her, too?" Silk asked.

"Hardly," laughed Fran. "My job was to deliver you to your assigned vessel, then it's off to Seattle for me."

Silk tried to see her eyes behind the sea-sprayed glasses and suddenly burst out laughing.

"Glasses are a real nuisance at sea, aren't they?" he said. "You can't keep the damn spray off them."

"Better on the glasses than on my mascara, ducks," Fran smiled.

"Fran, I appreciate your getting me into the NMFS observer program," Silk said. "Without you and Cowboy I doubt if I would ever have made it."

"Actually, you'll do rather nicely as an observer I would imagine," she said. "If the bloody task brings about a bit of a setback now and again, press on. And keep that cool, logical head of yours. Almost every bloke is successful. Joel Scott never loses a bob."

Silk smiled at the thought of Fran and Cowboy together. They were the most incongruous couple he could imagine. And neither of them fully realized what the other was after. Fran thought Cowboy was a good catch, and Cowboy wasn't about to be caught.

"Now what brings about that amused smile of yours?" Fran studied Silk's face with a dubious expression. "Something I said?" "Just thinking about Cowboy," said Silk. "I miss him." "Cowboy is a love," said Fran. "But the bloke is an absolute mystery to me. One moment I fancy I have his attention, and the next moment he's absorbed in some rubbish about that beastly boat of his."

"Cowboy's a died in the wool fisherman, Fran," Silk said. "I guess I am too."

Silk hadn't heard from Cowboy or anyone else in Port Bane since the morning he got on the bus over three weeks ago and headed for Seattle. He wondered about the mushroom hunt and if his van was still intact. He wondered if Lillian and the Locals had been caught by some narc.

Silk's mind was no longer on the observer school, or Seattle, or Port Bane for that matter. The hull of the colossus MURMANSK hovered above them. The immense vessel dwarfed the two-hundred foot catcher boat that he and Fran were standing on. He could see the great red spots of peeling rust and smell the fish. Tons of fish guts and heads dumped overboard mingled with the reeking odor of smoke coming from the smoke stack high above the gurney. The factory below the decks was canning fish meal. The stench of the whole process was almost overpowering to a fisherman used to clean air.

Silk began to gather his luggage together. One bag held the sampling equipment, the survival suit, his rain gear, his rubber boots, his gloves, his hard hat, his work clothes, his heavy wool socks, his warm jacket, and his knit wool cap. In another bag he had packed his better clothing and shoes which would be needed for dining in the wardrooms with the ships' officers. The transfer boat, which would deliver him to the MURMANSK, was drawing near.

Fran handed him a third bag he did not recognize.

"This doesn't belong to me," he said.

"A small gift which will do you rather nicely. You'll see," she said. "American liquor and Levi's. Those Ruskie tars think your American Levi's are smashing."

"Fran," Silk said. "You picked me up at the bus station and

61

delivered me to NMFS. You introduced me to Blackburn and set me up with the observer program. You found me a place to stay near the NOAH complex. You drove me to the doctor for my physical. You wined and dined me and took me to the theatre. You escorted me to the Russian ship, and now you give me a gift. Thank you."

"You might have thanked me in a much nicer way, ducks." Fran cocked her pretty head to one side and raised herself to her tiptoes. She put her red lips close to Silk's ear. "You might have given me a jolly good romp once, at least bloody once."

Silk put his hands on her shoulders and looked directly into her eyes.

"Fran, you are beautiful."

"Your skiff is alongside, Mr. Silk," said the first mate coming to the rail. "Please have your gear ready to board her."

"Off you go," Fran said.

Silk handed his luggage to the Russian officer and smiled at Fran.

"Goodbye, Fran," he said and climbed down to the skiff.

"Goodbye, John Silk." Fran smiled mischievously. "Give my love to your friend, Joel Scott."

CHAPTER

8

THE MURMANSK

The moon's shimmering rays turned the tops of the undulating swells to long shards of silver. A huge shadow, its top brightly lit by spotlight eyes, moved eerily across the horizon. The MURMANSK seemed unaffected by the sea's force. Unlike the small trollers Silk was used to, the Russian mothership neither shuddered nor groaned as she plowed her way forward, her huge turbines turning heavy, steady.

John Silk stood at the rail on the officer's exercise deck smoking his first cigarette of the day. Except for a swooping seagull or two, he was alone. He didn't know when the Russians found time to exercise, but they certainly weren't here this morning. In the distance he could see the lights of an approaching stern trawler, his first cod-end transfer of the day.

After a quick breakfast of black tea, homemade bread spread thick with real butter, and a slice of cold lamb, he picked up his gaff and took his position on the trawl deck near the forward shute. He watched the large net from the trawler below creep up the stern ramp and rise above him. Soviet fishermen in hard hats stood nearby with scoop shovels to shovel the fish spilled on deck into the well which led directly to the conveyor belts below. In a minute they would be swimming in dead fish.

The large cod-end snapped open, and every kind of fish in the Pacific ocean slid down the shute and covered the deck. There were rock cod, ling cod, hake, crab, starfish, brown bombers, canaries, flat fish, salmon, and halibut. Silk gaffed aside three illegal coho, a Chinook and a large halibut, but he had seen more go down the shute to the processing room. As soon as the last fish had emptied from the cod-end, Silk ran below to estimate the number of illegal fish that had slid past his gaff.

The stench of the processing room was almost overpowering.

63

Eight Russian workers stood on either side of the conveyor belt. The first pair slit the bellies open; the second pair sucked out the guts with vacuum hoses; the third pair washed the slime from the skin, gills, and cavities with salt water jets, and the last pair vacuumed the fish a final time and laid the trimmed and dressed fish on a belt moving toward the freezers. Silk and the workers were continually splashed with blood, scales and fish guts. He estimated the catch size, the density of fish pack and measured the fifteen ton bins. He checked the fillets being processed and estimated the freezer weights. By midnight, he fell into his bunk exhausted.

The transfer itself looked like a ballet, especially at night when the giant floodlights exposed every cog moving in unison. Each movement of man and machine was coordinated for the single purpose of hauling in and emptying the net. Every man in the unloading process seemed trained especially for his job. Below, the cutters who headed and gutted the fish moved with quick efficiency, seldom speaking except during their thirty minute tea breaks.

When Silk had first begun his job as an observer, the Russian fishermen were especially co-operative. Although none of the workers understood English, when he pointed out an illegal fish, both the deck crews and the cleaning staff smiled and quickly set the fish aside. Then one day he had involved himself in an argument with the trawlmaster who gestured angrily at an illegal halibut, a big one, and kept repeating "many rubles".

After that experience the illegal fish above and below decks became more and more scarce. In fact, now when he ran below to check the conveyor belts he could swear that the Russians were using the same small salmon over and over for him to find. And instead of smiles when he came below, he was greeted with scowls and silence. He knew the waste should run at least one percent, but now it wasn't even one percent of one percent. But he could never prove that the Russians aboard the ship were hiding illegal fish.

Of course, he had written his observations in his reports to NMFS, but as yet he had received no word from headquarters. He was beginning to realize that being a foreign fisheries observer for NMFS was a paper job. There was no way he could be in two places at once. There should be at least three expert observers on each foreign ship. He had written that observation in his reports, too.

Silk thought he knew the problem. He was the problem! Everytime he found an illegal fish the Russians lost money, and they were not used to losing money. That was an obvious fact. Foreign fishery observers were put on ships to observe, not to find illegal fish. They had thought he would grow tired of his job, so they were co-operative the first few days. Now that they knew he meant business, they had become sullen and uncooperative.

After three weeks on board the MURMANSK, Silk understood why certain observers were selected and why they cooperated. Working as an observor was a better life than most of the young recruits had ever lived. Their facilities were almost as exotic as one would find on a pleasure cruise. The officer's cabin was about the size of a small motel room with a closet, adequate bunk, and port holes through which one could always see the ocean. The food was excellent; meat at every meal, either lamb, beef tongue, steak, or roast beef, all deliciously prepared with rice and sour cream as condiments. Every Sunday there was a chicken dinner served with rice and fresh vegetables. Chunks of raw onion and whole pods of raw garlic were served for snacks, and black tea or coffee was available twenty-four hours a day.

As with all observers, Silk dined in the mess hall with the officers. The majority of the officers appeared to be friendly but quietly reserved, and most of them knew the English language when they decided to talk. The commissar was the least liked on board ship. Most of the time he seemed to be watching the men individually and at a distance. Dr. Dimetriov, the fleet's surgeon, looked like a mad scientist with his steel-rimmed glasses, and white, electrified hair. He was willing to communicate, except that his English was so limited he could only discuss music or literature with questions. "Do you like," he would say, then use nouns like STEINBECK, CHOPIN, or perhaps Rock and Roll. But when Silk tried to explain how he felt, Dr. Dimetriov nodded his head wisely as if he understood every word.

Silk had met only once with Captain Alexi Semyonov. The tall man with the mustache had stared directly into Silk's eyes like he wanted to tell him something but couldn't.

It wasn't until an American joint venture boat ran over his own cod-end during a transfer that Silk felt secure enough with the men to begin asking questions. The hundred foot ship was hopelessly entangled in its own cod-end, and the main cable attached to the net and the drum on the MURMANSK had to be cut. There was some discussion and fear between the welder and the Russian fisherman about the dangers of such a task. At last Silk had come forward and taken the welding torch from the welder's hand. As a hundred of the crewmen watched at a safe distance, he had cut the cable. There was a snap like a cannon shot, and one hundred feet of cable flew back across the deck and wrapped around the wench drum. The heavy line missed Silk by an eyelash.

If the steel cable had hit him, it would have cut him in half. As he turned around, shaking at his near death, a cheer went up from all the Russian fishermen. While a diver was sent below to cut the net from around the catcher boat's props, Silk was toasted with glass after glass of vodka in the officer's mess hall. He thought most of them were talking about his exploit with the torch until he heard one English sentence. "The American captain will pay $5,000 American dollars for new net and lose $3,000 fish." the smiling fourth mate said.

One by one he began to question the officer's of the MURMANSK whom he knew had some command of English. He knew they understood his questions; yet, every time he asked about the possibility of a Russian ship striking a fishing boat, the answer was the same. "No. No," they would answer. "That is not possible."

"But with so many boats in the vicinity, isn't it possible that you could have such an accident?" Silk pursued. "Especially if there were fog."

"No," they would answer solemnly. "We have constant surveillance by all ships in the fleet. And we have good radar. No, it is not possible. If we had struck a boat, everyone would know."

By the time he had questioned the last officer, Silk knew that he would not find out about the CRACKANOON on board the MURMANSK.

He began to notice a distinct change of attitude toward him. When he was around the fishermen, they would turn their backs on him and talk to each other in Russian as if he weren't present. And now the officers avoided him. When he came into the mess hall to eat, they would excuse themselves one by one. He decided that they were afraid to talk to him, or to sit near him, and would tell him nothing even if they knew.

Silk decided to question the American joint venture representative, Roberta Potts. Roberta was short, stocky, and very popular among the men who seemed to revere her as some kind of American beauty queen. They couldn't pronounce her name, so they affectionally called her Rob. She spoke perfect Russian and had been with the fleet on one ship or another for three years. Her short-clipped blonde hair and her authoritative voice gave her the demeanor of a trained military officer. Silk had watched her standing in her tight Levi's and seaman's sweater between an American catcher boat captain and the management representative of the Russian ship translating back and forth. The haggling over fish prices always followed the Soviet purchase of the American catch. At the end of each discussion a price was finally agreed upon, and Roberta invariably came out of the verbal tiff a good ambassador on both sides. She was well respected and admired by all the officers whom Silk thought secretly wanted to get her into a bed. He thought she would surely know, with her connections, whether any ship in the fleet had hit a small American troller.

One evening Silk found himself alone with Roberta in the officer's mess. This was a good opportunity to ask the question. With three hundred and sixty Russians aboard, it was difficult to talk to her alone, unless, of course, that person went to her cabin. Silk had never been invited. In fact, Roberta to this point had ignored him.

"Roberta," Silk introduced himself. "I'm John Silk. I'm the new foreign fisheries observer, and since I'm also American, I thought it was time we got acquainted."

"I'm called Rob on this ship," she said curtly. "And I know who you are. There's an American catcher boat due here any moment, and I have to be on deck."

"I understand that," said Silk. "But before you go, I'd like to ask you a question."

"Go ahead and ask," said Roberta, impatiently standing up and slipping on her heavy pea jacket.

"I'd like to know whether or not you've heard of any collision between a Russian ship and an American troller recently?"

To Silk's surprise the woman suddenly sat back down and slammed her clipboard on the table.

"Let me give you a piece of advice, mister," she said vehemently. "You're on a Russian ship now, and your role is to be an observer and an ambassador. You're hired to do a necessary job and not to ask questions. These people don't like questions, or haven't you noticed? Russian ships running down American fishing boats? The answer is no. I haven't heard of any incidents like that, and I wouldn't tell you if I had. If I were you and knew about such an incident, I wouldn't mention it. If you keep prying into matters that don't concern you, I can promise that you won't be an observer long."

After Roberta had left the mess hall, Silk decided he didn't like the woman at all. Like most people who work for someone else, she was more concerned about her job than the truth. However, she had served one purpose. She had warned him about asking questions overtly, and from now on he would be more cautious.

He decided to keep his mouth shut and his eyes open and prepare to finish out his tour. That was when he noticed another presence always nearby; around the corner of a bulkhead; near a Russian fisherman; across the table in the officer's mess; and even on the observation deck when the cod-ends were being transferred from the catcher boats, Sergi Malenkov, the Soviet commissar and a known member of the KGB, was shadowing him.

'So it's all out,' thought Silk to himself. 'They're on to me, and they have reported to the commissar. I'll never be left alone again, and if old Malenkov doesn't shoot me himself, I'll probably be dumped overboard. But this is definitely the end of my sleuthing.'

Silk had been alone on the ocean for five days at a time, but he had never been so cut off from humanity. Now he was treated like the worst kind of traitor and shunned by every Russian fisherman on the ship. On top of that his longevity looked a little shaky. He had asked the wrong questions, and he was at their mercy.

Malenkov was omnipresent, always standing nearby with his hands in the pockets of his furlined jacket, his eyes staring directly ahead in case Silk happened to look up. The atmosphere aboard the MURMANSK was heavy and threatening.

One night after a particularly long day Silk was standing on the observation deck having a final breath of ocean air before he fell into his bunk. Suddenly there was a Russian fisherman standing next to him smoking a cigarette. He had a knit cap stretched down low over his forehead, and his pea jacket collar was turned up over his chin, so that Silk couldn't see the man's face.

"The captain requests you come to his quarters. Now, please,"

68

the man whispered in broken English. Then he disappeared in the darkness beneath the base of one of the giant floodlights.

'That does it,' thought Silk. 'I have been reported to NMFS headquarters. Now I'm going to get my walking papers.'

He located the captain's quarters on the first level beneath the stern bridge and knocked on the door. He didn't know what to expect, but if this was to be the end of his trip, then he intended to confront the Russian captain head on and ask him whether or not he had any knowledge of the sinking of the CRACKANOON. The least the man could do was to deny the incident like all the others had done.

"Please come in," the voice spoke in accented but excellent English.

As Silk opened the cabin door and stepped inside, he was amazed at the captain's quarters. The room was sparsely furnished but elegant. An antique Cossack rug lay on the floor in front of a teak desk and leather chair. Rich, tapestry hangings covered the walls. On the bed was a coverlet of blue silk with a gold, hand-tied fringe.

Silk figured that this was probably the entire sum of the captain's worldly possessions. Memorabelia that he had collected over the years; things he could transport from ship to ship.

The captain sat on the end of the bed smoking a Russian cigarette. His clothes were wrinkled. His sockless feet were stuck into old cloth slippers. His graying black hair was disheveled, and a three day growth of black shadow on his face was a strange contrast to the thick mustache that drooped on either side of his thin mouth. For a moment there was silence while the man smoked.

Silk stood inside the door, waiting for whatever was to happen at this midnight meeting. The captain seemed totally preoccupied. As he smoked, he stared fixedly at the rug, his brow furrowed like he was in deep concentration. Finally he looked up and stared directly into Silk's eyes.

"Sit, please," he indicated a straight-backed chair next to the desk.

Silk sat down and waited through another long period of silence, broken by several deep sighs from the captain. At last Silk spoke.

"I believe you sent for me, captain," he said. "I am John Silk, the foreign fisheries observer assigned to your ship."

"I know who you are," said the captain. "I know everything that happens on the MURMANSK, and I know why you are here. You are here as a spy for the United States government to find out about the sinking of an American fishing boat."

"So you know about that," Silk was surprised at the captain's candor.

"I have ears," said the captain squashing his cigarette in an alabaster ashtray. "You have been asking questions of the wrong people. You should have come to me first if that is your mission."

"I'm not on any mission for the United States," said Silk. "I don't work for any organization, and I'm not a spy for the CIA or the FBI. I'm a fisherman from Port Bane, Oregon. About a month or so ago I was out here fishing when one of your ships came right through the Port Bane fleet. That night two of my friends disappeared and neither they nor their troller has ever been seen again. I think that a ship ran over them, and I also think that this isn't the first such incident. I just wanted to see for myself what you people were doing out here. So far, I have found that you are fishing inside the American 200-mile limit and catching a lot of fish. I don't agree with the policy, but since the United States permits it, and the joint venture ships support it, there is nothing for me to say. But why have you invited me to your quarters this time of night?"

Captain Semyonov lit another cigarette, took a deep drag and exhaled.

"As an American fisheries observer, you have done your work admirably," he said. "But, you see, it is your questions about the fishing boat that have caused this summons. Most observers do not ask questions. Russian fishermen are by nature suspicious, and they don't trust people who pry into matters which do not concern them. Such practices threaten their livelihood. Their living is the tons of fish they bring in and process. They are not paid to think, only to work. Matters that do not concern them cause them to worry."

"If I have frightened your crew," said Silk, "I apologize. I didn't know who else to ask. Obviously, the workers know nothing about the missing troller. But now the officers will not talk to me, and I seem to have picked up a shadow. Your commissar, Malenkov, follows me wherever I go. I can't even take a piss without his waiting outside the door."

The captain stood up and paced the floor. He twirled the large world globe standing in one corner of the room and looked long at the silk map of Soviet Russia which hung on the wall behind his desk.

"I must explain to you about Comrade Sergei Malenkov. To understand the role of the political officer one has to be a Russian. Malenkov is a KGB officer appointed to report to the Soviet government any violation of regulations. He is officially subordinate to the party committee, and after every trip he must issue a political report to the Maritime Division of the KGB listing any irregularities in the crew's performance of their duties. His main job is political indoctrination of the crew, but he has the title of commissar which is equal to that of first mate." The captain sat down again and faced Silk.

"You see, the political officer is literally the scourge of Soviet seamen. In the merchant marine the sailors have termed the official title of POMPOLIT into the insulting nickname POMPA, the Russian word for "pump", implying that their task of ideological indoctrination is like pumping propaganda into the sailor's resistant heads. In the fishing fleet, they are called "commissars", with an overtone of contempt, and sometimes "invalids" because they are so totally incompetent at any seaman-like work.

With rare exceptions, they are seamen with no understanding of the sea, and no sense of responsibility for the safety of the ship and the lives of the crew. A commissar is only duty bound to report openly at meetings of the party committee, to which he is officially subordinate, and to report secretly to the Maritime Division of the KGB. He is also required to write so-called political reports. He can even veto my orders. All this gives him practically unlimited power over the crew. That is why the fishermen are afraid. Often the captains themselves go in fear of the political officer."

"I can see that the Port Bane fishermen are not the only fishermen that have a piece of hell," said Silk.

"We have had our ups and downs," said the captain. "Malenkov is a blockheaded gorilla with no more than five years of schooling, and even that is doubtful. You see, I had Malenkov thrown off my ship once for drunkenness, and he never got over it. He tried to throw his weight about, and I waited until he overstepped the mark. Then I reported him to the party committee. I had him taken off my ship. That is why he is after me, to get even for this humiliation. So, when the assignments were being passed out at the beginning of this tour, Malenkov went to party headquarters and created doubt in their minds about my capabilities as captain and about my loyalty as a party member. Then he requested to serve aboard the MURMANSK in order to watch my activities. He is to report back to the committee his findings. He is planning to find some irregularity in my command to discredit me, and if he can't find one, he will likely invent one."

Silk listened. He was amazed at the vehemence in the captain's voice. It was evident that he hated commissars, especially Malenkov.

"Do you think he may be eavesdropping on our conversation now, or could he have wired your quarters?" Silk asked.

"I doubt he would go that far, but I have heard of stranger activities on Russian ships," said the captain. "Tonight I sent him to his cabin to write out an official report on you, so I do not think he will bother us for a while. Yet, you never know."

"When you sent word for me to come to your quarters, I was certain that it was the end of my job," Silk said. "That you wanted me to get the hell off your ship."

"That is precisely the request I make," said the captain nodding his head. "You must leave this ship as soon as possible. But that is not the only reason I sent for you. I sent for you to face you with a confession and I pray that you do not despise me for it." He took a deep breath. "You see," he said, "I am responsible for the sinking of the American fishing boat."

The admission of guilt by the captain stunned Silk, and for a brief moment he hated the man. This sonofabitch was responsibile for the death of Cap and Trevor, and now he had admitted it.

Silk felt like tearing into the man. But Captain Semyonov's quiet admission conveyed another emotion, a kind of plea for understanding, or a final confession that purged the spirit, like a criminal who finally admits to the officer that he was the one who committed the crimes.

"How did it happen?" Silk asked.

"The incident should not have happened," said the captain looking directly at Silk with his intense blue eyes. "I had not slept for twenty-four hours, and I thought I was losing my effectiveness. I thought if I slept just two hours, I would be alert again. I gave charge of the ship to my first mate Vladimir Usyenko. I gave him the proper co-ordinates to follow and he followed my instructions. However, he failed to allow for the drifting of your American boats. In the fog he could not see and neither could your American captain.

I had just come back on the bridge when I saw the shadow of a small vessel. I ran to the wheelhouse to stop engines. The collision occurred, I felt the slight impact, and I knew we had not struck a natural object. I feared the worst, but I did not know until you began to ask questions that it was the MURMANSK which killed your friends."

"Shit," said Silk in frustration. "You knew you had hit a boat, so why didn't you report it?"

"It was not that I am a coward," said the captain. "It was best not to report it. You must understand something about the Soviet system. You see, we are only allowed to function as fishermen because of the Soviet government. Their only interest is the amount of fish we catch and process. If an accident is outside the process of fishing, we are not allowed to involve ourselves. We cannot even help our own ships if such a maneuver interferes with our mission.

The Soviet government only requires that we fulfill our quotas; no other activities are tolerated even if those activities result in loss of life. Reporting that we hit an American boat would have brought attention to the Russian fleet. There would have been an investigation. That investigation would have taken time, and we could not have fulfilled our quota. Such a circumstance would have cost every man aboard ship his job, and my first mate, Usyenko would have gone to prison. He was in charge of the ship when the boat was struck, but I was responsible even though I was not on the bridge. I could do nothing to save the boat. Nothing. Only Usyenko and I understood what had happened, so I kept silent to protect the innocent fishermen. Usyenko is loyal and will say nothing."

"But I will say something," said Silk. "I don't understand why you are telling me this when you know full well I intend to report it."

"I was hoping that you would not," said the captain. "I have a plan and I need your help to carry out this plan. If you will listen, I will tell you about myself, then maybe you will understand. Usually when one commits evil, he is an evil man to begin with. But for me the case is different. It was not until I had been a fisherman for many years that I understood the full extent of my crime. Following this observation, I then realized it was too late for me to change my di-

73

rection. Now, I understand. After this latest catastrophe I must pay. I must make retribution."

"What the hell are you talking about?" asked Silk lighting another cigarette.

"I have been a captain in the Soviet fishing fleet for over thirty years," the captain said. "I do not like what I have seen nor what I have had to do. I, myself, have been responsibile for much destruction of the world's fishing grounds.

For some time now I have been thinking of what I must do to make, how do you say it, make amends for the crimes I have committed. Always in the past when I have tried to save lives or prevent wasteful practices by our fleet, I have been stopped by the KGB or the very structure of the fishing strategy of the Soviet government. They do not care what lives are lost or what fishing grounds are destroyed as long as we meet the quotas they assign.

After the night of the tragic accident with the American boat, I made a decision which was very painful to me. In order to live with my conscience, I must give up my position as captain of the MURMANSK, and I must leave the ship and my homeland, which means leaving Russia.

But how shall I accomplish this task? In Russia when you achieve the rank of captain, you cannot just resign your commission and take a new job. You work directly for the government, and your position is for a lifetime unless Soviet rules are violated, then they remove you. Likely they will go through your files and find past crimes, then you are sent to prison.

I did not relish that choice. Because of previous transgressions against the KGB and making my own decisions, I was already under suspicion and assigned a watchdog. So what should I do? The MURMANSK will be at sea for at least another year, and I could not just walk off the ship.

Then you came aboard and began asking questions. At first I thought you had found out about the collision, and I was afraid you would report me. Then, it became obvious that you were merely seeking information, and I made a decision.

I could not just walk off the ship without Malenkov's knowledge. He might report me to the authorities in the U.S., and having little knowledge of your country and no friends, the CIA would soon apprehend me and deport me back to Russia. As I have stated, Comrade Malenkov is a KGB agent assigned to this ship especially to keep an eye on me because I am already under suspicion. I must be very careful. But with your help there might be a possibility for me to escape. Therefore, I must know whether you will report me or help me."

As Semyonov was speaking, Silk's mind was in a turmoil. He had come aboard the MURMANSK to find out the truth about old Cap and Trevor. He had all but given up hope and resigned himself to the fact that his mission had failed and that he would be dismissed for asking the wrong questions. Now, suddenly, Captain Semyonov had come forth and confessed to being responsibile for the sinking of the CRACKANOON and the deaths of Cap and Trevor. All he had to do was to report the information to the proper authorities, and the matter would be settled. The captain would be punished, and he could go back to Port Bane with the information.

But what would that accomplish? So everyone knew what had happened to the CRACKANOON. What would the people in Port Bane do about it? What had the fishermen ever done to protect their fishing rights? Nothing. They would continue trying to scratch out a living on the sea; and Captain Semyonov would be sent to prison. Nothing would change, and reporting the captain would not bring old Cap and Trevor back. The captain more than anyone seemed stricken over the tragic accident, and the man wanted to do something to ease his conscience. Maybe, if he helped the man to defect, he could really make an impact on the NMFS and the American fishing policies. He was sure if he got the captain off the ship, this man could help him.

After a long period of silence Silk looked at the captain's proffered hand.

"Captain," he said, "I will help you get off the ship, but I'll expect a favor in return."

"Ask the favor," said the captain.

"I'll expect some help in exposing what the Russian ships are doing off our coast," he said. "At some time in the future I shall ask you to tell what you know about the Soviet fishing operation."

"Thank you," said Captain Semyonov with a sudden smile. "I shall not disappoint you. But first, we must plan our escape. Remember, Comrade Malenkov is already suspicious, so the task will not be easy. We have talked too long. Go now. We must meet later. Wait for my signal."

CHAPTER

9

MALENKOV

Silk was sitting at the end of the long table in the officer's mess with a copy of a Russian magazine SOVIET SPORT before him. He couldn't read Russian, and he wasn't interested in Soviet athletics, but he was hoping that one of the officers off the second shift would slip him a message from Captain Semyonov.

Seven days had passed since he had talked to the captain, and he had no proof that the conversation had even taken place. During that time he worked harder than ever, stretching his schedule up to sixteen hours a day in hopes of regaining some rapport with the Russian fishermen. But Comrade Malenkov's continued presence had made him a pariah on the ship. The workers were afraid of the Pompolit, as he was entitled, and whatever surveillance system he had going aboard the MURMANSK.

Silk's thoughts were interrupted by the very man he was thinking about. Commissar Malenkov suddenly walked through the door and sat down next to the officers. The officers kept their eyes on their food while Malenkov's little black eyes, piercing through the heavy folds of skin, surveyed the men before they focused their gaze directly on him. The stare was churlish.

'That politico bullet-head hates my guts,' he thought. 'I'll bet he was waiting outside that door all the time to see if one of the officers talked to me. If Semyonov and I are going to get off this ship without his knowledge, we'd better have a plan that's foolproof. How are the two of us going to hide from three hundred and sixty frightened seamen. He's probably already threatened every man on board the ship. One slip and we'll have both the Russian and American authorities on us.'

Silk slowly got to his feet and excused himself. There was no doubt about it; he was a marked man. But how was he to extricate himself from this dilemma? He hadn't counted on a captain wanting

76

his help to defect when he came on board. All he was after was the truth about the disappearance of the CRACKANOON and other trollers which had mysteriously dropped out of sight. It was time for the Port Bane fishermen to put a stop to such foreign misconduct. Time to act before it killed every fisherman and sunk every boat in port. He was already earmarked as a trouble maker; the Russian's wouldn't come near him and the KGB was on his neck.

As he walked through the hatch door, he was met with a summons to the radio room for a message. He was surprised when the radio man handed him a computer letter written in English.

"Attention: John Silk

This is to notify you that your job aboard the MARMANSK will be cut short. We have received a request from Dr. Hubert Huxley of the Marine Science Department of Zuma, Arizona, to serve as an observor on board a Russian ship in order to gather hands on information for a study he is doing on joint venture. Since we have only the MURMANSK available, we are replacing you as soon as transportation for Dr. Huxley can be arranged. I'm sure you can understand our position. We thank you for your dedication to our service.

Yours sincerely,
Dr. C. F. Blackburn
Director NMFS"

Silk smiled. Well, here it was. He was fired for doing too good a job. He knew for a fact that there were many foreign ships without an observer where Dr. Huxley could have done his "study". He wondered if the radio man had understood the message. The best thing to do was to take the letter to his cabin and place it somewhere out of sight.

When Silk stepped into his own cabin, he felt the hackles on his neck rise. At first he couldn't tell whether it was his imagination or his distrust of the KGB officer, but something about his room felt different. The door had been locked, but certainly that wouldn't stop someone like Malenkov from entering, whether to plant a bug or to search the room. Hadn't Semyonov said he wouldn't put it above the KGB to hide a recording device in his own cabin? Carefully, Silk searched the room for any kind of hidden recorder. The lampshade above his bunk, the closet, the mattress, the small desk, and even his oil skins and boots turned up nothing. Then he looked behind the photograph of himself standing next to the CRACKANOON. A small corner of the frame had been peeled back with a razor or a very slender knife. There was still a splotch of damp glue on the corner of the frame. Malenkov, or someone sent by him, had searched his room. He had nothing to hide, and no foreign substance like dope. What did they think he could possibly conceal behind a small picture frame? Had Captain Semyonov lied to him and set him up for the KGB?

He felt the doubt of Semyonov turn into a pang of fear. If he couldn't trust the captain, then whom could he trust? The whole meeting might have been a set-up to get him off the MURMANSK. But then why did Semyonov confess to sinking the troller if he knew Silk could report him?

Images of his being thrown overboard seeped through his mind. On the ocean, if his fishing troller ran into trouble, there were usually steps he could take to save himself. He could fix a fuel line, repair a hydraulic system, or radio for help. But on a foreign ship his problems were more subtle. Now the KGB, the NMFS, and possibly the Russian captain had thrown him a curve. What options were left for him?

It was past 3:00 A.M. that same morning when Silk got his answer to the question; Malenkov was standing on the stern bridge watching him.

The MURMANSK had unloaded three Russian trawlers in a driving wind, but when the fourth maneuvered alongside, the sea was obviously too rough, even for the Russians to keep working. The man was still standing there when the operation was called off, and Silk walked off the deck for a cup of black tea before going to bed. If he were going to survive this trip, something had to be done about the KGB officer shadowing him.

He had just decided to go directly to Captain Semyonov, regardless of Malenkov, when he heard a soft knock on the door. He was out of bed and across the room in three steps. He switched on the light and waited, holding his breath. That was when he noticed a slip of paper on the floor. The note written in neat black ink read, "Stern bridge. Immediately! Semyonov."

Quickly, Silk slipped on his Levi's, a heavy sweater, his peajacket, and a wool cap. In less than five minutes he was top side in a

78

forty-knot wind and driving rain. The entire deck beneath the bright flood lights seemed deserted. Then he saw the dark figure standing against the outside rail of the stern bridge.

"So you came!" the captain said staring straight ahead into the solid wall of white spray being swept off the top of the waves.

"Damn right I came," said Silk as loudly as he dared. "Where the hell have you been the last seven days?"

"I have been trying to escape the shadow of Comrade Malenkov," said the captain. "Step back against the stern bridge so we will be unobserved."

"Where is Malenkov now?" Silk faced Semyonov.

"He is in his cabin, so we are safe here," said the captain. "I am having him watched."

"Who is watching him?" asked Silk.

"The only one I can trust, Usyenko. His future is at stake so he will do as he is told," said the captain. "Have you given the matter we discussed some thought?"

"I have thought of little else since that night," said Silk. "But time may be running out for such a plan. I received a radiogram this morning from NMFS to the effect that I have been dismissed as an observer on the MURMANSK. I figure we have only a few days before they take me off. So, if we are to get either of us off this ship, it had better be quick."

"I realize the probabilities of the success of such action is small," said Semyonov. "With Malenkov constantly at my heels, only with your help can we succeed. What are your ideas?"

"I have been looking at several of your lifeboats," said Silk. "They look like something out of a JULES VERNE novel."

"You mean the old ones," said the captain.

"The ones that are shaped like a bullet with a hatch in the center," said Silk. "You could get inside to run the diesel engine, and I could stick my head out to steer. Once you were inside, anyone looking would only see me and think I was escaping. I doubt that the crew would try to stop me if I was by myself. Once we reached shore I have friends who would help us."

"Those boats are not trustworthy," said Semyonov. "They are outdated, and very few of the engines run. Besides, who would lower us?"

"You mean we're out here without lifeboats?" asked Silk.

79

"You do not understand the Russian philosophy," said the captain smiling. "The government does not wish to spend rubles to protect fishermen, and besides, a refrigerator/factory ship the size of the MURMANSK has never sunk. Why waste money on lifeboats which will never be used? No, I'm afraid your idea will not work."

"Well, what have you come up with?" asked Silk.

"I, too, have given the problem much thought," said Semyonov, "but I have not as yet reached a solution. I am well known throughout the fleet, so there is very little chance that we could make our escape on a Russian catcher boat or one of the supply ships. I know for sure that Comrade Malenkov is preparing a report to the KGB that you have asked questions and talked to me. That is grounds enough for me to lose my license as captain, since I was reported once before by another KGB commissar of being an American sympathizer. I am sure that he has alerted every KGB man in the fleet, so no one on any Russian ship is likely to help us. If I should radio the United States Coast Guard, Comrade Malenkov will have me arrested before I can talk to the American authorities. If he knew for sure about the American fishing troller, he would already have arrested me."

"It seems to me," said Silk, "that Malenkov is the main problem. I'm sure he has searched my cabin."

"Oh, yes," said Semyonov. "Comrade Malenkov is very thorough. He is watching both of us constantly."

"Is he armed?" asked Silk. "I wouldn't put it past him to shoot me in the back or have me thrown overboard."

"That is not the way of the KGB," said Semyonov. "He would prefer to watch us until we make a mistake. Then he will arrest me and have you turned over to your government. But in answer to your question. I doubt that he carries a weapon. However, should he need arms, he has access. There is an armory aboard ship. We must be very careful and not make a mistake. Our time is almost up. I told Usyenko fifteen to twenty minutes, then he is to report back to the bridge."

"What am I supposed to do?"

"Do nothing except your job, and talk to no one. Wait for me to contact you again. Good night."

The second contact with Captain Semyonov occurred three days later, much sooner and under different circumstances than Silk had anticipated. He had just checked the last cod-end transfer from a Russian trawler, soaked to the skin and dog tired.

As he had come off the trawl deck on his way to the officer's mess, he noticed a familar figure in oil skins following him. He ordered black tea and lingered over the cup for some time, hoping the man would come into the officer's mess. As he sipped his tea, he watched a couple of junior officers with whom he was familiar. No one else made an appearance. At last he gave up and walked out of the room to the passageway that led to his quarters.

He had just unlocked the door to his cabin when the man in oil skins rushed down the passageway, stepped in front of him, and entered the room. Silk closed the door and turned to face his intruder.

"Alright," said Silk to the man who now was pulling off the oil skins. "Who are you and what do you want?"

When the man raised his head, Silk realized why he had thought the fisherman seemed familiar. It was Captain Semyonov himself.

"I did not mean to frighten you," Semyonov whispered excitedly. "But circumstances have changed. Malenkov knows about the American fishing boat. He knows about my meeting with you. Usyenko, under pressure, has confessed. I must leave the MURMANSK immediately, or I will be arrested and sent back to face charges in Russia. Malenkov has won."

For a moment Silk was stunned. Then the coldness of a mind coping with the impossibility of dealing with the vicissitudes of the ocean set in. There was a problem, but it was not insoluble. The captain was still free, and if the proper decisions were made, he could get him off the ship. He just didn't know how.

"Sit down," he said. "I'll get out of these wet clothes, then we'll make a plan. Drink?"

"Vodka," said Semyonov.

Silk went to his footlocker and picked out a bottle of Russian vodka and set it before the captain. He had traded a fifth of good American scotch for the bottle, but he didn't know the quality of the Russian vodka he had received in return. Semyonov didn't seem to mind, however. He uncorked the bottle and poured himself an eight ounce glassful. By the time Silk had undressed and put himself in a dry pair of Levi's and sweatshirt, the captain had finished half the bottle.

"Now," said Silk pouring himself a shot of scotch. "Tell me

exactly what happened, and then we'll figure out what moves we can make."

"This predicament is my fault," said Semyonov grimacing over a large swallow. "I underestimated Comrade Malenkov. I did not foresee that he would go so far as to threaten Usyenko. I became suspicious when two other KGB agents came aboard to talk with Comrade Malenkov. About what I did not know. I only knew that they were having a meeting. Then, when Usyenko did not show up in the wheelhouse to relieve me, I knew there would be trouble. I left the ship in command of the second mate, even though he is young and inexperienced, and went in search of Usyenko. In all my years that I have served with the man, I have never known him to be a minute late for his duty.

I sent my communications officer to Usyenko's quarters, and he informed me that Usyenko was under arrest, and that there was a guard stationed at his door. That was when I went on deck to mingle with the crew while they were working. Word travels fast among Russian fishermen. It seems that the steward took tea into Usyenko's quarters while they were questioning him. I guess it didn't take them long to break him down.

The KGB has efficient methods to extract information from even the toughest man. All they had to do was to threaten the family of officer Usyenko, and he no doubt told them about striking the American fishing boat and my irresponsibility in the matter. While he was serving the tea, the steward overheard Comrade Malenkov give the order for my arrest. When I heard these rumors, I knew I had very little time if I was to get off the ship. That is all I know. I waited until I saw you leave the trawl deck, and I followed you to your room. Soon Malenkov will be here with the other agents, and even you will not be safe if I am caught in your room. There is no place on the ship where I will be safe. It is a matter of time."

"But, I don't understand," said Silk. "Why would Malenkov arrest you? Usyenko was in command of the ship when it struck the CRACKANOON? No one has any idea that you are planning to leave the MURMANSK, do they?"

"No," said Semyonov thoughtfully. "No one knows except the two of us. But that does not alter the circumstances. Like I mentioned before, I have had dealings with Comrade Malenkov in the past. I know the man! He will stoop to any lie in order to destroy my creditability. In reality, I have no recourse but to leave the ship, if that is at all possible."

"I understand," said Silk. "Is there anyone on board that you can trust; anyone who might help us?"

"No. There is no one. This is a good crew. The men are hardworking and trustworthy, but they are also afraid for themselves. They will not interfere, but neither will they help us. Only you are willing to help me, but perhaps it is already too late."

"I have been thinking while you've been talking, and I have an idea. Since you cannot go back to your quarters and you can't stay here, we must have a place to hide you, and I think I know the place. Remember when I mentioned those strange bullet-shaped lifeboats where the pilot must stick his head out of the hatch to steer? It seems to me that one of those lifeboats would be the perfect place to hide you for a period of time. No one can look inside them unless they actually climb through the hatch. This is a large ship, and it's going to take Malenkov some time to search it thoroughly. I doubt that he'll think the captain would be hiding inside a lifeboat."

"But as I mentioned, those boats are not seaworthy," said Semyonov. "Besides, we would be seen if we tried to lower it."

"I didn't say we would try to launch it," said Silk. "The lifeboat is only a hiding place for a short period of time until I can figure a way to get us off this ship. Certainly we're too far out to jump overboard."

"Very well," said Semyonov," I will manage to hide in a lifeboat, but what will you do?"

"I have another idea," continued Silk. "Are you familiar with the joint venture catcher boat called the OTTER?"

"No," said Semyonov. "I have worked with numerous American boats, but I am not familiar with that one."

"Well, she's out of Port Bane, and I know her captain. He told me before I left for Seattle that he would be fishing off the coast of Oregon. If I could somehow contact him, I know he would give us a hand. If we can't get the Russian fishermen to help us, perhaps we can call on an American."

"That is good," agreed Semyonov, his enthusiasm beginning to show through his depression. "But how could you contact the American except by radio? You would have to have access to the radio room, and I know that by now Malenkov has one of his agents waiting there."

"You leave that to me," said Silk. "They're waiting for you, not for me. I'll simply think of an excuse and insist on making a call. After all, until they take me off, I am an official representative of the NMFS. Since I am an observer, they can't refuse my using the radio without calling attention to the fact that I'm an American trying to do my job, and they're attempting to prevent it. If I am refused, I'll tell them this breach of contract will be mentioned in my next report, and that I'll demand an investigation of the whole Soviet fleet. That ought to get their attention."

"It might work," said Semyonov. "At least it's the only chance we have."

"It'll work," said Silk beginning to draw on his rain gear. "But first, we have to get you inside one of those lifeboats while it's still dark. Since it's raining and the wind is up, and there are no boats to unload, I doubt there'll be anyone on deck. However, I'm afraid you won't be able to pack."

"The vodka is enough," said Semyonov. "I have been in conditions much worse."

"Well, here's some more vodka," said Silk handing him an unopened bottle. "It'll keep you warm while you are waiting."

Silk opened the door, checked the passageway both ways, and motioned Semyonov to follow him. Using the same procedure, with Silk leading the way, the two men reached the main deck without seeing a single officer or seaman. It seemed like every crewmember on board the MURMANSK was taking advantage of the crescendoing storm to rest; unless, maybe Malenkov was searching in another part of the ship. Silk doubted that the mate in the wheelhouse could tell the difference on deck between one man and another.

They were lucky. Through the giant floodlights Silk could see the outline of the lifeboat he wanted far forward. That would be the one to use if they were to escape unseen. The OTTER would transfer the cod-end, then move ahead past the MURMANSK. While the crew was working with the cod-end transfer, they could climb over the side.

Quickly the two men made their way across the deck to the lifeboat. Silk figured the force of the winds from the sou'wester at about forty-five knots and increasing. The sea was definitely too rough for the Russian fishermen to manage a transfer. Semyonov would have to remain in the lifeboat until the winds diminished, even if it were three days.

Like a cat, Semyonov climbed on top of the boat, unsnapped the hatch cover, and disappeared inside. A moment later his head and shoulders emerged.

"You'll have to stay in that thing until Cowboy gets here," said Silk. "If it's too long, I'll try to bring you some food and water."

"Go," the captain said and he disappeared pulling the hatch cover shut above him.

Silk returned to his cabin, hung his raingear on the hook outside his closet door, undressed, and dropped into bed. He needed a few hours rest before he attempted to radio Cowboy. Certainly it wouldn't do to show up so early in the morning and demand to radio a call.

He had been lying awake staring into the darkness when there was a firm knock on his door.

"Damn," he whispered as he switched on the light and sat up in bed. "I've only been here fifteen minutes and Semyonov's already in trouble."

When Silk opened the door, he was unprepared for the scowling features of Comrade Malenkov. Without speaking the broad man in a heavy leather coat lined with fur stepped past him into the room.

"May I come in?" he said in a grunting accent.

"You're already in, aren't you?" said Silk. "What can I do for you?"

"I am trying to locate Captain Semyonov. Perhaps you have seen him?" Malenkov's small cold eyes quickly glanced around the room.

"No, I have not," said Silk evenly, hoping the man couldn't detect the lie in his voice.

"The captain has not been seen since he came off duty," continued Malenkov. "He is not in his cabin or in the plant. It is important that I find him immediately. It is a matter of grave urgency."

"I suggest you look elsewhere. As you can see he is not in my cabin. You have interrupted my sleep," said Silk. He held his cabin door open to indicate that Malenkov vacate his quarters immediately. "If I should see the captain, I will give him your message."

"I wish you good evening," said Malenkov, "and I would not advise your going on deck again until the storm is over." Silk heaved

85

a sigh of relief as the big man toddled down the narrow corridor.

The steady oscillating of the huge ship in the heavy seas gradually lulled Silk to sleep, even though, for an hour after Malenkov had left, he thought sleep would be impossible.

The Russian political officer had cast a malefic spell over everyone on board the ship. Silk felt that evil spirit, and he didn't like the feeling. He didn't know what the KGB man could do to him, an American citizen, but he knew if the fat bastard apprehended Semyonov, the captain would no doubt spend the rest of his life in a Russian prison. He felt sorry for Semyonov, huddled inside the cold hull of the lifeboat while he was lying comfortable in his own bed. But the captain had a bottle of vodka to keep him warm and some hope of getting out of his dilemma now. One night of discomfort was certainly better than being arrested by Malenkov. If they managed to get off the MURMANSK, and that was a big if, he would see that Semyonov had every opportunity to begin a new life. The thing he had to do was to get to the radio room and use the radio.

CHAPTER

10

THE ESCAPE

Silk awoke with a start and stared into the darkness of his cabin. He could tell by the steady roll of the MURMANSK that the seas were still too rough for cod-end transfers; the catcher boats would not be able to unload their fish. If the catcher boats could not unload their fish, then there would be no need for ship to ship transmissions by radio and no activity going on near the radio room. Silk rolled over and reached for the light above his bunk. His mental clock had been working. If he were lucky, he would catch Ivan, the Russian night operator, who would still be on duty. He was almost certain that Ivan understood very few English words. This was his opportunity to try to contact Cowboy without creating suspicion. It might be his last opportunity. The problem lay with Cowboy. If the OTTER was in the vicinity, then Cowboy would help him. But if Cowboy had changed his mind and gone back to Alaska, then Silk's whole plan was shot. He dared not think about this contingency.

Silk brushed his teeth, dressed, and slipped into his fishing boots and rain gear. He thought it best to go prepared for work in order to make his behavior seem as natural as possible.

After a quick breakfast of sweet rolls and tea, Silk headed directly toward the radio room, just like he was going to make a call to an observer on another ship or to radio the NMFS office in Seattle. This was normal procedure for any observer, and he had made such calls in the past.

The driving rain whipped across the deck in gusting sheets. In the distance Silk could see the wind tearing the tops off twenty foot waves, churning them into whitecaps.

He ducked his head against the storm, and felt his way toward the radio room which was located on the bridge.

The bridge was deserted except for a large man in oil skins leaning against the outside wall of the navigation room. He seemed

87

to have no purpose for being there, and Silk knew the Russian fishermen were always kept busy at some task. Obviously, the man was one of the KGB agents who was waiting for Captain Semyonov. Silk hoped the agent wouldn't stop him from making his call to Cowboy. He could do nothing except go through with his original plan.

He waved casually at the KGB agent, who ignored him, opened the door to the radio room, and walked in calmly like this communication was part of his everyday activity.

"Good morning," Silk said with relief. This was the radio officer, Ivan, whom he had hoped to find on duty. The short-haired Russian was in the process of lighting a cigarette. "The ocean is too rough for a transfer this morning," Silk continued, "but I have to make a radio call to an American catcher boat." He gestured toward the large CB. "Will you put my call through?"

The Russian handed Silk a clipboard.

"O.K." he said smiling through his one day growth of black stubble. "You write."

Silk took the clipboard and wrote CHANNEL SIXTEEN, and handed it back to the officer. The stocky Russian punched the necessary switches before he handed Silk the mike. Then he resumed his smoking. Silk turned his back on the Russian before he spoke into the mike.

"Hello! OTTER! This is the Russian factory ship MURMANSK calling the OTTER. Do you read me, Cowboy? O.K."

"This is the OTTER," the voice came back over the radio. "But this isn't Cowboy, it's Mike. Cowboy's out on deck. Hang on, and I'll get him. O.K."

For several minutes Silk fidgeted uneasily while he kept the channel open. A glance over his shoulder at the Russian told him the man was not paying any attention to the call; either that, or the fellow was a hell of a good actor. Then Cowboy's voice drawled over the loud speaker, and Silk relaxed.

"Good morning, MURMANSK. This is Captain Joel Scott. What can I do for you? I have a broken hydraulic line that needs some fixin', so make it snappy. Over."

"Cowboy, this is John Silk. I'm on the Russian factory ship, MURMANSK, off the coast of Oregon. Are you in the vicinity, and are you fishing? O.K."

"Well, I'll be dogged," said Cowboy. "So you made it ol' buddy.

Fran must a done her job right. Yeah, we're fishin'. Fact is, we're headin' down your way in a few days. Our company just signed a new contract with them Ruskies, and we're gonna be sellin', probably to that same ship. Soon as the weather eases up and we get a full load, we'll be headin' down that direction. Yeah. We're just a little ways north of you right now. O.K."

"That's perfect, Cowboy," Silk said. "I'll be glad to see you, but right now I have a little problem, and I need some help. The problem is between you and me, and I don't want the rest of the fleet to listen in, so switch to our old channel we used to use when we were fishing together. How bad is your leak, Cowboy? O.K."

"Nothin' serious," Cowboy's voice was calm and easy. "Just a little crack in the hydraulic line, and we about got her fixed. She'll hold 'til we get through talkin'. I'm switchin' to 'fish on the way.' O.K."

Silk glanced at the radio man who was now engrossed in a Russian magazine. Since the man was paying no attention, he switched to channel 23 just as Cowboy's "do you read me Silk" came over the speaker. This was the channel he and Cowboy had used before when one of them had come upon a heavy school of salmon and didn't want to share that information with the rest of the fleet.

"I read you, Cowboy," said Silk quietly. "Now listen carefully to what I have to say. I have a radio officer here with me who may or may not understand English so I have to guard my language. Just remember how we used to talk about catches and see if you can fit the pieces together. O.K."

"Real clear," said Cowboy. "I got you, Silk. Say on. O.K."

"I have a big fish," said Silk. "I would like to bring that fish aboard the OTTER, as a gift let's say. The problem is that I can get no one to lower a basket or an inflatable life raft to carry my fish in. You have to help me transfer my fish to your boat. If I could manage to lower a Jacob's ladder over the bow of the ship, do you think you could have a boat of some kind underneath waiting for me and my fish? O.K."

"Hell, Silk, I don't see why not," Cowboy said. "But if I got to pick anythin' in the world I'd want right now, you can bet your bottom dollar the last one I'd pick would be a big fish! This one must be somethin' special. O.K."

"Yeah," Silk said. "We'll go into that later. What about the boat? Can you manage it without any problems? O.K."

"I just happen to have the perfect set-up for the job," Cowboy said, "a little inflatable Zodiac with a outboard motor. I can have her waitin' underneath you the minute you drop that ladder, but how're we gonna co-ordinate the time,old buddy? O.K."

"That's another problem," said Silk in the same tone he used in talking to another observer or NMFS headquarters. "We have to co-ordinate this visit now because I'm going to be busy, and I may not have another chance to use this radio. Where are you in relation to the MURMANSK, and how soon can you get here? O.K."

"I'm inside and up hill from you, about a hundred miles, I guess. But there ain't no way we're gonna unload in this weather. When that wind dies down, I figure it'll take me about twelve hours just to get there, and right now I'm only a little over three quarters full. O.K."

"We'll do this then," said Silk knowing Cowboy was now on to the plan. "As soon as the weather calms down, you radio the MURMANSK that you wish to transfer your fish. Try to time it so you get here just about dark. I'll be on the trawl deck working the transfer. The minute you have the money aboard the OTTER, move out. We can catch the OTTER in the Zodiac if we have to run for it. I'll be hanging at the bottom of a ladder on the starboard side of the bow. Have your deckhand pull underneath me. I'll have my fish with me or I won't be there. Understood? O.K."

"I got you, Silk," said Cowboy. "But it's a hell of a shot just to come aboard for a visit. O.K."

"It'll be worth it," said Silk. "This'll be the greatest haul you've ever made, but I doubt that there will be much money in it. Any questions? O.K."

"Nope," said Cowboy. "We'll be right there ready and waitin'. O.K."

"One last thing," said Silk. "This could be a little dangerous. You know, weather conditions and all. O.K."

"What are friends for," answered Cowboy. "Lookin' forward to seein' you. Over and out."

Silk placed the receiver back on the hook and smiled at the radio officer.

"Thank you," he said. "Catcher boat."

"Catch...er boat," the Russian repeated.

90

"No fish yet," Silk said.

"No fish nyet." The Russian was into it now.

"Too big waves," Silk said. He gestured with his hands in rhythm with the rolling ship.

"Too big," said the Russian repeating Silk's movement.

"Goodbye. Thank you," said Silk.

"Goodbye. Thank you," said the Russian.

Silk left the radio room, again smiling at the agent outside the door, and climbed down the ladder onto the trawl deck. He glanced over his shoulder just in time to see the bulky Malenkov enter the radio room. The bastard had almost caught him, but at least he knew where the commissar was for the moment. It would take him some time to question Ivan about the radio call he had just put through. This was his opportunity to approach the lifeboat in which Captain Semyonov was hiding. He ran toward the bow of the ship, climbed the ladder to the lifeboat, clambered up on top, unfastened the hatch, and stared down into the darkness.

"Are you all right?" he asked in a voice somewhere between a shout and a whisper. For some reason he thought the captain would be gone.

"Yes, I am fine," Semyonov's voice came from somewhere inside the hole. "Have you a plan?"

"Yes, I do," said Silk into the hollow emptiness of the boat. "I called my friend Cowboy Joel Scott on the OTTER. As soon as this storm calms down, he will come to pick us up. He will unload his catch first, then I will throw a Jacob's ladder over the starboard side of the bow below this lifeboat. Somehow we have to be hanging from that ladder when Cowboy's Zodiac picks us up. I will leave the trawl deck and signal you by knocking on the side of the lifeboat. But you still have to get out of this boat and down the ladder without the men seeing you."

"The men are not a problem," muttered Semyonov from the bow end of the boat. "It is Malenkov who must not see me. The men will be busy and pay no attention, but if Malenkov or the other two KGB men spot me, I will not have a chance."

"I'll try to divert Malenkov," said Silk. "But I have to go now before that fat bastard comes out of the radio room. He went in there immediately after I made my call. Does the radio officer without the beard understand English?"

91

"No, he does not," answered Semyonov. "You do not have to worry about him. Besides his ignorance of English, he is what you call...not too swift. He will simply tell Malenkov you made a call to another observer about weights and measurements. That is the stock answer. Malenkov might suspect, but he will never find out from Orloskovich. But you must be careful; Malenkov is probably having you watched."

"I'm certain he is," said Silk. "He came to my room last night asking questions about you, but I told him nothing. If I get a chance later, I'll try to bring you something to eat."

"No," said the captain. "It is not worth the risk. Remember, do not underestimate Malenkov. He is well trained and efficient. Now go, and I'll wait for your signal."

Silk went directly to the officer's mess for a cup of black tea. He was about half finished when Malenkov entered the room and sat at one end of his table. Silk nodded a greeting and the heavy-browed officer nodded his head in return. A moment later Silk stood up and left the room. Malenkov remained sitting, seemingly unconcerned.

When he reached his quarters, Silk discovered a man he didn't recognize lounging in the corridor. The man was dressed like a fisherman, in a heavy wool sweater and dungarees, but he had all of the earmarks of a prison guard. Probably one of Malenkov's boys. No wonder the commissar had not followed him. He didn't have to. Semyonov was right. Malenkov was thorough. From now on he would not be left alone, even for a second. The KGB troops were here.

The storm did not die down until well after midnight. Silk made several trips topside to check the weather and went to the mess hall on schedule for lunch and dinner. He did not have an opportunity to get near the lifeboat where Captain Semyonov was hiding even though he knew the man had neither food nor water. Everytime he left his cabin there was either the KGB agent in dungarees or Malenkov himself waiting around in the corridors. Even when he purposefully made a trip down into the cavernous freezer hold under the pretense of taking weight and density measurements, the man followed him. At last, in resignation he confined himself to his quarters to finish up his reports.

During the night the steady gale ceased, and the ocean calmed herself. By daybreak there were only sporadic gusts with heavy periods of rain. The MURMANSK was now riding easily on the heavy swells. The Russian fishermen, who had unloaded ships under conditions much worse, thought this was good weather for work.

By dusk, the crew of the MURMANSK had unloaded six Russian catcher boats, and Silk had measured every one of them, supervising the unloading on the trawl deck then going below to watch the heading and gutting and to take his measurements. The day's routine had been like any other, hard and repetitive, except that now he was obviously being watched by Malenkov on the observation deck and the other KGB agent below. No questions had been asked, and as far as he knew, the captain had not yet been discovered. The truth would come out when the OTTER arrived.

There was heavy rain again, yet the sea was still calm enough for a transfer. The conditions were perfect for Semyonov to slip out of the lifeboat and over the side without being seen. The problem was Malenkov who was still standing on the observation deck. He could see the whole ship. Something would have to be done about him if the captain were to have a chance.

Darkness had come swiftly, and the big deck floodlights were turned on before the lights of the OTTER were sighted. Silk, staring up into the streaks of rain, could see the shadow of Malenkov standing on the observation deck. Did the gorilla never rest? He knew the other KGB man was below deck waiting, probably having a cup of tea with the filleters before the work began. He would have to make his move the minute the OTTER'S last fish was in the chute. He had already managed to kick a Jacob's ladder to the spot where it would be thrown over the railing. None of the fishermen had spotted it, or if they had, they hadn't replaced it on the rack.

The OTTER closed on the MURMANSK, pulling a net only partially full, and soon the two ships were busy with the transfer. The cod-end was pulled on deck, and from that moment on Silk was on the trawl deck, busy with his gaff sorting fish as they poured down the chute to the cutters below.

The spotlights were trained on Silk; full, bright lights; not a shadow of sanctuary anyplace. He knew he was trapped. Stalked like an animal. No way out. He glanced up toward the observation deck once more. Malenkov was still silhouetted like a stone statue with his binoculars trained on him.

Silk began to feel panic and then a sense of madness took over. Fuck Malenkov! Survival was what counted. A man will do

93

anything to survive. For the first time in his life Silk understood the full meaning of that statement. He realized he would even kill this time. The circumstances were that desperate.

Not only did he have to save the captain, but he also had to save himself. And if he and Semyonov were to ever leave the MURMANSK alive, it had to be now while Cowboy was here to help them. His only recourse was to, somehow, take Malenkov out. At this moment he frankly didn't give a damn how he did it. The bastard had to go.

Still holding the heavy wooden gaff, Silk moved toward the gangway as if he were going down to the cutters in the processing room. As soon as he reached the shadows, he bent low and headed toward the observation deck where Malenkov was standing. He had little time left before the KGB agent would expect him below to begin his sorting on the conveyor belt.

Silk located the outside ladder that led up to the shadowed side of the observation deck and quietly climbed the steel rungs to a point where his head was level with the deck. He peered over the edge. He could see Malenkov leaning against the rail, his eyes focused intently on the trawl deck operation. Quietly, Silk pulled himself over the deck edge and sidled along the wheelhouse. Now, if only no one appeared to interrupt him.

He took three quiet steps to the leaning man, and swung the heavy gaff at Malenkov's head.

The blow behind the ear would have killed a horse, but it only stunned Malenkov. The KGB man threw up an arm, grunted from the pain, and turned to face his assailant. Silk struck him again directly between the eyes, and he felt warm blood splatter on his face. The big man went backwards against the rail, then fell face forward. Silk caught the heavy body under the arms and dragged him toward the shadows. He felt the blood that had spurted from Malenkov's head running down his cheeks, or maybe it was only the rain. He only hoped he hadn't killed him. Murder, a Russian's murder at that, was a serious offense. He lay the unconscious agent flat on his back, and climbed quickly down the ladder to the deck that led to the bow of the ship where Semyonov was hiding.

He tapped three times on the outside of the lifeboat with the gaff handle before he hurled the weapon overboard. He heard an answering tap from inside. Seconds later the hatch cover opened, and Semyonov slid over the side and down beside him. None of the crew seemed to have noticed. Both men were dressed in boots and heavy raingear, so it would be difficult for an observer to tell them from the rest of the crew.

"Malenkov?" asked the captain.

"I had to hit him with a gaff," said Silk. "He's unconscious on the observation deck."

"I only hope you struck the pig hard enough," said the captain. "Where is the ladder?"

"It's lying on the starboard side near the bow," said Silk. "But we have to move quickly. Cowboy is already unloaded. I'll go first and drop the Jacob's ladder."

Silk left the captain and located the rolled up ladder hidden in the dark next to the starboard bow. He affixed the hooks to the railing and dropped the heavy manilla rungs over the side of the ship. He slipped over the rail and waited. Out of the corner of his eye he could see a tall figure slide over the rail above him and the ladder began to jerk as Semyonov followed him down.

Long minutes passed, then out of the darkness, a pitching shape came into view. With a sigh of relief, Silk dropped the last few feet into the Zodiac followed a moment later by the captain. The rugged outboard engine roared, and the tiny craft began to move away from the black ship through the swirling water. Soon it was riding smoothly over and between the heavy swells in the opposite direction of the giant factory ship. At a distance Silk could see the staring faces of the Russian fishermen gathered along the rail. They probably wondered what a small craft like the Zodiac was doing this time of night over a hundred miles out in the ocean. Silk was relieved when the MURMANSK didn't change course.

The deckhand steering the Zodiac busied himself with the motor, and in a short time they pulled alongside the OTTER. The small boat was rising on the swells, one minute towering over the OTTER, and then the next dropping swiftly beneath. Someone on board threw a line to the steersman, and then another. He attached the first one to the rounded bow of the Zodiac and the second one to the stern.

"Now," he said. "When we come even with the deck or just a bit above it, jump."

Silk leaped across the gap between the catcher boat and the

95

Zodiac, and the captain followed him a wave or two later. As he touched the deck, Cowboy enveloped him in a bear hug.

"Silk, you old sea dog, you," he said. "What a way to come callin' on a friend a hundred miles off shore. What the hell's goin' on, and where is that damned fish you was mentionin'? I couldn't understand what on earth you was talkin' about, so I just followed orders."

"You were great, Cowboy," said Silk stretching to assist Captain Semyonov as he slipped momentarily on the wet deck. "I'd like you to meet Alexi Semyonov, captain of the MURMANSK, and recently a member of the Russian fishing fleet. He has decided to become an American."

"Well, I'll be damned," exclaimed Cowboy shaking Semyonov's hand. "So that's what all that hocus pocus was about. Welcome aboard, captain. Now come on inside and sit down. We'll have a drink to celebrate, then we'll eat somethin'. I want to hear the whole story."

"Do you have a privy, er... how do you say, a restroom on this ship?" asked Semyonov.

"Down in the fo'c'sle and to your left," said Cowboy.

Silk and Cowboy stood staring at one another as Semyonov fled into the cabin. The look on Cowboy's face was a combination of chagrin and amazement. At last he spoke.

"Well, Silk, looks like you really got your ass in a sling this time. You not only jumped ship, but you took the Russian captain with you. That must be some kind of a international crime or somethin'. You'll be lucky if somebody don't shoot the hell out of you. But, now that you got him, what're you goin' to do with him?"

"What about Claude's place?" Silk asked. "Claude has plenty of room. He'd be the perfect man to hide the captain."

"What if Claude won't take him?" asked Cowboy. "That's askin' an awful lot."

"Oh, Claude will take him alright," answered Silk. "I know the man. He'll want to help in this case. Trust me."

"Come to think of it," said Cowboy. "You're probably right. Him and Claude'll get along like peas in a pod. Let's do it."

CHAPTER

11

CLAUDE

Most of the people on the bay front thought Claude was just a big, good-natured guy, but they didn't understand the real Claude. Claude was some kind of eccentric genius. There was very little about diving or the salvage business the man didn't know, but these two pursuits were just the tip of the iceberg so to speak. Claude had imagination, and all one had to say to set his mind in motion was, "this job is impossible."

Claude lived over on South Beach in a house he had designed himself. The structure sat on a knoll overlooking its own private beach on Bane Bay. The plan was simple. The house was one large wooden hexagon with a flat roof.

Claude's business was as unobtrusive as its owner. From the road frontage an observer could see only a small square building that looked like a bum's shack covered with unpainted cedar siding. In front of the building stood a tall wooden ship's mast planted in the ground from which swung a hand-painted sign which read CLAUDE'S DEEP SEA DIVING AND SALVAGE.

The building was larger than it looked, and inside there was every kind of diving gear one could imagine. Besides this portion of his business, Claude himself was available as a diver. He could be hired to retrieve anything that had been dropped into the bay, whether it was a pair of prescription glasses, a gold watch, a woman's necklace or purse, or a bicycle. He was also available for larger projects like raising fishing boats which had sunk or salvaging valuable objects from large ships that littered the ocean bottom near the mouth of Bane Bay. Claude's dock extended out into the bay so he would have clear access to the jaws and the ocean and still be protected from the incoming and outgoing sea traffic.

Behind his retail business lay two long, one-story barracks which Claude had purchased from the U.S. Army in Fort Lewis,

Washington. He located the two barracks on either side of the dock where he tied-up his salvage barge and crane. Though it was impossible by definition, Claude had somehow managed to maneuver these two large buildings past the business and down the hill close to the water.

The barrack next to the dock Claude remodeled into a marine workshop. In the second barrack Claude stored the objects closest to his heart: strange antiques that he thought worthy of collecting and bizarre building materials. There were odd piles of lumber, electrical conduit, grand pianos, antique cars—an old LaSalle and an ancient Packard were his favorites—chests of drawers, and pioneer baby cribs. These items were piled the length of the whole barrack, helter-skelter with no visible method of arrangement. Whenever he was frustrated and didn't feel like donning his diving gear, Claude retired to this building by himself. Here, he sat at one of the antique grand pianos and played CHOPIN and DEBUSSY in the dark. There were no music sheets because Claude could not read a note, and he had never had a music lesson. He simply played by ear. As he said, "Any music that the waves inspire, I can play."

When Silk, Cowboy, and Semyonov reached South Beach, they found Claude's house to be as dark as a deserted tomb. Silk pounded heavily on the door with his fist, and the three men waited. The rain was pouring now, the kind of Port Bane rain that got you wet in thirty seconds. Silk felt a trickle of water run down the crack of his ass and drip dispassionately into his right boot. He was tired and cold. He and the captain hadn't slept much during the last four days.

The men shuffled about in the rain waiting for a sign of life inside, but there was no light and no movement. Silk pounded on the door again.

"There seems to be no one in the house," said Semyonov at last.

"He's not here, Silk," said Cowboy. "Now we're really into it, and it's three or five miles back to my boat."

"Quiet! Listen!" Silk said and motioned for silence.

Above the beating of the raindrops there wafted the faint tinkling of a piano. It seemed to be coming from the bay area.

"Claude's here," breathed Silk at last. "He's down in one of his barracks."

"Can't be Claude," said Cowboy. "Claude plays that classical stuff. That's rock, man! I heard it on the radio the other day. It's that SWEET ROCK AND ROLL song by that fella, SOREN."

"You never know about Claude," said Silk. " He's likely to be doing anything at any time. Let's take a look."

The men sidled down the hill to the first barrack. Again there were no lights, but the sounds of the piano were definitely coming from somewhere inside the long building. Again Silk pounded on the door. This time the music stopped, and they saw the flickering flash of a light showing through one of the windows.

"What the fuck do you want?" boomed a base voice behind the barrack's door.

"It's Silk, Claude!" he shouted. "Open the damned door."

There was the sliding of a heavy bolt and the door opened a crack so the men could peer inside. There was Claude, standing barefoot in a pair of long flannel underwear holding an antique oil lamp in his hand.

"Well, I'll be damned," he said. "John Silk. I thought you went to hell and the sharks ate you. It's good to see you, man. Well, come on in. I got some beer in the refrig, and I just scored some mighty good bud. I get in the mood to play and smoke once in a while."

"It's damn decent of you, Claude," said Silk calmly. "But we don't have time tonight. We're in deep trouble, and we need some help."

"It won't take any longer to have a beer while we talk, will it?" Claude grinned.

Claude passed out a six-pack of Corona and sat down on the grand piano bench.

"Alright. Let's have it."

"It's a long story, Claude," said Silk. "And I can't tell you all of it. This is Captain Semyonov. He's decided to become an American and I'm helping him, or, I'm trying to. We need a place to hide for awhile."

"Well at least you came to the right place," said Claude. "You can hide here! No one could find you here in a million years."

"What do you think, Semyonov?" asked Silk.

"I think you have a very fine friend," said Semyonov.

"Sounds good to me," said Cowboy.

99

"Well, It looks like we're going to take you up on your offer, Claude," said Silk. "It's mighty nice of you."

"No problem," said Claude. "She's not much to look at, but she's warm and dry. No electricity, but she has imported hurricane lamps that burn kerosene. I've been collecting them off ships for years. And there's a wood stove, plenty of wood cut, bunks, a CB radio with extra batteries, and plenty of blankets. Cowboy can use my jeep to bring in supplies when you need them, and I guarantee no one will find you if you don't want to be found."

"I think this is exactly what we're looking for," said Silk. "What do you think, Semyonov?"

"Excellent," said the captain. It reminds me of a place I lived in Russia when I was young."

Claude grinned happily at the captain and started a fire in the wood stove.

"Looks like we're going to have some time on our hands, Semyonov." Silk said.

"Just worry about doing what you set out to do, Silk," said Claude. "Hide."

As soon as Claude and Cowboy took off, Silk and Semyonov stretched out on the two bunks and slept the sleep of the dead. They did not awaken until late in the afternoon when they heard the sound of the jeep returning. Cowboy pushed open the door with a sack of groceries in each arm.

"I bought everythin' I could think of you might need for three weeks," he announced. "I'm figurin' on stayin' out on the OTTER for about a month, just so the trip'll look natural. If anybody's lookin' for you, I'll let Claude know on the CB radio. And if you find out you need somethin' else before I get back, Claude said to give him a shout. He's wantin' you to call tonight to make sure the CB works. Said to use channel 16."

As the men helped carry in the groceries, a feeling of well-being crept over them. They had done the impossible. They had escaped from the MURMANSK and were safely hidden where no one could find them for as long as they wanted to stay. With Cowboy's expert purchasing, they would not lack anything to eat or drink. For the captain he had purchased a case of vodka and for Silk a case of scotch. Last he carried in cartons of Camel cigarettes.

Silk immediately opened a carton and pealed the top off a

pack. He licked the length of a cigarette and lit up.

"Damn good," he inhaled deeply.

"You mean you haven't had no Camels since you left?" asked Cowboy. "I figured Fran would fix you up."

"Fran did alright," said Silk.

"Say," said Cowboy smiling. "Did I steer you right about Fran, or did you get a chance to find out."

"You were right, Cowboy," said Silk glancing at Semyonov. Semyonov was beginning to look a little green around the gills. "Let's get these steaks on, Cowboy. I want to see if they taste as good as I remember."

After the dinner was over, the three men gathered about the table smoking cigarettes and sipping cognac out of kitchen glasses. The conversation ranged from each man's account of the escape to fishing problems on the west coast.

"I don't know why I should continue to concern myself with fishing since I no longer have the SEAGULL," Silk said.

"I guess you didn't hear what happened to the SEAGULL while you was gone," said Cowboy.

"No," said Silk. "What?"

"After old Plank fired you that day, he come roarin' down to the docks and sunk her hisself," laughed Cowboy. "Right down to the old bottom she went. Course, the old bastard didn't mean to do it, and that made it funnier than hell. Plank wanted Claude to raise her up, but Claude told old Plank to find somebody else. He was busy. I guess Plank finally got a salvage company in Astoria to come down and take a look at her, but it took them weeks to get around to it. He may as well kiss the SEAGULL goodbye. She won't be worth nothin' now."

"No matter," said Silk feeling a twinge of regret for the SEAGULL. "I'm beginning to think commercial trolling is a thing of the past."

"You got to be enterprising," said Cowboy. "The problem with them little boats is you got to come in ever' five days to sell your fish to them American fish plants. That's why I went into joint venture in the first place, you can fish seven days a week."

Semyonov, who had been silent while the two American skippers talked, poured himself another glass of cognac.

"I would like to tell you what I know about your problem," he said politely.

"Say on," said Cowboy.

"You Americans do not understand the huge damage the Russian fishing authorities have wrought upon the oceans of the world," he began in a quite voice. "Silk has an inkling of understanding; but Cowboy, you do not understand. I am an old deepwater fisherman, a captain in the Soviet fishing fleet for more than thirty years. I have fished on every kind of Russian trawler, and I have served for many years on large Soviet factory ships like the MURMANSK. With these tools I have helped deplete most of the world's fishing grounds, and I am partially responsible, you see, for I knew what I was doing."

There was silence for a moment. Both Silk and Cowboy were caught up in the captain's story.

"Perhaps I should not speak of this," the captain said. "Am I offending anyone?"

"No, no," said Cowboy. "Me and Silk have been takin' sides on this stuff for years. Now I want to hear what you have to say about it."

"The story is very sad," Semyonov took a deep breath and a large swallow of vodka. "But it must be told. For nearly a decade and a half I worked as a trawler captain in the Barents Sea. We fished alongside Norwegian, French, Spanish, and British trawlers. Our production targets always included a proportion of cod steaks and filleted cod. Cod steaks, of course are not sold in the Soviet Union. They are exported to other nations in order for Russia to earn hard currency. Both cod steaks and fillets can only be obtained from mature cod that are at least two and one half feet in length. The British love cod, so being intelligent fishermen, we always kept the British trawlers in sight because we knew they would be trawling in areas where there were concentrations of fully grown cod or halibut.

So you see, for many years we had a reliable indicator of where to find those large, fully-grown, commercially valuable fish. At this point I must tell you that we had the complete support of the Russian government who understood our methods.

But catching mature, adult fish was not enough for the Soviet authorities. Instead of allowing the species of cod and halibut to re-establish themselves naturally, we took the insane step of wiping out the small species of fish on which the cod fed. Can you imagine a fleet of huge, deepwater super-trawlers, capable of freezing seventy

102

tons of fish per day, and each of them catching ten thousand tons of fish a year? These fleets were pledged by the Soviet authorities to raise the production of edible-grade fish by thirty percent. The results were deplorable. We destroyed the very fish which we needed.

Of course, when the cod were gone, we turned our attention to catching moiva. Moiva are not found here off your west coast, but there were many of them there. Moiva is another name for small fry known to you as whiting. In their immature stage they are approximately four inches in length. These tiny fish are the chief food of the cod, the halibut, and the other large fish of the Barents Sea. Our huge engines of destruction concentrated on these small fish. We in charge knew what we were doing, but we had no choice. The Soviet fishing authorities had made their demands.

We knew that less than thirty percent of our catch reached the consumer as food. The rest was ground into fishmeal for fertilizer. Fifty percent of the total catch, due to the inefficiency of Soviet fishing methods, spoiled while it was still in the trawl-net. This spoiled fish had to be thrown overboard. Since fish will not feed on dead matter, the chances of cod and other valuable species re-establishing themselves in the Barents Sea was impossible.

After we had destroyed the Barents Sea cod and the moiva, we moved to the shores of the American continent. We fished right up to the Nantucket light at the entrance of New York Harbor. Your government was willing. We fished there, then we moved on up to the famous George's Bank where your fishing fleets fished.

At one point in time ninty percent of the fishing vessels on the George's Bank were Soviet. Of course, the fishing policies of the Soviet authorities were no surprise to me, but I have never understood how your American government has allowed us to destroy your own fishing grounds. At times our great flotillas made it almost impossible for your ships to pass through, but there was never a complaint. It was almost as though your government welcomed the destruction of its resources.

Our government dispatched fishing ships to your shores for six months a year: large trawler factory ships, medium-sized trawlers, herring trawlers, and seiners. They sent along tankers to refuel them and transfer ships to take off the fish. The aim of the Soviet government at this time was to catch any species of fish in any waters in any quantity and at any price.

103

On the George's Banks our herring trawlers could take up to seventy tons of fish apiece. Our large stern trawlers could catch up to forty tons of fish in twenty minutes trawling. Of this catch eight tons would be sent down for freezing, and five tons would be shovelled into the bunkers of the fish-meal machines.

The next trawl would start at once, and it also would yield another forty tons of fish in twenty minutes time. In the meantime the remaining twenty tons of fish that had been caught in the first trawl, and not yet processed, would be thrown overboard where it would sink to the bottom to decay. With this practice not only were the fish themselves destroyed, but the sea-bottom was ruined as a breeding ground. Because the layer of dead fish on the bottom rots and gives off hydrogen sulfide, other fish cannot spawn in the poisoned area.

But this poisoning is only part of the destruction. When a fleet of between four hundred and five hundred vessels simultaneously drag their trawls across an area, the bottom is scored. It is rammed down like steamrollers have been rolled across it. There is nothing left alive for the fish to eat.

And as the years passed, there were no large cod and few hake to be caught on the George's Banks, so with your government's cooperation we moved our Soviet fleet to the Flemish Cape Bank where we annihilated the sea perch. At about the same time we began catching young hake on the southwestern slope of the Newfoundland Bank.

We had opened the door, so it was an easy step to move our fleets to Alaska and both coasts of America. Our fleets now ranged from the southernmost tip of Africa to the Indian Ocean. If the Argentine government had not seen the danger signals in time and extended its territorial waters to the two hundred mile zone, no doubt we would have destroyed all the stocks of hake in the south Atlantic."

"I got a question," Cowboy interrupted. He had been concentrating on Semyonov's story so intently that he had left his first glass of vodka untouched.

"Yes?" asked the Captain.

"Do you mean to tell me that you're doin' the same thing to the ocean bottom off our Oregon Coast?"

"Yes, that is true, my friend," said Semyonov. "The regulations are more strict now, but the methods are the same. The destruction of the fish will just take a little longer."

"Damn me," said Cowboy leaping to his feet, working his mouth up and down so his mustache twitched. "Am I helpin' to do this here destruction to my own country?"

"Yes," said Semyonov, "you are. Without the American joint venture companies, our Russian fleet would not be allowed to fish here."

"Damn a mighty!" said Cowboy drinking half his glass of vodka in a single gulp. "That's terrible!"

"Yes, it is terrible," said Semyonov.

"Don't nobody know or care about what your fleet's doin' to the ocean? I mean, if you're doin' it, then all the rest of them dern foreigners out there's about as bad. And I've sold fish to almost ever' one of them bastards."

"Of course," answered Semyonov. "Your government and educational institutions know what we are doing, but I should not speak for America. Only Russia. I know for a certainty that the Soviet authorities are aware of these facts and have been for years. They have used this knowledge for their own purposes. The Soviet Union has more fishery research vessels than any other country including your own. When an area is finished, they simply change fishing grounds."

"Ain't that awful cold blooded?" Cowboy's temper was rising. "Hell's bells! What's the world comin' to? What's it goin' to take to satisfy that blasted government of yours? Ain't they never goin' to get enough?"

"No," said Semyonov. "The Russian government will never be satisfied until the oceans are dead. We were never satisfied just to destroy our own fishing stocks.

For years our huge fishing fleets have set off from the Maritime Province. We have moved further and further out from Kamchatka and Sakhalin. We have moved into the Pacific and Indian Oceans. We fish off the west coasts of Canada and America; we fish the Aleutian Islands; we surround Alaska; we fish in the waters of Indonesia and the Philippines; we fish around the islands of the South Seas. At times we even sneak into Japanese waters, despite the fact that Japan suffers from a shortage of fish."

"Holy Moses!" cried Cowboy. "You boys is destroyin' the whole world."

"Not the world," said Semyonov. "Just the fishing grounds of the world. Russian fishing vessels are now at work from the Arctic to the Antarctic. Everywhere we are decimating the world's fish."

"You've known this all along, haven't you, Semyonov?" Cowboy said in an accusing tone. "I've been sellin' fish to you bastards for years, and there ain't nobody ever told me about this part of the deal!"

"You do not understand," said Semyonov. "What I have told you tonight is a well-kept secret. Only those of us in command know. The joint venture people do not know. The Russian fishermen and factory workers do not know. Only governments and politicians and captains know. I have kept my mouth shut for over seventeen years, but I cannot continue to be part of such a system. That is why I am here in this lonesome place."

Semyonov took a deep breath and sucked at his glass. It was empty. He glanced at his watch and heaved a deep sigh of relief.

"I am sorry," he said to Silk. "I have told the truth, but I have talked much too long. I must sleep now."

"You go on to bed," Silk said. "Cowboy and I have to make a call to Claude to check out the radio."

"You check out the fuckin' radio," said Cowboy. "I got to go outside and have me a smoke."

While Semyonov stretched out on his bunk, Silk cranked up the CB and made his call to Claude. The reception was clear and Claude had heard no word of any pursuit of him or Semyonov. He plunked down the receiver and went outside to see about Cowboy.

Silk found Cowboy in the beam of the flashlight, with the water running off his head and drooping mustache. He was staring into the distance with a perplexed frown on his face.

"What are you doing out here in the rain?" asked Silk. "It's dry inside, you know."

"I'm thinkin'," said Cowboy. "Leave me the hell alone."

"About what?" asked Silk trying to humor the man.

"About what the captain said," murmured the Cowboy.

"Just what did he say that you've not heard before," asked Silk. "I've been telling the same thing to you for years."

"Yep," said Cowboy. "You damn well sure have. But I just ain't been listenin', have I? You know that man makes a lotta sense."

"I know," said Silk. "He was telling the truth as best he could."

106

"Yeah," agreed Cowboy. "But I've been mighty stupid. I shoulda listened to you, Silk."

"But I didn't know everything Semyonov said," said Silk quietly. "I've only suspected."

"But you was right," said Cowboy. "Seems like you're always right. I was the one that's wrong."

"You were just trying to survive, like everybody else," Silk said.

"Sure. I survived," said Cowboy. "Damn right I survived. And I sure made a whale of a lot a money doin' it. I just didn't know what I was doin'."

"Maybe you were right, and I was wrong," said Silk. "You do have your boat and money; I have neither."

"No, Silk," said Cowboy heading for the jeep. "You was right. I shoulda listened better."

"Get back here, Cowboy," said Silk. "Where do you think you're going? Don't you know we were expecting you to spend the night?"

"I gotta go," said Cowboy over his shoulder. "You got everythin' you'll be needin' for awhile?"

"Yes," said Silk. "You've brought enough supplies to last for months. But I don't think you should leave."

"I gotta, Silk," said Cowboy. "I got me some thinkin' to do."

The weeks went by before Cowboy returned.

"Where the hell have you been, Cowboy?" Silk asked.

"Ever'wheres," said Cowboy handing boxes to Silk and Semyonov. "Port Bane, Coos Bay, Corvallis, Portland, Seattle. Member when I was here last time, and you told us about them Russian fishin' ships destroyin' the oceans? Well, that kinda got next to me and got me to thinkin' about myself. So, I went and sold the OTTER the other day. That's why I was gone so long. A guy in Portland wanted her. Got good money for her, too."

"Cowboy," you're crazy," Silk said. "I thought you loved the OTTER. Now neither of us have a boat to fish this season."

"Gettin' rid of the OTTER wasn't the worst part of the deal," said Cowboy. "It was firin' the crew that hurt. They'd been with me for a long time. But after the captain told all them things about Russian fishin', hell, I got pissed off at myself. I'd been helpin' them

Russians for years. I was worse then they was 'cause I was a American. If my old daddy down in Texas had a known what I was doin', he'd a kicked the shit out of me. Honest to God, Silk, I never figured I was helpin' to ruin the ocean!

Maybe I'll just quit the fishin' business and buy me a ranch somewhere. But tonight, I'm goin' to get dog drunk."

Cowboy spent the night and left the next morning with the promise he would return as soon as he collected his money from selling the OTTER.

One week passed, and Silk was on the verge of dying his two-month growth of beard and hair, changing his name, hitchhiking to Port Bane, and getting on a boat to Alaska when Cowboy returned in the jeep. He had collected his money from the sale, and he had two interesting pieces of news. The PBFA had decided to strike this season for higher salmon prices, and Hattie at the Albatross had received a call from a man named Ziggy Searles in Ashland, Oregon. The man had ordered enough salmon to feed an army.

"You ought to go into Port Bane and handle the deal, Silk," Cowboy said. "This might be a good thing for the fishermen. There ain't nobody lookin for you now anyway, and the captain can go on stayin' here for a few days by hisself, 'til you get a feel of the situation."

"What do you think, Semyonov?" asked Silk. "Would you be alright here by yourself?"

"Yes, of course," said Semyonov. "If you think you can help the fishermen, then you must go."

"It's settled then," said Silk happily. "To tell the truth I've been ready to get out of this place for at least a month. I'm ready to leave."

PART II

SILVIA CASTLETON

CHAPTER

12

THE STORM

Old Cid had been hanging around the Port of Astoria for as long as anyone could remember. Once a fisherman in the days before boats had engines, he now contented himself with watching over the port and reliving his fishing experiences on most of the world's waters.

Cid was a tall, spare man in his nineties. His parched face consisted of a thousand wrinkles which ran in deep rivulets from his white hairline to somewhere beneath his shirt collar. Because of these wrinkles and a pair of unblinking watery-blue eyes, Cid often shocked a newcomer to the docks with his appearance. Without the Egyptian wrappings, he looked exactly like a mummy.

He sat on his wooden bench at the top of the ramp watching a young flatlander and his wife, or girlfriend, load the white troller for a fishing trip. Obviously, the man didn't know his ass from a hole in the ground, and the woman knew even less. He only hoped they didn't plan to leave the port today. Today was the wrong time to be out on the ocean. The fishermen who knew their business had sensed the quiet atmosphere and remained in port. There hadn't been a fishing vessel across the bar in twenty-four hours. Only he understood the phenomenon that was causing the calm, and he hadn't seen it in thirty years. Sometime today the sun dog would appear in the Oregon sky; two suns at the same time, and you couldn't tell which one was real. Both were deadly; both bit in silence. There was no telling what such a sign would bring.

Cid had just placed the tiny butt of his hand-rolled cigarette between his cracked lips when he felt the weight on the other end of his bench. He turned his head to see who the sweet-smelling intruder was.

She was all in white, and her green eyes and longflowing dark hair reminded him of a mermaid he had once seen on a ship's bow.

The young woman, sensing the old man's awareness, turned to meet his gaze.

"Good morning," she said smiling. "Do you mind if I sit here for a moment? I'm trying to stay out of Matt's way. Matt is my friend on the boat down at the end of the dock." She pointed. "He's getting it ready for us to embark on a deep-sea fishing trip."

"Yer not plannin' on going out today!" the old man said in alarm.

"Matt says we are," she said. "We have everything loaded, and I'm just waiting for him to finish arranging the supplies. See, he already has the engine running."

The old man finished his cigarette butt and ground it beneath his heel. He stared out at the broad expanse of the Columbia bar which was now beginning to churn white foam as the tide reversed herself. This was his opportunity to warn the young woman. If she listened to him, perhaps he could save her.

"You ought not to go out today, lass," he said. "Wait two or three days until the storm's spent. Ye notice the fishing boats docked down there? They're still here. They know better'n to go out. There's big combers runnin' out there, else they'd a gone before dawn."

"I beg your pardon?" she said.

"The old sun dog's comin' out today, and whoever he gets his evil eye on is goin to the bottom of the sea. Don't go out. I'm a warnin' ye. Don't go out."

The young woman seemed fascinated by the old man's deep, trance-like voice, but she found his words alarming.

"What in heaven's name is a sun dog?"

"Two suns instead of one," the old man said in a sing-song voice. "Can't tell which is the real one. If he shows his ugly head, you're as good as gone, and your ship's as good as sunk. Don't go. Let your friend go by hisself. Let him die by hisself."

"You are trying to scare me," she said indignantly and stood up to leave. "You have no right to do that. Who are you anyway?"

"I'm trying to help you, love. I am called Cid."

"Are you a fisherman?" she asked.

"Aye, a long time ago," said the old man. "Before you were born."

"Then how do you know what's going to happen when you're just sitting here?" she asked. "I have to go now."

She was halfway down the ramp when she heard the old man's last words.

"I'm warning ye," he said. "The sun dog is out to get ye."

Old Cid sat silently watching the woman stride toward the boat. If she didn't listen, there was nothing he could do.

The handsome man she called Matt offered his strong right arm and helped the pretty lady on board the COLUMBIA PRINCE. "Did the old man up there on the bench say something to upset you, Silvia?" he asked.

"Nothing that I could understand," Silvia looked at Matt. Suddenly, she felt frightened. "He said there will be a sun dog today, Matt. He said there would be a storm. He said we would die."

"That is ridiculous!" said Matt. "I've already made a call to the coast guard weather station, and they say there is virtually no wind and the weather will hold. It's ideal for our first trip on the COLUMBIA PRINCE. That old fellow probably doesn't know anything about the ocean. He's just full of the old wives' tales that he tells to tourists. Maybe that's his job! He certainly doesn't understand about weather forecasting."

Silvia stepped into the cabin and gazed admiringly around the pilot house.

"She is very handsome, isn't she, Matt?" Silvia said.

"That she is," he said proudly. "For an older boat she's in excellent condition. All those gadgets you see in front of you are man's latest inventions to catch the illusive salmon. We have a wheelhouse, a galley, and a toilet and shower. We have enough electronic equipment to find our way anywhere. See? Three radios, a depth finder, a compass, a paper machine, a loran, and an autopilot which you can also run from the pit in the back. We have a 335 horsepower Detroit diesel engine, and the hull is almost as deep as the boat is tall. The man who sold her to me said she's virtually unsinkable, the safest troller we could buy."

"That's wonderful, Matt," said Silvia. "I feel so good knowing we're safe. Now, what can I do?"

"Just stand here beside me while I back out of the slip. After we get underway, you can go below in the fo'c'sle, select your bunk, and put your clothes away."

111

Thirty minutes later old Cid, still sitting on his bench, watched the white troller leave the port and push her way toward mid-channel of the Columbia River.

"Damned fool," he muttered to himself. "Flatlanders never listen. She's a mighty pretty boat, but she's never comin' back. That one's goin' to the bottom."

Neither Silvia nor Matt expected the waves to be so large on the Columbia River, but they didn't understand that because of the persistent ground swells which roll in from the Pacific Ocean, the west coast bars are among the worst waters in the world.

The Columbia River bar where the troller was laboring is called The Graveyard of Ships. The troller was heading into six foot breakers with a force behind them of two tons per square inch.

The sound of the big diesel engine was comforting to Silvia, particularly when it strained its three hundred and thirty-five horsepower against an especially large wave. The water was rough, but the troller had little trouble against the incoming tide.

As Matt steered the COLUMBIA PRINCE down the middle of the channel toward open sea, he kept up a running dialogue of information.

"The front of the boat is fore, and the back is aft," he said. "When you're facing forward, the left hand side is called port and the right side starboard. Understand?"

"I think so," said Silvia. "Front, fore; rear, aft; left, port; right, starboard. I feel kind of funny, Matt. I can hardly stand up; it's this rocking from side to side."

"Don't think about the motion of the boat," said Matt. "That feeling is just your body trying to compensate for not having dry land under it. You'll get your sea legs under you as soon as we get out of this channel. Everyone feels the same way their first time out. Concentrate on what I am saying, and try not to think about the waves."

Silvia listened intently to Matt's explanation of the five pieces of electronic equipment in the cabin: the compass, C.B. and VHF radios, depth finder, paper machine, and loran. She pushed the red electronic buttons that turned the autopilot on and off, then she followed Matt out on the deck. There was a brisk headwind that blew her long hair back from her face. For a moment she forgot her giddy feeling.

"How did you learn about all that equipment," she asked clinging onto Matt's arm. "I thought you were a San Francisco businessman, not an experienced captain."

"Sheer genius," he said. "I've read books for years, taken a few courses, and I've been out on lots of charter boats."

Silvia took a step backward and looked into Matt's dark eyes. She was properly impressed. Standing on deck Matt looked tall and strong. His black curly hair peeked unruly beneath his wool seaman's cap. She wondered, as he put his arm around her and gestured toward the back of the boat, if she really loved him.

"That square receptacle back there at the very end of the deck is where you stand when you fish. It's called the pit. From there you can control the autopilot if you're fishing by yourself, but I don't have to worry about that, I've got you. As soon as we hit the open sea, I'll let you steer while I put out the lines. Those strange looking contraptions on either side are called gurdies, and they're hydraulically controlled. Those round lead weights are cannonballs, and they hold the lines down at different depths. We're almost there, Silvia. Do you see how those chops are smoothing out. I'm so damned happy. We'll have our first salmon by noon. Did you know that by tradition you're supposed to cook and eat the first salmon you catch? Yes, sir, we're going to eat salmon tonight."

The thought of eating salmon was more than Silvia's stomach could take. She suddenly jerked Matt's arm from around her waist and staggered to the rail. Her whole breakfast came up, sour and ugly, and the wind blew it back in her face, spattering the front of her white wool sweater with the vile mess. Her nose was full, and she couldn't breathe.

"Matt," she gasped. "Matt, I'm sick."

Matt look at Silvia clinging helplessly to the rail with one hand, her dark hair matted against her face, her free hand covering her mouth.

"You're really seasick," he said. "Turn aft with your back to the wind so the stuff doesn't blow in your face."

Silvia felt embarrased, but she didn't have time to dwell on it. Somehow, she managed to turn around, but not before she got another dousing. She hung her head over the side and felt the cool spray on her face. She only hoped she wouldn't fall overboard. She had never felt so sick in her life.

"Is there anything I can do to help?" Matt asked laying his hand on her shoulder.

"No, nothing," she gasped. "Go away."

As Matt stepped into the cabin to check the automatic pilot, she was sick again. This time nothing came up except the bottom of her stomach. The COLUMBIA PRINCE plowed steadily into the chopping bar waves.

Then the sea turned an oily calm, and the COLUMBIA PRINCE was cutting through the almost flat surface. The rocking, rolling motion had ceased as the Columbia bar spilled her contents into the open sea.

Matt pointed the troller due south and pushed the button for the automatic pilot. He helped Silvia inside, and fed her soda crackers and hot chowder from a thermos. While he lowered the trolling poles and baited the hooks, Silvia went below to the fo'c'sle to wash and change.

The day, indeed, had turned beautiful. Silvia had survived the Columbia bar and her sickness; now she was totally captivated by the beautiful ocean. The swells were rising and falling rhythmically, matching her own breathing. There seemed nothing solid on which to fix her eyes; blue sky and horizon had become one with the ocean. Gradually the energy of the sea dominated her own, and she became a mere pinpoint in a universe of blue. She watched the slight ripples where the lines entered the water. As they rose and fell with the waves, the small droplets seemed filled with minature rainbows. Occasionally, she glanced at the stainless steel springs through which the lines were fed. According to Matt they would jerk and expand like a shock absorber when a salmon struck somewhere in the dark fathoms below. She tried to think like a salmon seeing the moving herring, but somehow she couldn't concentrate. Something seemed to lurk in the unknown depths, or maybe it was drifting on a horizon that wasn't there. Where were the birds? Weren't there birds when she first came on deck, flying on both sides, moving with the boat? And where were the other fishing boats?

"Matt," she said as he was trying to extract a hook from a flopping fish. "Have you noticed that there aren't any birds around?"

"I haven't had time," he said. "It's all I can do to keep the hooks baited. This is the second salmon today."

"And there aren't any other fishing boats," she continued. "The old man said they weren't going out because of the big combers, but the sea seems perfectly smooth."

"I told you not to listen to him, dear," said Matt. "Obviously, he doesn't know what he's talking about. The coast guard reports heavy squalls northwest of here, but they're not supposed to strike the coast until tomorrow. We'll be safe in port by then." Matt tossed the salmon into a box and began rebaiting the hook.

Silvia thought back to the old man called Cid with his azure-blue eyes.

"Have you ever seen a sun dog?" she called to Matt.

"I've never even heard of one," Matt answered. "Just relax and enjoy yourself. A sun dog is probably just a figment of the imagination, or maybe some kind of story from Greek mythology. It doesn't have a thing to do with catching salmon, and that's what we're here for."

Sensing Matt's agitation at her fears of the old man, Silvia concentrated on the gentle motion of the boat and the silence of the open sea. She was almost asleep on her feet when she heard Matt's roar of approval as one of the lines began to jerk under the strain of another fish.

"Got one," he said. "And by the looks of that line, he's a big one."

Silvia glanced upward at the spring to see if it was acting like a shock absorber, then she gasped as a sudden chill ran up the back of her legs. There it was. Just like the old man had said. A double sun, the sun dog. At first she thought her eyes were playing tricks on her, but it was real! It never occurred to her that she was one of very few who had ever seen a sun dog on the west coast.

"Look at the sun, Matt," she called to the laboring man who was standing in the pit at the stern of the boat playing a big salmon. "It's like the old man said. There are two suns."

Matt glanced up at the phenomenon then turned his attentions back to the gaff.

"Yep," he said. "It appears to be two alright."

Silvia felt a premonition as she watched Matt club the big salmon unconscious and try to extract the hook. "The old sun dog has got him. Let him go by hisself. Ye don't have to die with him."

"Doesn't that mean a storm is coming?" she asked hesitat-

115

ingly, not wanting to incur Matt's wrath. "Shouldn't we prepare the boat, or go back? That old man said a sun dog meant..."

"For God's sake, darling, enjoy yourself instead of worrying," Matt interrupted. "I checked the weather. If anything were going to happen right a way, I would know about it. That old wharf-rat doesn't know about the sea. He's still sitting back in port on his bench trying to scare another tourist. We're into salmon now. Just look at this one. He'll go twenty pounds or my name's not Matt Butler."

Silvia tried to concentrate on the large fish and keep the image of the old man out of her mind. There was only one sun now, and the sea was as calm as ever. Matt was probably right. Certainly he knew much more about fishing than she did. She had always had an active imagination, but Matt had seen that double sun too. How could the old man have predicted they would see a double sun? Maybe it was just an illusion that happened all the time.

True to Matt's word the sea remained perfectly calm, and she felt at peace as the red sun, now a Chinese lantern, sank beyond the horizon, capping the ocean's surface in vermillion.

It was just dark when the salmon quit biting. Silvia felt chilled and made her way to the wheelhouse for a jacket. That was when it started. She thought she noticed an imperceptible change in the atmosphere. As she clung to the rail, the first gusts of wind began to stir up from the northwest. Matt was just beginning to pull in the gear, and the red and green running lights were shining eerily into a dark something that had silently appeared behind them.

"What is that dark spot back there?" she asked Matt. "It wasn't there when I went below."

"Probably just an ocean squall," he answered staring in the darkness at the black horizon. "They come up quite suddenly, but there's nothing to worry about. Just a little rain and wind. Put on those oil skins hanging on the cabin door."

The gusts of wind were now becoming more violent, and the sea had become a mass of six foot chops topped with dirty foam.

"There is going to be a storm," she whispered. "The old man was telling the truth."

Matt seemed to be working unusually fast now as he pulled in another line on the back deck. He must have noticed the ocean's change, she thought. He had just got the third line on board when the first eight-foot swell rolled underneath the COLUMBIA PRINCE. A

116

chest holding their food suddenly careened from one side of the deck to the other. Matt lost his balance and fell against the side of the pit. Silvia grasped an overhead wire and somehow managed to keep her footing.

In less than forty minutes a calm benevolent sea had turned into a raging monster. Thrashing her tail like a dragon, she was now shipping up fifteen foot waves before her fifty knot winds.

"Release the automatic pilot," Matt shouted into the wind. "Like I showed you. Push the left button on the panel. Shove the throttle forward and try to steer into the waves."

"I'll try!" Silvia screamed into the darkness.

Somehow she pushed the right button and pressed the throttle forward. She grasped the wheel with both hands and felt a strange sense of relief as the big diesel took hold. Out of the corner of her eye she could see something black and huge rolling toward them, too fast. Before the COLUMBIA PRINCE could reverse directions, the giant wave was upon them. It slammed into the troller broadside. The thrashing water deluged the boat. The wheel was wrenched from her grasp, and she was thrown against the cabin wall. She quickly managed to regain her footing, but the sudden violence and snapping sound she had heard when the wave struck terrified her.

"Oh my God," she thought. "Matt!"

Then she heard his voice out of the darkness from the back deck.

"I said into the waves, Goddamn it! Before the next one hits us. We've lost our salmon, and the lines are scattered all over the deck. Hold on to the wheel. I'm going to try to cut this gear loose."

Silvia struggled to turn the wheel, but in the trough between the waves, the prow of the COLUMBIA PRINCE shifted. This time it was more like a roller-coaster ride. The troller climbed to the top of the wave, then crabbed down the other side. Silvia felt a terrible sense of doom as she watched the foaming tops of the waves move up behind them in the lights like giant specters brought suddenly alive by some unknown force.

"Oh God," she said to herself as she watched Matt work his way forward to the cabin door. "Thank God you are here. Are you alright?"

"Yes, I'm fine," he said. "I'm wetter than hell, but it's the equipment I'm worried about. One of our trolling poles has snapped

117

off, and we've lost all our gear. At least the lines didn't get fouled up in the prop."

"I'm afraid I'm not much help," she said above the roar.

"You're doing just fine, dear," he said putting an arm around her. "I should have been paying more attention to the weather instead of the fish. I don't understand where those big waves are coming from, and so suddenly."

Silvia had formed the words "sun dog" on her lips, but she didn't say the words. Something inside warned her about angry gods or the old man. She prayed he wasn't right.

For the next hour she stood beside Matt staring helplessly ahead into the blackness, unable to think of anything but the terror of the deafening crashes and the mountains of water pounding the COLUMBIA PRINCE. But the COLUMBIA PRINCE seemed to be riding sturdily as she rose high and then fell into the abyss below.

"Don't worry," Matt shouted above the belting storm. "This is a good boat, unsinkable so the man told me. We'll be out of these squalls soon, and I'll make it up to you, sweetheart. Do you think you could manage to get me some coffee?"

As Silvia stumbled toward the thermos of coffee, a chilling, overwhelming sense of uneasiness shook her. It was like a veiled, disguised, sinister secret from a sealed book of the unknown was about to be revealed. Inadvertantly, she turned her eyes toward Matt who seemed paralyzed. There was a wave, monsterous and black, a moving monolith rising behind them. Matt jerked his head back to see it's top, but the curl of the wave above the troller was so high, he could see only eternal blackness.

"What is it, Matt?" she whispered.

"I don't know," said Matt. "I don't by God know, but it's something I don't understand."

At first the bow of the COLUMBIA PRINCE began to lift over the wave. Silvia thought the whole boat was going to roll over backward. Then the avalance of water plunged over them. Antenna's snapped and glass shattered. Silvia was thrown back against the wall, suddenly drenched, unable to move. The water was two feet deep. Matt lay on the floor near the loran. Silvia thought she saw blood. She crawled on her hands and knees toward the motionless man. Then the second wave struck the COLUMBIA PRINCE, more violently than the first. For long moments the troller was submerged

118

before she finally stuck her bow through the cascading water.

The cabin was a mass of wreckage. The loran was demolished; sparks were shooting out of one VCR; and the whole cabin was awash with water and debris. Again, Silvia crawled toward Matt, half paddling in the deep water. He lay motionless, face up, a long spiral of broken glass protruding from his neck. Somehow she reached him. She cradled his head in one arm and grasped the sharp glass with her other hand. She jerked, and it slid out, followed by red blood in sporatic spurts. Another wave hit and water again submerged them. In panic, she struggled to keep Matt's head above the water. When the troller righted herself again, his eyes flickered open, and he smiled slightly.

"Those people put plate glass in," he said. "A mistake. I'm sorry."

"Don't talk," she said. "Can you move?"

"No," he answered. "Something's broken. I think I'm hurt pretty bad, darling. Don't panic. Do what I tell you. We'll get out of this yet. Take the wheel, now!"

Slowly Silvia staggered to her feet and waded to the wheel. She grasped it with both hands.

"I have it. What do I do now?" she asked.

"Turn it," he said. "Do you feel any response?"

"No," she answered.

"Then call the coast guard! Do you remember how I taught you to use the radios? Take the mike from the small one."

"I think I remember," said Silvia, taking down the mike.

"Now say MAYDAY! Say MAYDAY!" This time his voice was a whisper.

"I am," she said. "There's no answer! There's nothing! No one is listening! Matt!" She dropped the receiver and waded back toward him. His head had slipped beneath the water. "Matt, my God!" she screamed.

She had almost reached him when the third giant wave struck the troller broadside. There was a blinding flash as the radar and the rest of the antennas were swept away. Torrents of water crashed through the broken windows and submerged her. She struggled through the seawater to reach the wheel. She must pull herself up! She must make it work! She turned the wheel, first left and then right. Nothing happened. But she heard the throbbing of the bilge

119

pump still pumping, and she heard the engine still running! Why wouldn't it work? They were rising to the top of another wave, broadside, then falling. She must hold on to the wheel. She must not let go. Why wouldn't it work? What was it Matt had said? The rudder, yes the rudder. It was sheared off.

She switched on the radios one at a time, frantically calling over the mikes like Matt had showed her.

"MAYDAY!" she shouted. "MAYDAY!" There was no response. She tried again. "This is Silvia Castleton, COLUMBIA PRINCE, please help us." There was no response.

She tried the remaining VHF. She remembered that Matt had said Channel 16.

"MAYDAY!" she screamed. pushing the button for transmit. "MAYDAY...COLUMBIA PRINCE... Help us! For God's sake, help us!" Three more times she tried without response. She was at the ocean's mercy now and Matt was dead.

The troller, freeboarding in the strong northwest gale, pitched helplessly from wave to wave like a small toy. Silvia climbed onto the short bunk and pulled her knees up next to her chest. She had come into the world in this fetal position, and now she would die in the same position, a victim of the sea. The old man had been right. She would never leave this bunk alive. She would die like Matt. There was nothing left to do but wait for the final crashing wave. She must not sleep. She must be awake when it happened. "Stay awake." she whispered. "Stay awake."

Suddenly, there was a grinding crash, and the boat was lying on her side. Another wave smashed against the foundering troller, and Silvia's body dropped silently into the dirty white foam.

CHAPTER

13

GRAVEYARD

A particularly strong gust of wind rattled the front windows of the Albatross Tavern. Except for the sound of the driving rain beating against the windows there was almost total silence. At the door one of the dogs started to howl in discomfort, and Rat went outside to comfort the animal. Silk stared at the front door as he lit another cigarette.

'Dogs and fishermen go together,' he thought. Almost every fisherman had a dog or wanted one, and there were always three or four large, mangy mongrels lying around the door outside the Albatross. Silk couldn't remember if he had ever seen a pure bred dog on the bay front. 'Fishermen are just like their dogs,' he mused. 'all mongrels.' Yet, the fishermen had a choice and the damn dogs didn't. They could always go back into a decent job in the society, or most of them could, but the dogs were stuck. He, of all men, shouldn't be sitting here in this smokey tavern, uncomfortable, hoping he could get enough money together so he could buy his own troller again and work himself to death on it. 'Stupid,' he thought, 'stupid.' At the same time he knew that given the chance, neither he nor any commercial fisherman here, drunk or sober, would return to that insane asylum he had escaped from called the straight world. Fishing and the magic of the ocean had gotten them all. He doubted if there was a man in the room who could hold down an eight hour office job; yet, they were a working bunch of sonsabitches. They would work on a boat for thirty hours at a stretch without sleep if that's what it took. "Poor bastards," he muttered.

"What'd you say?" Cowboy leaned across the table toward

Silk.

"I said, give us another round," said Silk. He looked at his empty pitcher, then up and down the table. "We're out."

"Silk, you can't afford another round," whispered Cowboy.

"The hell I can't," said Silk. "Bring us another round."

"Well, suit yourself," said Cowboy. "I'm not your keeper. But if you don't start savin' your money, you're never goin' to get another boat."

Silk was aware of the pride of each of the men around the table, although he knew it was a foolish pride. Beer was all in the world they still had left, especially the damned Locals. There wasn't a man among them that would ask for a drink. Hell, it wasn't their fault that commercial fishing out of Port Bane was almost a thing of the past.

"Thank's, Silk," said Doc Ox as the beers arrived. "I'll get the next round."

"Nope," said Cowboy. "The next round's on me."

'Too damn bad,' Silk thought. 'If half the world had the ability to understand how things really were, the world would be a decent place to live in. We could put some salmon back in the ocean, and these guys would be able to work again.

Graveyard was holding forth, attempting to keep the party going as long as the beer held out.

"In the early days," he said, "when a lot of us come to Port Bane, there was a lot of artist type people here. It was a liberal refuge from ever'one ever'where else, and that's why we come here. We didn't want to live in the reality of cities or deal with the straight families. We was all outcast, fugitives in our own right, from them other civilized forms of communities. We all fit together and felt comfortable, so we stayed.

To us, Port Bane become like a home of homes. You're all sittin' here feelin' like hell 'cause of the storm, but look at me! Where do I go from here? I got no boat. I have a seventh grade education; I got no formal skills other than self-taught. I can't join the union. I don't have a certification for anythin' that I do know, so it don't do me no good. So what do I have to work for the rest of my life? Why, $4.00 an hour! No matter what I do, I have to do something basically illegal to survive. That's why I have a marijuana farm. No matter what kind of a occupation I get, I still have to do somethin' dishonest

122

to make up for it 'cause I can't make enough money in the work force of the real world to survive. I've worked, I've always worked. I'm not afraid to do whatever it takes to make money to survive. That's why I don't take welfare or food stamps. I don't need 'em. As long as I'm physically able to go out and get money, I don't need the government to give me nothin'. It don't owe me nothin'. If anythin', I owe my country. I'm a American. I'm not ashamed or afraid to admit I was in Viet Nam, that it was my patriotic responsibility to be there 'cause I was a American. I would not go to Canada. I would not do anythin' like that. In my opinion, America is the best country that there is to live in.

My dad was in the Korean war. He got out on a hardship discharge and everythin', and he told me I ought to have went to Canada when I was bein' drafted. Hell, I joined the Marine Corps instead. I was ready to fight."

"Tell them about when you was young," said Rat nudging Graveyard's elbow.

"Well, when I was a kid, I was raised under radiation. I lived 120 miles from Jackass Flats where they blew all them bombs up. My dad put in all the air-conditioning in the ground zero buildings from 1952 until 1959. I woke up at three or four years old; I seen the mushroom plumes. I was born there and lived there til I was 14. Jackass Flats is where they want to put the waste right now. Yep, I've done a lot of livin' in this short thirty-six years. It blows some people away. I done a lot of things, but what has it done but teach me how to survive in the world. I'm a person who will never starve. I'll always figure out how to have a dollar or two in my pocket."

"Do you know how to captain a boat?" Claude was trying to change the subject and shut Graveyard up.

"Ship rights? As far as rights go, I can do woodworkin', take a motor apart and put it back together again. Anythin' physical I've never had any trouble with. When I was a kid, my dad was very mechanical. We always had dune-buggies, motorcycles, airplanes. I flew air shows, did acrobats. Dad had an AT-6 World War II plane. We had a P-51 Mustang and old Ryan biplanes. From the time I was a tot, I was raised with that type of mechanical skills.

My old man was a welder, steam fitter, and pipe fitter. He had a room in his house that was wall to wall radios. He's talked to ever' country in the whole world. Right now, he teaches ham radio in Vegas, and he's got exclusive highest radio clearance and satellite equipment. He's talked to Challenger; he's talked to different astronauts; he's worked on their radios. So, I've always been around that type of mechanics, and never got interested in anythin' else. I wish I had now, but that's in the past. I know where my limitations are.

Anyway, when I was a kid we flew civil air patrol. I had a junior certificate 'til I went into the service. Air Search and Rescue throughout the state of Nevada. Air races in Reno, Vegas.

Hell, I've had jump trainin'. I could go over here and steal a airplane, take off and land it without a second thought, but I'm not certified. When I went into the military, I didn't want them to know I had any air jump trainin' because they would of made me a paratrooper, and there was no way I wanted to be floatin' down in a parachute and have somebody shootin' at me. So I avoided that whole reality and ended up in the air wing group anyway. I analyzed and tested fuels. I could tell what gas was mixed with other gas. After the war I come back to Nevada, but it was all the same old shit. I didn't want to stay there so I ran away from home. One of the rebellious trips, you know. That's when I come to Port Bane..."

Esteban had become more and more excited as Graveyard talked about his early flying experiences. Here was another flyer with whom he could communicate. He jumped to his feet and said,

"I make for to fly thees airplane weeth two wings too! En Los Angeles I walk thees wing; theen I jump off. The parachute she open and..."

"We all know about you jumpin' out of that biplane, Esteban," Graveyard said briskly, not wanting to relinquish his position as chief story teller. "Like I was sayin', I come to Port Bane. What a experience.

One night in 1975 I'd eaten about a 150 mushrooms and drunk a fifth of Jack Daniels, and I 'member the night I was up on the hill from Port Dock 3, and this lady was going to get her ears pierced. She was young, maybe twenty or twenty-one years old, and she said, 'does it hurt,' and she was squeamish about the situation, and I said, 'here, give me that damn needle.' And I pulled my nose up, and I shoved the needle through it, and it had a string fixed on it, and I tied the needle off, and I grinned at her and said 'I didn't feel a thing.' I said, 'here let me pierce your ear now.' And I put it through her ear and said, 'see you didn't feel a thing!' I tell you it was mind over matter.

I woke up the next morning, and I was real thirsty. But I had forgot all about that I had pierced my nose. So I went to wipe my face and I hit this string. 'Man, what is this?' I thought. I went to look in the mirror, and it was all the way through my nose. So I was workin' this string back and forth, and I said 'shit, it's not too bad, might as well leave it in for a while.' So a jeweler here put a ring in my nose, and I wore it for two or three years. When we first met, my ex-wife had a ring in her nose too. That's how we got acquainted.

When I first saw her, I thought, 'well, that's kind of neat.' But that was kind of hippyism. I wore gold rings on all my fingers. I wore silver and gold bands all up and down my arms.

I got into a fight one evenin' and somebody tried to grab the ring in my nose. And I said to that guy, 'If you ever get a hold of that ring, I'll follow you to the end of my days.' After that fight I took the ring out! If that guy would a got ahold of it, man, that would have been my worst reality. In them days my beard was down to the middle of my chest, and my hair was almost to my waist. But down through the years, I've altered myself to where I am now. Now, I'm down right conservative lookin'. No one from back in them days would ever know me now."

"They might not recognize you now, Graveyard, but you're still not the Marlboro Man," said Silk smiling at the image of the big man.

"Well, my philosophy of life is simple," said Rat. "I intend to hurt until I die, and if somebody don't hurt me, why, I'm just going to hurt myself. I think we're put here mostly to suffer anyway."

Bent suddenly sprang to life.

"They spark off on this philosophy stuff, then all of a sudden they think they're some kind of a authority. What are they talking about anyway? Heaven or something? Where do they all get off? They talk about world peace too. Well, hell! How are they going to do that? All I know is you put two of us out there in a tree, and the next thing you know, you're out on a limb, and that other guy's right behind you with a saw."

"There is no beginning and there is no end," said Claude who claimed at times to be a metaphysicist of sorts, especially when he was contemplating while he was sitting on the bottom of the bay in his diving suit.

125

"All the energy transforms to a new state of energy. Molecules do not die. A table is just as much alive now as it was when it was a tree. Just look into a microscope and you can see that the molecules are still moving. Just because life exists as we know it doesn't mean that it won't continue in a new state. Supernatural powers, everything is energy. Energy doesn't die, it just transforms into a new state of energy, an endless circle.

Evolution is just an energy field continually changing. As far as we know we can be just a bacteria in a giant scale. We can't be the only life form there is. I have experienced minimal realities. I've seen objects levitate without touching them. One time in California I remember physically leaving my body, turning around and looking at myself, a form of telepathy. I remember leaning against the wall on an Indian reservation at the foot of Mt. Palomar. We lived in a place called Maggie's Farm in 1970; twenty-three of us lived on this farm. These people met me when I lived in Big Sur and sort of adopted me. We got to getting into these drug tests. We ate LSD, and these people would sit with a clipboard and write down everything we did. Acid test. I am one of the experimental pieces in that story and was written up in Time magazine."

"Hey! That's enough about religion and philosophy," said Bent. "Let's change the topic of conversation to women."

"A woman in all her fury is hell on wheels, is all I know," said Cowboy.

"You've got to have beauty before you can capture beauty." Doc Ox chimed in. The giant man had been silent all evening, lost in some thoughts of his own, but here was a subject on which he could wax poetic.

"This group is getting drunk," said Silk. "This storm has us talking nonsense."

"Storms," Graveyard said in a loud voice, regaining the control of the conversation. "If you think this storm is bad, you should a been here when we had that big one in the sixties. She come in from the northwest with winds over a hundred twenty knots. She blew out the front windows here and all the power was out for more than a week. Nobody could get in or out to repair the damn power lines, and there was whole sides of mountains left without a tree standin'. That old bay out there boiled like a caldron, and most of the boats on the port docks broke loose from their moorin's. There must a been half a

126

dozen that went down right out in the bay. Hell, we chased loose boats for two weeks after that storm. She was a real freak and did some funny things.

To storms," he said, and slugged the whole pitcher down.

"That was last call. Time to go," said Hattie. "See you gentlemen in the morning."

The men put on their rain gear, and two or three at a time meandered out into the rain. As Silk and Cowboy climbed the hill to the parking lot, they were overtaken by Bent and Graveyard. Each had a six pack under his arm.

"I thought that was you, Silk," said Graveyard. "Me and Bent's goin' to walk out to the jetty and take a look at the storm. We'll be back in the mornin', and if she shows a sign of quittin', you'll be the first to know."

"Thanks," said Silk, "but this is a hell of a night to go out there. The wind's so bad you won't be able to see a thing, and those rocks are treacherous even in the light. Are you sure I can't give you a ride someplace else?"

"Na!" said Graveyard. "We already made up our minds, and that's where we want to go. I don't feel like sleepin' tonight anyways. 'Sides, I'm pretty drunk and the walk'll sober me up."

"Yes," said Bent. "We haven't been out to the ocean in awhile."

"Thanks for telling us," said Silk. "I hope you boys have a good night. And be careful that you don't get swept off the jetty. We'll see you in the morning."

Graveyard and Bent cracked open their first two cans of beer as they started their long walk toward the jetty. The wind almost blew them backwards with sudden swift gusts, and the rain in the street lights was a single glaring streak. The water ran down their faces as they tipped up their beers.

"That Silk is a hell of a nice fellow," said Bent leaning his long legs into the wind. "Offering us a ride like that."

"Yep," said Graveyard. "He's some kind of a genius or somethin'. Anyways, you can tell he's educated 'cause of the way he talks."

There was silence while the two men sloshed through the water that lay in the deep potholes of the paved road.

"That wind's a whistling bitch tonight," said Bent. "Those breakers on the jetty have to be something awesome."

You know, Bent. I think we timed her just right tonight. By the

127

time we get to the jetty, it'll be almost dawn, and we can tell what's happenin' a lot better in the light. We'll be able to see the horizon and see what's comin'. Then it'll be almost time for the Albatross to open, and we'll go back with some news and warm up. If the storm's quietin' down, everybody'll want to know. We'll be ahead of the weather reports this time."

By the time the two men had reached south jetty, a distance of at least two miles, each had consumed three beers. But when they had climbed the thirty feet to the top of the rock wall, it was apparent that they weren't going to drink and walk over the rugged rocks at the same time. Graveyard had slipped twice and almost fallen, and Bent did fall flat on his back. He saved himself by flinging his hand over a rock, smashing a beer can, and dropping the remainder of the six pack. Several minutes went by while they retrieved the two beers which had lodged in a crevice halfway down the slope. When they finally managed to stand up, they had to bend forward to keep their balance. Here, unprotected by the shore line, the water seethed and raged. The wind seemed determined to blow them off the top as they struggled against her force.

"I think I broke my damned hand when I fell back there," shouted Bent into Graveyard's ear. "And worse, I spilt half that can of beer. Let's go back. We ain't goin' to find out nothin' new out there. This is a regular hurricane."

"Now wait a minute, Bent," Graveyard shouted back. "I didn't ask you to come with me on this here expedition. You volunteered, and your damned hand ain't broke. It just feels like it. You know as well as I do we can't turn back now. Our reputations is at stake, both with the Locals and Silk and Cowboy. I ain't no damned coward, and I ain't a liar. I'm goin' all the way out to the end and see what's happenin'. If you're goin' to chickenshit out, that's O.K. by me. I'll go by myself."

"Shit," snorted Bent. "You're not goin' without me. But let's slide down off the top a bit and finish these beers first. We're not goin' to be able to walk in this wind and drink at the same time. Look at my hand, it's bleedin'."

"O.K.," said Graveyard, sliding down a huge boulder until he was beneath the top of the jetty. "I think you're right. We'll finish them two beers, and then we'll go...but we got to go!"

The two men hunkered on the steep slope and finished their

beer in silence, then they climbed to the top again. Bending low into the wind, they slowly worked their way, slipping and sliding over the treacherous boulders until they had worked their way to the end of the jetty. Now there was nothing but a fifty foot slope to separate them from the violence of the ocean. In the dark the great waves, frothing foam, began somewhere out in the darkness and swept in, seeming to grow in size as they moved. The spray shot up in white sheets high above them.

"Hot Damn!" screamed Bent. "This is one hell of a storm. I didn't know she could get this bad."

"She's a brute. Look out there Bent," said Graveyard. "Ain't that light? Yes, by damn, it's gettin' light. Now just stare out there until we see if there's any break in them clouds." The two men strained their eyes, hanging on to each other with one hand and drinking beer with the other. The two remaining cans had been slipped into pockets beneath their rain gear.

As the sky became lighter, they could see the outlines of the monstrous rocks that lined the jetty. And further out, where there should be nothing but deep water, there was another rock moving toward them. Bent blinked the rain out of his eyes and stared into the semi-light.

"I'll be damned!" he exclaimed. "There's a big rock out there, and it's moving toward us."

Graveyard wiped his hand across his eyes and squinted. When he got drunk and went to the cemetary, he had often seen images of long dead fishermen rising from their graves, and he had talked to them. But this was different. He had never seen a huge moving rock before. Slowly his vision focused, and he could see the outline of a boat moving toward the jetty sideways.

"That's no rock, you damned fool," he shouted. "That's a boat and she's freeboarding. She ain't got no rudder."

They watched spellbound as the breakers swept over the COLUMBIA PRINCE, sweeping her helplessly toward them.

"She's a goner," said Graveyard. "And there ain't nothin' we can do but watch. Sure hope there ain't nobody aboard her."

The two men ground their teeth as the troller crashed bow first into the first rock. She careened off, sideways, until another breaker struck her. She flipped over on her side, and a large section of jagged rock pierced her wood hull. There she hung. To the two

129

men the scene had been like watching a beautiful animal die. They stood silently in sorrow watching the breakers smash against the speared troller. Suddenly, they saw something light come sliding across the deck, hang for a moment, then drop into the surf.

"Did you see that?" shouted Graveyard. "Did you see that?"

"I sure did!" yelled Bent. "That was a body, but I didn't see it until she turned over. How'd it get on deck?"

"Don't know," said Graveyard. "I didn't see it neither. We'd better try to fish it out."

The two men worked their way down to the water and slid into the violent surf until it came up to their waists. Slowly they worked their way sideways until they were directly parallel with the troller. At the same time they stared into the heavy breakers trying to see the body. The work was difficult, and the men found it almost impossible to keep their feet on the slippery rocks.

They were about ready to give up and crawl back to the top of the jetty when something struck Graveyard's arm. It was floating and it had long hair. He grasped the hair with both hands and pulled. At the same time he stumbled backwards over the rough rocks. Somehow he managed to get the body halfway out of the water, but he could move no further without using his hands which were now clinched tightly around the limp form. He stood there helplessly until Bent had worked his way along the steep embankment in order to help him.

Straining and cursing, the two men carried the limp body to the top of the jetty. For a moment they lay there exhausted, then they rolled the form over on it's back.

"Lord have mercy," said Bent. "It's a woman!"

Graveyard nodded in agreement, then he placed his big head on her chest and at the same time felt for the heart beat in her neck.

"She's still alive," he said, "but her pulse is damned weak. We got to get the water out of her lungs if she's goin' to make it."

The two men gently rolled the woman back on her stomach, then Graveyard strattled her and began pressing his huge hands rhythmically on her back.

"Looks like she's hurt bad," said Bent.

"That don't matter none right now," Graveyard said never breaking rhythm. "It's the water she's swallered, and shock we got to worry about. We got to get her out of this weather. Bent, you go for

130

some help. Run like hell to the Albatross and get Silk. Tell him to bring some blankets and somethin' to carry her off of this here jetty. I'll try to keep her alive 'till you get back."

Bent was a tall man who stooped continually forward resembling a tall spider standing on his hind legs. Tonight, as he stood over the kneeling man, he realized this was an emergency, and he must run.

As Graveyard resumed his C.P.R, over the striken woman, Bent began to work his way across the jagged boulders. The moment he climbed off the jetty and the road leveled out, he began running, his long legs striding smoothly over the road. He never stopped until he stumbled gasping into the Albatross and collapsed on one of the long wooden tables.

The fishermen at the counter continued eating large plates of Hattie's delicious sausage, bacon, pancakes, eggs, bisquits with gravy, and in the case that anyone had any money, steak.

"Where's Graveyard?" asked Silk who was sitting at the counter with Cowboy having their morning coffee. "Did you men make it to the jetty last night?"

"There's a wrecked troller out there, just at the end of the south jetty," said Bent breathlessly. "Graveyard and me was standin' out there tryin' to see the first light so we could see if there was any break in the clouds. Right out of nowhere she came planin' in and hit that first big rock at the end. She bounced off that and a big breaker caught her midship and tipped her onto that pointed middle rock. She's hangin' there now like a beached whale."

"Is she one of ours?" asked Silk.

"No," said Bent. "Neither of us have ever seen her before. But that's not all! Just after she tipped, we saw a body come sailin' off her deck. We fished her out, and she was a woman. She's alive, Silk, but she's hurt bad. Graveyard's still out there with her. He sent me for help. He said to get you and get your van. And we need a stretcher too if we can find one."

"Hey, everyone," Silk's voice rose above the murmur of talking men. "Bent says there's a troller on the rocks off the south jetty and an injured woman who needs some help. Rat, you run up and get Doc Ox. Tell him to bring his bag and I'll pick you both up on the way." Silk was standing now and giving orders.

"Hattie, do you have some blankets?"

"You bet," said Hattie. "There's a pile in the back room."

"Does anyone here know where we can get a stretcher?"

"There's some three-quarter inch marine plywood down on the OTTER," volunteered Cowboy. "We can use half a sheet of that for a stretcher. And I got a light tarp to keep the water off her."

Everyone in the Albatross was now on their feet. A troller was in trouble in their port, and a fisherman was injured. In short order the rescue party was organized. Cars and pickups were volunteered, and the fishermen quickly loaded up the rescue equipment: thermoses of chowder and hot coffee, blankets, extra rain gear and warm clothes.

The vehicles arrived at the south jetty. Led by Bent, Silk and Cowboy, the men moved rapidly down the rugged path until they came to Graveyard who was kneeling beside the injured woman trying to protect her face from the wind and rain. He was clad only in his underwear, shirt, and dungarees. His rain gear and old coat lay over the woman. Doc Ox knelt by the woman, took a stethoscope from his black bag and began to undo her clothes.

"Give me a couple of blankets to put over her," he ordered.

While he examined the woman, Silk held a flashlight. The other men turned their backs politely and huddled together against the driving rain. Whether she lived or died, she deserved her privacy as a woman. Silk stared at her face which was framed against the dark ground. Even though she looked gray in the light, he could tell hers was a face he would never forget. He thought he would choke from the lump in his throat. She was too beautiful to die. Now that he had seen her again, she had better not die...

At last Doc Ox raised his head and said, "She'll live. I want to make sure there are no broken bones, then we'll get her off these rocks and somewhere warm."

Silk breathed a sigh of relief, and he handed Graveyard the flashlight. He joined the huddled men and turned his attention to the pinned troller.

Except for her name, the troller might have been a twin to Cap's CRACKANOON. To Silk she was alive and trapped like a wounded animal. Like the woman, she didn't deserve to die either.

"The COLUMBIA PRINCE," Silk shouted. "Any of you men heard of her? Looks like she's hanging there pretty good. I wonder if there is anyone else aboard her."

132

"It don't matter none," said Cowboy. "There ain't no way a man can find out in this storm. Gettin' on that deck would be like committin' suicide."

"Damn," said Silk. "We'd better call the coast guard. Maybe they can do something."

"If they were worth a shit, they'd already be here," said Bent.

"Well, we have to try something," said Silk. "Is there anything we can do to keep her on that rock?"

"Not a thing," said Cowboy. "Those waves poundin' her will break her to pieces in no time. If the storm quit right now, we might be able to hold her on that rock, but that's a mighty big if."

"How?" Silk shouted.

"You see them big rocks further out on either side?" Cowboy pointed to the rocks. "We could wrap a cable around each of them and tie the other ends to the bow and stern of the boat. Then we could hang another cable to the jetty and fasten it midship. Three lines would hold her steady on the rock. Then we could pull her off gradually with a tug. If her pumps was still runnin', she might hold 'til a diver could get inside. But the way she is now... Hell, Silk, that boat won't last a hour under the beatin' she's takin'."

"I sure hate to see her lost when there might be somebody else on board," said Silk. "Let's run into port and get some cables and..."

At that moment a large wave broke completely over the foundered troller, and the backwash pulled her off the rock. For a moment she floated in a deep trough where she moved a short distance off the jetty before she sank, stern first, beneath the waves.

Silk felt sick. He was sure the woman wasn't aboard the boat by herself, but there was nothing he could have done.

"She's gone, she's sunk," he said, the grief showing in his voice.

"Well, that's that," said Cowboy philosophically. "If there was anybody else on board her, he's gone now."

"Damn it, Cowboy," cursed Silk. "Somebody might have died because we couldn't help him. Damn it all to hell."

"I know it's terrible, Silk," said Cowboy. "But we got to look at the best side of things. Maybe at least we could save the boat."

"How can we do that?" asked Silk.

"Well, as she is now, them waves on the bottom will grind

her to pieces," said Cowboy. "We got to move fast. We got to get a diver and some truck innertubes. Right now she's sittin' safe in ten to fifteen fathom's of water. We stuff her full of innertubes and float her to the top. Then we have her towed up to Tristars, haul her out of the water, and patch her up right."

"Then we should contact Claude immediately," said Silk switching his mind to the new problem. "I'm going with Doc Ox. You take the rest of the men, go into town, and start rounding up all the truck innertubes you can find. Tell Claude to get his barge and compressor and his diving gear ready. As soon as this storm dies down, we're going to try to raise that boat."

"Alright men," shouted Doc Ox. "We've got her on the plywood and wrapped up warmly. I don't dare try to examine her any more. She's suffering from shock and exposure, and we better get her out to Jodie's quick or she isn't going to make it."

"Let's get her up to the van," Silk ordered. "We need one man on each corner of the stretcher."

The men cursed and slid over the slick rocks, but they never ceased following Doc Ox's instructions, and that was to keep the motionless body level. At last they reached the vehicles at the end of the jetty.

As soon as the injured woman was safely loaded, the vehicles of the rescue crew pulled out in different directions. One pickup to contact Claude and his diving crew, a second to let the coast guard station know there was a sunken troller near the mouth of the jaws so they could ascertain if there were other passengers on board, and a third to the strip to collect innertubes in preparation for raising the boat.

Silk, with Doc Ox sitting beside him, discovered his mind was in a turmoil. Images of the sinking troller, Cap and Trevor, Semyonov, and the old SEAGULL kept appearing and disappearing in the headlights. Over them all was the beautiful face of the injured woman.

"Do you think she's hurt badly enough that we need to take her to the hospital?" Silk glanced imploringly at the big man.

"No," said Doc Ox. "She's out of the weather and warm. I think I can treat her better in a private facility. Since she came off a fishing troller, and we don't know her, we can guess she doesn't want to waste money on no damn hospital unless it's absolutely necessary. We'll take her to Jodie's house."

CHAPTER

14

JODIE'S HOUSE

About a mile east of the Albatrose Tavern, on Bay Front Road, there is a telephone pole with a tin can lid nailed to each side. Here, an almost imperceptible dirt road turns off the pavement to the right. This is the road to Jodie's house, and the tin can lids are nailed on the pole to help a caller know where to turn. The dirt road meanders leisurely up a slight incline until it comes to an old two story house sitting on a bluff overlooking Bane Bay. At night the lights from the front room window serve as a beacon to fishing boats navigating the curving shoreline from the furthermost boat docks to the end of the bay. If a fisherman is injured and cannot afford a doctor or a hospital, it is to Jodie's house the unfortunate victim is driven until he recovers from his malady. Jodie's house is a legend in Port Bane, and Jodie herself is the guardian saint of the fishermen.

Silvia Castleton lay motionless on the narrow makeshift bed in Jodie's front room. She was not beautiful in a model sense of the word, but she was striking. A tall slim woman, she had high cheekbones, a straight nose, thin aristocratic lips and olive green eyes. Her hair was a curious shade between brown and black and hung straight down her back to just below her waist.

Silvia's father, Jerome Castleton, was a SHAKESPEAREAN actor born in London and educated at the East 15 Acting School and the London Academy of Dramatic Arts. Unable to make a living in that fair English city, Jerome came to Hollywood to make his fortune as an American movie star. Despite his large stature and good looks, however, Hollywood wasn't interested in SHAKESPEAREAN actors from England. He soon found himself starving and drinking and traveling from one repertory theatre across America to another, and it wasn't until he arrived in Ashland that he found his niche in American theatre.

No one who witnessed his first performance as SIR JOHN in Ashland would ever forget it. Years of debauchery had expanded his girth and jowled his cheeks so that he looked the part of FALSTAFF, and with his false white beard and a half fifth of scotch whiskey in his gut, he acted the part to perfection. He was the epitome of the "fat-kidney'd rascal, fat guts, woolsack, whoreson round man, fat paunch, horse-back-breaker, huge hill of flesh, and cup o'sack" that made SHAKESPEARE'S character so famous. As he listened to the music of the standing ovation on the fifth curtain call in the Elizabethan Theatre, he breathed the fresh air of this Siskiyou paradise and vowed never to go on the road again. He had found a home.

The character of SIR JOHN FALSTAFF became magic for Jerome Castleton and synonymous with the annual SHAKESPEAREAN Festival in Ashland. Although he played many other characters in the plays, whenever the part of SIR JOHN FALSTAFF came up for tryouts, he got it. And when he acted SIR JOHN, the patrons of the SHAKESPEAREAN Festival, which included famous people from as far west as San Diego, California, and as far east as New York City, flocked, much as they had done in SHAKESPEARE'S England, to see the fat man walk the stage in KING HENRY IV, PARTS I AND II and THE MERRY WIVES OF WINDSOR. Although Jerome Castleton hadn't made it big in Hollywood or on Broadway, he had become a celebrity in his own right. He was the one and only true American SIR JOHN FALSTAFF.

Jerome was very happy playing SHAKESPEARE. He was called SIR JOHN in every tavern from Ashland to Medford, and the more he ate and drank, the more he resembled in girth and boisterous revelry the character he had become. And, like SHAKESPEARE'S SIR JOHN, he became a victim of his own bad habits, and his downfall was a woman.

Instead of DOLL TEARSHEET in KING HENRY THE IV PART II, Jerome fell in love with a local singer named Leslie Onslott who sang bawdy songs at the Boar's Head Tavern in Ashland. She was a beautiful girl with long black hair and light blue eyes, but except for her perfect figure and seductive smile, she didn't have much talent. They were married in a week, and about nine months later a baby girl was born.

When Silvia was born Jerome was delighted with the tiny sprite

136

and promptly named her after SHAKESPEARE'S virtuous heroine in TWO GENTLEMEN FROM VERONA. But Leslie was already dissatisfied with the thought of being a mother. And she hated Jerome's heavy weight and continual poverty. She could see old age creeping up on her without dame fortune.

One morning, when Jerome was feeding the baby, Leslie walked out the door and Jerome never heard from her again.

Jerome continued his role as the knight about town and watched from the wings as Silvia grew into a striking young woman. When Silvia was eighteen, he shipped her off to college and after that to the American Conservatory School of Theatre in San Francisco to become a professional actress.

It was during Silvia's last semester at the conservatory school in San Francisco when Jerome's lights went out. One night after a particularly heavy imbibing of ale, his mind suddenly went blank. He couldn't remember whether he was Jerome Castleton or SIR JOHN FALLSTAFF. Old age and years of drinking had taken their toll, and the effort to solve this dilemma caused a blood vessel to break in the left side of his head. A stream of spittle ran from his mouth onto his gray-white beard. Through the haze, or fog, or whatever, walked the giant SIR JOHN FALLSTAFF, sword in hand, dressed in full armour, searching for just one more cup o'sack.

Silvia had never told Jerome she didn't want to be an actress. Now her father was gone. She was alone! What was she to do?

Then Matt Butler had come into her life. She had become enamoured with the handsome businessman while she was a first year drama student at San Francisco State University. After a passionate beginning, their romance had heated and cooled for years, depending upon her rehearsal schedules and his frequent trips as a business consultant for the Butler Stock and Bond Corporation. Matt was rich, handsome, and possessed a promising future in his father's brokerage, but every time he proposed marriage, she could never fully interpret her inner feelings enough to say, yes. Sometimes she suspected she loved him; at other times she felt a strange pressure, like she must wait before making that final decision.

But now, that decision had been made for her. Matt was dead.

Silvia opened her eyes and stared at a strange white ceiling. There was another light coming from somewhere. She turned her head and gazed through a large picture window toward a clear sky. A gray hummingbird darted from it's feeder and sipped from a honeysuckle blossom. Blue water rippled on a wide bay, gently rocking a fishing boat in concert with CHOPIN'S POLONAISE which was playing softly somewhere in the background. She felt strangely peaceful, but she knew she was not in her own bed. She turned her head back toward the ceiling and shut her eyes. Suddenly she heard a door open followed by heavy footsteps. She sensed the presence of someone standing beside her. And she felt a hand on her wrist then a slight pressure. She could hear her own heartbeat furiously pumping its blood. She was terrified and opened her eyes.

There was an enormous man towering above her. His face and head were covered with wirey white hair. His bright blue eyes were gazing at her. He put long rubber tubes into his ears before he hovered over her and placed something cold against her chest.

"Help," Silvia whispered hoarsely as she tried to push his hand away.

"Jodie!" the big man called loudly. "She's awake! Bring that damn broth in here. Hurry! She's got to eat something before she shuts her eyes again, or she's not going to make it. She's too damned weak!"

Silvia saw a small woman appear from the shadows. She was dressed like an Indian maiden; yet, her long hair was golden brown and her skin was fair. The woman came closer and stared at her with calm, brown eyes. She smiled as she placed a wooden bowl close to Silvia's mouth and dipped out a spoonful of the bowl's contents.

Silvia felt the pressure against her lips and the warm liquid run into her mouth and down her throat. She choked and swallowed as the light in the room turned black.

When Silvia awoke the second time, the man stood up quickly and grasped her wrist firmly in his huge hand and looked steadily at his watch. Then he nodded to the woman standing beside the bed.

"Now, open your mouth," the woman said."We want to give you some food to make you stronger."

Who were these strange people? Why was she with them? What did they mean, 'make her stronger'? Her mind was racing in circles now as she took spoonful after spoonful of the most delicious broth she had ever tasted. She felt warmth and strength flow back into her body, and she tried to sit up. But her chest hurt, and every bone felt like it had been twisted or torn apart. She put her hand on her left side and discovered a heavy bandage was strapped tightly across her chest. Why was she bandaged? Why did she hurt all over? With a supreme effort she raised her head and looked at herself. She was stretched out on a small bed with a pink sheet over her.

"Oh God," she whispered. "It isn't a dream! What has happened to me? Where am I?"

"You're in Port Bane, in my house," said the woman. "You fell off a fishing troller. Some of the men rescued you from the surf. You're safe now."

"Am I going to die?" Silvia faltered.

"You'll be just fine as soon as the bruises heal and a few cracked ribs mend," said the big man. "But you're going to be mighty sore for awhile."

"Matt. Where is Matt? Is he here?" Silvia asked.

The woman took Silvia's hand gently and held it in both of hers.

"Your Matt is dead."

"Oh, No!" Silvia cried and shook her head from side to side. "Now I remember. A terrible storm. Matt under the water, staring. Glass stuck in his neck. He tried to save me. He told me what to do."

"Let's not talk about that right now," the man said soothingly. "I'm Doc Ox, and this is Jodie. Can you tell us your name?"

"My name is Silvia," she whispered. "Silvia Castleton."

"Can you tell us where you live, Silvia?" asked Doc Ox. "Is there anyone we can call?"

Silvia slowly closed her eyes. She felt extremely tired. She needed to rest, a long, long rest. Perhaps when she awoke she would be with Matt in a familiar world.

One morning she saw a tall young man standing at the foot of her bed. He was clean shaven with a partially bald head, and he was dressed in a naval uniform.

"I wasn't going to let him in," explained Jodie, "but he insisted. He's from the U.S. Coast Guard Station, and he says he has to ask you some questions about the wreck. If you don't feel like talking to him, I'll throw him out."

"I don't understand," said Silvia. "Why should he wish to see me?"

"I am Lieutenant Rasmussan," said the young officer. "I be-

139

lieve you were aboard the COLUMBIA PRINCE when she foundered. Is that correct?"

"Yes," whispered Silvia.

"Was there anybody else on board besides a Matt Butler?" he
asked looking at his notepad.

"No," answered Silvia in a low voice. "There was no one else."

"Can you tell me which fishing port was your home port?"

"We left from Astoria, Oregon," Silvia said, not quite certain
what the term 'home port' meant.

"And, where were you bound?" he asked.

"We were just fishing," she said. "Somewhere south of the
Columbia River bar."

"We have located and notified Mr. Bulter's parents," continued the lieutenant. "Is there someone whom you would like us to
notify?"

"No," she answered. "There is no one."

"Can you tell me how you got way down here on Bane Bay?"

"I don't know," said Silvia.

"Would you like to tell me what happened?" asked the officer.

"I can't," Silvia said.

"Who in the hell let you in!" roared Doc Ox as he came through
the front doorway. "I gave orders that nobody was to disturb my
patient. She's too sick to answer a bunch of damn fool questions."

"Who are you?" asked the lieutenant, not as yet intimidated
by the presence of this huge fierce man.

"I'm Doc Ox, and this is my patient, and no one is to disturb
her without my orders."

"If she's that sick, she should be in a hospital," said the officer. "What's she doing here? You aren't even a doctor, are you?"

"Damned right I'm a doctor. I've been taking care of fishermen in this port for years. Fishermen can't afford the damned hospital in this town, and this woman's a fisherman. Where were you when
she needed you? You're never around when a boat's sinkin'! This
woman's too sick to answer your bullshit questions. You get the hell
out of here, or I'll throw you out. You can just tell your captain, or
whomever it is you report to, that I'll call you when I feel she's up to
talking, and not before. You savvy?"

"I'm going to report you, mister." said the lieutenant shaking
his finger in Doc's face.

140

"Report me to whomever you wish," Doc Ox roared as he backed the lieutenant toward the front door. "And bring the whole damned coast guard, and the police department too. I'd like that." His big fist clenched, and his voice dropped to a dead quiet whisper as he moved closer to the lieutenant's face. "But you'd better get the hell out of here right now, mister coast guard, or you're going to be lying on the floor with no teeth."

Silvia shut her eyes and feigned sleep until she was certain that the lieutenant and Doc Ox had left the room.

"That man is violent!" she said to Jodie as she raised herself on one elbow.

"No, he's not violent to you, and not to me and the fishermen. He just hates the coast guard, and he probably has a good reason. He told me once that they were a bunch of useless gin rummy players who piddle around when they should be answering distress signals. Maybe it's true.

Doc Ox is a special man. I've even seen him fix a bird's wing and feed the little thing until it could fly again. The more you're around him, the more you'll realize how special he is."

"I realize it already, Jodie," Silvia said. "He saved my life. My problem is that I don't know how I will be able to pay him. In fact, I don't know how I can pay you for my room and board. Everything I owned was on that boat, and I haven't a cent. I'm so grateful for your care, Jodie, but I guess you'll just have to put it on an account until I'm well enough to work."

"You don't have to worry about paying anything," said Jodie. "Doc Ox wouldn't take your money anyway. And as for me? To tell you the truth, it's kind of nice having a woman around here for a change, somebody I can talk to. There's nothing much you can say to a fisherman except 'how's fishin'!' Wait'll you meet them, then you'll understand."

The first fisherman Silvia met, however, was not the stereotype she had expected from Jodie's description. Several days had passed, and Silvia was beginning to limp about the room when there was a knock and Jodie opened the front door. Jodie hugged the tall man and they whispered a moment before they entered the room.

"Silvia, this is my favorite fisherman," said Jodie proudly. "He has some things he wants to talk over with you. If you need me, I'll be in the kitchen."

Silk looked at Silvia with steady gray eyes and felt embarrassed. He had been staring at her. He took a deep breath and got control of his thoughts.

"My name is John Silk, Miss Castleton," he said. "I see you are coming along. I'm here to ask about your boat, the COLUMBIA PRINCE."

"My boat? You are mistaken," said Silvia. "The COLUMBIA PRINCE wasn't my boat. It was Matt's. I was only on a fishing trip with him."

"I know this is going to be painful," Silk said. "Are you certain you feel well enough to go into this right now?"

"What is there to go into?" Silvia was confused.

"Well," said Silk. "The night the boat sank the fishermen thought there was a chance she could be saved. They hated the idea that she would be ground to pieces on the rocks. To make a long story short, we contacted Claude, our local salvage expert, collected some innertubes, and raised her as soon as the storm ended. Right now, she's down at Claude's place waiting to have her equipment cleaned up and the hole in her side patched.

I took the liberty of calling Mr. Butler's parents in San Francisco, and they were only interested in having their son's body shipped home for burial."

Silk stopped speaking when he saw the tears in her eyes.

"Please go on, Mr. Silk," she said dabbing at her tears.

"The parents were not at all interested in having the boat or the money to be made from selling her. They must think a lot of you though, because they insisted on turning the COLUMBIA PRINCE over to you. They've mailed an affidavit and the document replacement papers. All you have to do now is sign the documents. The men are already working on the boat, so I suspect she'll be ready for you to fish soon. I guess I came to tell you that you now own a fishing troller."

Silvia felt an unnatural queasiness in her stomach when she thought about Matt and the storm. Somehow, she managed to keep a straight face even though her voice trembled.

"The last thing I want is the COLUMBIA PRINCE. Please send the papers, or whatever it is you have, back to the Butlers."

"I'm sorry to talk about business so soon after the incident,"

142

said Silk sympathetically. "I think I know how you feel. I'd feel the same way myself if I had lost a loved one and almost died. But I hope you understand that the Butlers don't want anything to do with that boat either, and I don't want to get into an argument with them. What should I do about the problem? We have her down at Claude's, and I'd certainly hate to see her salvaged. She's a beautiful boat."

"Then, you take the boat," Silvia said. "I'll sign the necessary papers over to you."

"I can't do that," Silk said.

"But I know nothing about running a boat or fishing for that matter," she said, "and I have no desire to learn the trade. I also have no money to fix one up. So, I guess that means that it must be, as you put it, salvaged."

There was a tense silence for a moment, and both of them stared silently at one another. At last Silk spoke.

"I shouldn't have come to you so early with this problem," he said. "You weren't ready. But if its alright with you, I have a suggestion. I'll refurbish the boat with my money, or the money I borrow from Cowboy. I'll fish her this season on a thirty-five, sixty-five basis, just like I leased the boat from you. You get thirty-five percent off the top, and I will pay for everything she needs to keep her running. I'll bring you a check for your share of the profits after every trip. In the meantime, I'll try to put enough money aside to purchase the boat from you at the end of the season. How does that sound as a solution?"

"I suppose that will have to do," she said. "You seem to have a certain affinity for that boat. I do not! Please understand me on this point. I shall never set foot on that boat again or have anything to do with it."

"Thank you, Miss Castleton, for taking your time to talk with me," Silk said standing.

"Call me Silvia, please," she said.

"I guess, in a way, this makes us partners. I'll do my best for us," Silk said. "I'll be in touch."

Silk had no sooner shut the front door than Jodie came out of the kitchen.

"What did you think of Silk?" she asked excitedly.

"He seems quite polite," said Silvia. "Although, he doesn't strike me as a typical fisherman."

"You're right," said Jodie. "No one knows why Silk's fishing. The only thing I can tell you about Silk is that he watches out for everyone in the port."

"Perhaps he does," said Silvia. "But in my opinion, I think he likes fishing boats better than he likes people."

As Silvia's cracked ribs healed and her bruises diminished, Doc Ox came to the house less and less often. He arrived early in the morning, checked her pulse, inquired about her diet and left. Jodie explained that now Silvia was better, Doc Ox would only come when he thought it absolutely necessary. Until the next fisherman was injured, he would spend all his time pounding penny nails for his boat. Silvia hated to admit to herself that she missed the big man's company and his classical music. He was gruff certainly, but he was intelligent and always had something witty to say.

One morning he appeared, not by himself, but with a crew of men. When Jodie opened the door, they all trooped inside and stood uncomfortably about the room. Doc Ox came in last.

"You're well enough now to have company," he announced to Silvia, "so today I brought some friends of yours who want to see you. They've been concerned about your health."

"Friends of mine?" Silvia asked Doc in surprise. She stared at the rough-looking men. "I don't know anyone in Port Bane. I've met John Silk, and I know you and Jodie but..."

"Well, some friends of mine then," said Doc Ox proudly. "These men are past and present fishermen from Port Bane. We call them the Locals because they don't drift; they stick together and look out for each other like a family."

"Yea, we do that for sure," said a big man with a black beard.

"Some of them used to own boats," Doc Ox continued, but now they work as deck hands or mend nets or whatever, except for Claude there who owns a business. They wanted to meet you and see how you're feeling. They have a special interest in you because each had some part in saving your life and bringing you here."

Silvia stared at each of the men. They were eccentric looking. They all had long hair down to their shoulders, and two of the large men had full beards and earrings in one of their ears. One had sprouted a goatee, another a mustache, and a short man had a little patch of stringy cornsilk hair hanging from his lips. A small slender man was

144

obviously a Mexican and they were dressed in a variety of costumes. They wore old flannel shirts, dirty gray underwear tops, moth-eaten sweaters, and tattered Levi's over tennis shoes or fishing boots. One even stood uncomfortably in a wet suit. Each had a big knife on his belt and most had a fishing cap pulled low over his forehead.

"Men," Doc Ox said grandly. "I'd like you to meet Miss Silvia Castleton, the lady in question."

They cleared their throats and shuffled uncomfortably from one foot to another. The big man with the black beard thought to take off his cap. He raked it off quickly and stuck it under his arm. He mumbled something that sounded like "Gladtameetche." The rest of them simply stared at her like she was some kind of sacred statue.

"I'm glad to meet you," said Silvia simply. "And thank you for helping me."

The men were still silent, so Doc Ox continued trying to ease their discomfort.

"On my left here is Graveyard. He's one of the men who pulled you out of the water down at the jetty." Graveyard took one step forward, and Silvia could see where he got his name. He was a squared off, big man who looked like a granite tombstone. He would look quite at home in a graveyard she thought as she nodded to him.

"Next to Graveyard is Rat," said Doctor Ox. "Step forward Rat." Rat's ruddy complexion turned purplish, and he said, "Oh, hell", with a nasal twang and stepped forward. He looked at Silvia with his little beady black eyes and tried to cover his protruding teeth with his lips. His long nose began to twitch, and he sneezed and blew his nose into a wrinkled gray handkerchief.

"He always does that when he gets embarrassed," Silvia heard someone say.

"Your turn, Bent," said Doctor Ox. By now Silvia realized that each of the men had received his name from the way he looked, and she could almost name them without the introductions. Bent was a round-backed, bowed forward, skinny man. In fact, with his little round head barely covered with hair, and his long tooth-pick legs and arms that reached close to his knees, he could also be called "spider". She could just see him now crawling around the floor, or on the ceiling, or maybe up a steep hillside. Silvia smiled in spite of herself.

"Now, Bent here was with Graveyard when they found you,"

said Doc Ox, "and he did his part in getting you out of the surf."

Silvia smiled and nodded. Next in line was the man in the wet suit. He would either be called Brute or Hulk. She could see the thick neck and bulging muscles on his arms and shoulders.

"Claude here is our expert diver. He does most of our underwater work down at the docks."

"I'm very happy to meet you, Silvia," said Claude.

Silvia was surprised. Not only had she been wrong in her generalization about nick-names, this man seemed well educated.

"And this guy is a newcomer to the area," finished Doc Ox introducing the diminuative Mexican. "He's a trained pilot and an expert diesel mechanic."

"Esteban de Baca," the small man said bowing deeply. "Please accept thees little flowers."

"Thank you," Silvia took the flowers from the small man and smiled. "Thank you all for saving my life and coming to see about me." She didn't know what else to say and neither did the men.

Jodie, who had been standing silently in the background, took the flowers from Silvia's hand.

"I'll put them in a vase for you," she said. "They'll be beautiful on the fireplace mantle."

"Well that's that," said Doc Ox. "You men will have to leave now. You can see the patient's getting tired and needs her rest."

Graveyard was the last one out the door.

"You can call on us anytime," he said over his shoulder. "We're always around."

As soon as the men were gone, Doc Ox smiled broadly and rubbed his big hands together.

"Those men have been pestering me about your condition for a week," Doc Ox explained. "I finally had to bring them over and introduce you. You won't need me again, so I am dismissing you as my patient. I'll come to visit you now and then."

"Thank you, Doc Ox," said Silvia offering her hand. "I appreciate everything you've done for me. Please do come to see me."

CHAPTER

15

CRACKANOON II

Cowboy and Silk stood on Port Dock 3 studying the CRACK-ANOON II as she rocked gently in her slip. She was newly painted white with dark gray trim, and it would take a riveter's eye to see the new planks that covered the hole where the jagged rock had ripped open her hull. Silk felt proud; even the battering of the storm had been unable to destroy her. If the dead skipper had known what he was doing, she would never have had that hole bashed in her side and he would still be alive. Through some quirk of fate or fortune, he had become half owner of a fishing troller.

"She sure is some pretty boat," said Cowboy as he paced the length of the vessel. "Deep enough to ride real easy, and big enough to roam around on."

"She wasn't that pretty when she first came off the bottom," said Silk. "Claude knows his business."

As soon as Claude had heard about the sunken troller, he went into action. To most professional divers the task of raising the boat would have been impossible, but to Claude the job was another opportunity to prove his genius to the fishermen.

Cowboy and the Locals had collected, patched, and tested a large number of truck innertubes and stacked them in neat piles aboard the salvage barge. Claude had serviced the crane, filled the large compressor with gasoline, and assembled the divers' aquatic gear. Air hoses to fill the tubes, once they were placed inside the troller's fo'c'sle and cabin, were wound neatly on their hydraulic reels. Almost before the storm had died the wreck had been located and the barge towed and anchored in position above the sunken vessel.

Silk had watched intently as Claude and one of his partners went backward over the side, adjusted their face masks, and dived for the troller. As the work proceeded Silk paced the deck of the barge and smoked, frustrated because there seemed to be nothing he

147

could do but listen to the compressor engine and watch the bursting bubbles from the work below.

Then it happened. Claude had just switched on the bright spotlights when the COLUMBIA PRINCE rose slowly to the surface, lying on her side like a wounded white whale, water pouring from the jagged hole in her side and from off her deck. The men immediately towed the troller to Claude's dock, and Claude's crew went to work patching the hull. While the boat was being repaired, Silk had telephoned the Butlers in San Francisco and gone to see Silvia at Jodie's house. He had registered their names as the new owners with the Marine Safety Office in Portland and changed the name of the COLUMBIA PRINCE to the CRACKANOON II.

"Do you care if I ask you somethin'," asked Cowboy.

"What's bothering you, Cowboy?" Silk grinned.

"Nothin's botherin' me," said Cowboy. "I was just wonderin' why you wanted to call her the CRACKANOON II. That's all."

"Because she resembled Cap's old boat," said Silk. "They could be twins."

"Now that you got her up, what're you goin' to do with her next?" asked Cowboy biting off a chew of tobacco.

"She's looking pretty good on the outside, but you know as well as I do that the inside is a real mess. The storm destroyed most of her electronic gear, and even though I changed the oil, I know the engine's shot."

Cowboy switched jaws with his wad.

"You've been inside her," he said. "How much do you figure it would cost to get her fishin' again?"

"I'd say a boat like this deserves state of the art electronics," said Silk. "If I had unlimited funds, I'd put in a Furuno, L.P. 1000 plotter and an LC 80 for backup. That would take care of the lorans. She'd need a Furuno radar, new radios, an Icom big set, an Icom two meter VHF with scrambler, a Polaris radio direction finder, and a Bear Cat scanner. Of course she needs a new engine, probably a G.M.C. 671 diesel, and I'd want to put refrigeration in the ice hold in case I ever wanted to go tuna fishing."

"Sounds good to me," said Cowboy. "How much do you figure all that stuff would cost, in round numbers I mean?"

"Well," said Silk. "If I bought a used engine and used electronic gear, I could probably get by with $8000."

148

"How much if everything was new?" queried Cowboy.

"At least $15,000," said Silk realizing now what his friend was thinking.

Cowboy spit out his tobacco and reached in his back pocket for his checkbook. "Tell you what I'm goin' to do. I'm goin' to loan you $20,000 dollars to get you started. Since I sold the OTTER, I got more money now then I know what to do with. I'll keep lookin' for that ranch I want to buy while you do the fishin'. I figure with the right equipment you could pay me back in two seasons, and I'll even let you pay me interest."

After his loan from Cowboy, Silk had felt mighty grand about life in general. With his new boat and equipment, his luck had changed. After each trip he had been able to hand Silvia a fat envelope that contained sums in three figures. He had even put money in the bank, and could still pay for the ice and fuel for his next trip. At this rate he would be out of debt to Cowboy by the end of the season. But that was before the Fish and Wildlife had stepped into the act. Silk had readied the CRACKANOON II to head for the fishing grounds near the rock pile. Her fuel tanks were full of diesel; her fish hold filled with ice. Hoochies, spoons, and plugs were tied and arranged neatly in their racks behind the pit. Fresh cases of herring had been taken on board and lay snug in neat rows within their frozen packages. Stored carefully in chests lining the fish hold below, there was beer, coffee, and food for five days. The heavy diesel engine was throbbing deeply. He had just started to pull out of slip 14 when he heard the bad news about the new fishing regulation. The regulation itself had seemed innocuous in its simplicity. "For the remainder of the current commercial fishing season the ratio of adult salmon allowed will be two Chinook for every coho." The document gave a shortage of coho in certain fishing areas on the west coast of Oregon as a reason for the new rule.

For fishermen who were used to keeping all salmon, the ratio of Chinook to coho seemed insignificant. That is until this last trip. Every salmon on the line seemed to be a coho and had to be shaken off the hook and thrown overboard. He had fished for two days before he caught his first three Chinook, and the dead and dying coho stretched behind his slowly moving troller for as far as he could see. He had radioed several other fishing boats in the same vicinity and found they were experiencing the same difficulty. The carnage and waste of the beautiful coho made him sick. He couldn't imagine the warped minds in Washington D.C. who had spawned this terrible new law.

When he had come in to sell the fish, there was barely enough money to pay off Esteban and give Silvia her cut. He had tried to explain to the diminutive Mexican why he could no longer afford a deck hand, but he found himself at a loss for words. There was no logical explanation for the new scam. Esteban had seemed to understand, but Silk could read the disappointment in the dark eyes. The Mexican had waited patiently until the CRACKANOON II was ready to fish, and he had worked hard and learned fast. Now he was out of a job again. Not knowing what else to say, Silk had suggested the novice deck hand try to get a job up at Corners where Graveyard and Rat were splicing nets. He knew for a fact that Esteban had never seen a net in use let alone spliced one together. What a bitch. After last season's catastrophe with Plank, he had thought he was finally on his way to making a decent living. The world had turned his way. Then out of nowhere this new regulation and this sorry trip. He stared stonily at the flat envelope on the table. How was he going to explain to Silvia this pitiful offering? She wasn't interested in fishing, and she didn't understand fishing regulations. How was he going to explain that for the rest of the season the fishing outlook wouldn't be any better?

He thought of the previous season when he had tried to satisfy Mel Plank. He had never told anyone about how embarrassed he felt when that bastard terminated his lease and reclaimed the SEAGULL. Perhaps when Silvia saw the end of her money, she would terminate their agreement. Somehow he couldn't bear to think about that possibility. Cowboy had offered to lend him the money to purchase the new boat outright, but Silk had refused the loan. He owed Cowboy too much money already. Besides, he liked Silvia; he liked the excuse for seeing her at Jodie's, and he liked handing her envelopes filled with money. Now that the new law had put an end to this arrangement, he must face the music.

For a moment he looked ruefully at the flat envelope lying on the table. It had Silvia's name printed neatly on the outside, but inside it contained only thirty dollars, three crisp ten dollar bills, her share of the money for five full days fishing.

He picked up the envelope and walked off the boat still wearing his fishing boots and the clothes he had worn for five days. Silvia wouldn't care about his clothes or how he smelled as long as he brought his payments to her on time. She was going to be in for a surprise on this trip! Perhaps he could think of something positive to say on the way over.

In his frustration Silk beat on the front door of Jodie's house harder than he had intended. When Jodie opened the door, he strode directly past her into the front room looking for Silvia. When one was the bearer of bad news, he might as well get it over with as quickly as possible.

There, sitting nonchalantly on the front room couch was Silvia. She was breathtaking in new Levi's, a white shirt, and moccasins. Her slender arms protruded from the half-rolled up sleeves. Her face and swan-like neck was framed in flowing dark hair. Silk was almost across the room when she looked at him and smiled. He stopped in mid-stride and waved the envelope helplessly in front of him. He had forgotten the excuses he had prepared or even why he had come. Silvia took the envelope and handed it to Jodie before she spoke. He was suddenly aware of his own smell, and it was terrible. Why hadn't he showered and changed?

"Thank you, John," she said in a low voice. "You seem to deliver the money the moment you hit port."

Silk was flabbergasted. The envelope was gone. Jodie had taken it. He was still standing there with his hand outstretched except now there was no envelope in it. She hadn't even checked the money. She wasn't a Mel Plank after all, but what was he to say?

"I just brought you your payment for the last trip," he mumbled at last. He felt very much like a school boy who had just given an apple to his favorite teacher. "The amount doesn't measure up to your other payments, but it's exactly thirty-five percent of the fish I caught."

"Sit down, Silk," said Jodie. "I'll get some coffee. I'm sure Silvia and I would like to hear about your last trip."

Silk looked at the neat couch where Silvia was sitting and the only chair across from her. He was more than ever aware of the grease on his pants and the salmon blood smattered over his shirt and Levi's. God no, he couldn't sit down. He would ruin whatever he sat on except maybe the floor.

"Well, ah, I haven't the time," he said. "I have to go to the Albatross to talk to some of the fishermen. We have to decide on something we can do about this new fishing problem. That's why the payment is so small."

"What seems to be the problem?" asked Silvia. He wasn't sure that the statement was not in a contemptuous voice. "Has the boat broken down? Have you some unexpected repairs?"

"Well, no," said Silk. "I can't exactly explain it to you, but our U.S. Government has passed a new regulation. It prohibits us catching any coho without catching at least two Chinook."

"Can't the problem be resolved," asked Silvia.

"Hell! Who knows! Who knows what they may be up to next!" said Silk. "They may reverse the decision tomorrow, or they may continue the regulation for the duration of the season. I just hated to disappoint you with bad news."

"John," said Silvia. "Since we are partners in this venture, I think it only fair that I should share in your disappointments as well as your fortunes."

"Thank you for seeing it that way, Silvia," said Silk. "I just didn't want you to hear about the problem from anybody but me."

"I'm sure you can handle any problem that arises, John," she said.

"Did I hear you say you were going to the Albatross?" Jodie came out of the kitchen carrying a cup of coffee.

"Yes," said Silk. "I have to talk to some people."

"Well," said Jodie. "Since you can't stay, Silk, I wonder if you would take us with you. Silvia and I haven't been outside this house since she arrived, and I would like her to see the community and meet some of my friends. She may be going back to San Francisco soon."

Silk felt stunned as he stared into Silvia's clear green eyes.

"What about the boat? What about our agreement?"

"Our agreement doesn't have to change," said Silvia mildly. "Jodie can mail me my checks. Since I am better, I don't have any reason to stay in Port Bane. In San Francisco I can continue my acting career."

Jodie interrupted the conversation by coming back downstairs with a sweater for Silvia.

152

"Let's go," she said simply. "I know you're in a rush, Silk."

"You two go ahead," said Silvia. "I'll read."

"No. I won't hear of it," said Jodie. "You need a change of scene. You've been stuck in this old house too long already. You'll see. You will have a great time at the Albatross. And Silk's just the man to take us."

Silk drove directly to the Albatross Tavern cursing himself all the way. It was one thing to smell like a dead fish around the fishermen but with the ladies? Never! He vowed that from this day forward he would shower and dress appropriately before he went to Jodie's house.

He seated the two women on the back side of the long table at the rear of the tavern where they had a clear view of the entire room. His plan was simple. He would sit on the opposite side of the table and tell Silvia anecdotes about the various interesting customers who were either playing the arcade machines or sitting at the bar silently sipping their beer. That way, he would have something to talk about besides himself, and the stories would entertain and occupy Silvia. If she were interested in someone in particular, he could invite that person to the table. Today was a good day for such an activity. Claude was here, the Locals, and several interesting fishermen, a real kaleidoscope of the bay front inhabitants.

Silk walked to the bar and ordered a couple of wine coolers for the two ladies and a pitcher of beer for himself.

He realized he was speaking over the backs of two strangers so he moved to the next stool.

"Buy me a beer," said the seedy bearded man sitting next to a raw-boned, tall Indian with hair down to his shoulders.

Silk, usually an easy going character, didn't like the man's tone of voice. The statement was an order, and the guy already had a full pounder sitting in front of him. Silk decided to ignore the man. This was one day he didn't want any trouble. He was here for one purpose, and that was to impress Silvia.

"I said, buy me a beer," the man persisted.

"Look," said Silk. "I don't know you, and I don't like to be ordered around. You already have a full glass in front of you. If it were empty, I'd probably buy you a beer, but no one here begs without a reason."

There was silence until Hattie brought Silk's order, then the

153

Indian insolently shoved a quarter on top of Silk's money from a pile of change sitting on the bar. Silk took the man's quarter and put it back in the pile.

"Keep your damn money," he said in almost a whisper. "And don't start any trouble. I don't have time for it today."

Silk picked up his drinks and carried them over to the table. Silvia and Jodie had been joined by Graveyard and Claude.

"Who are the two strangers?" he asked carefully setting a wine cooler in front of Jodie and Silvia. "The two sitting at the end of the bar?"

"They're a couple of panhandlers," said Graveyard. "They been botherin' people in here for a couple of days. They hassle you, Silk?"

"No," said Silk. "But they're a couple of obnoxious characters. Why hasn't Hattie thrown them out?"

"The Indian's worse," said Graveyard, "But they ain't done nothin' so far but talk."

"Is this a bad place?" Silvia looked at Silk.

"This place is as peaceful as they come," said Silk trying to make Silvia feel at ease. "There are just a couple of people who shouldn't be in here. This is a fisherman's family bar, and they're not part of the family."

"There is never any trouble in here," agreed Jodie.

As if to punctuate his words, there was a loud crash, and Silk turned to see the big Indian with a head-lock on Rat. Silk groaned as the Indian threw the small man against the pool table.

"I'll be damned," said Graveyard. "Rat finally got tired of that Indian pushin' him around. I shoulda' warned him to leave the little guy alone."

Now the two men were on the floor rolling over one another, and the Indian had a choke hold.

"My God," gasped Silvia. "He's killing Rat. Somebody do something."

Somehow Rat squirmed out of the Indian's grasp, and while the man was still on his knees, grasped him by the long, black hair and kicked him in the face. Blood spattered everywhere. While the Indian was still stunned, Rat carefully laid his glasses on the bar then grabbed the Indian by the hair again. Backing around the pool table, just in front of the bench where Silvia was sitting, Rat kicked the staggering man in the face again and again, first with his knee, then with his boot. Blood from the Indian's face was now spattering the floor in gushes. Rat banged the Indian against their table so Silk stepped between the fight and Silvia. He shoved both men back in the aisle with his fishing boot.

"I think it's time we left." Silvia looked at Jodie.

"Better stay where you are," warned Silk. "Wait until it's over."

"Don't kick him any more, Rat," said Claude in a concerned voice. "You'll kill the fucker."

But Rat was really into it now. He continued to kick the Indian in the face. The big man's arms were dangling limply at his sides as he followed Rat helplessly around the table and up the length of the bar toward the front door. Every few steps the little man still kicked him viciously in the face. The fishermen at the bar respectfully made a path for Rat and his quarry.

"Call the cops," Rat said to Hattie as he dragged the Indian into the street.

Somewhere a mop had materialized in Graveyard's hands, and he was calmly mopping up the Indian's blood as Silk, Silvia, and Jodie left the bar, their drinks untouched.

"I should have warned that stupid bastard about Rat," said Graveyard again. "The little guy doesn't like to be pushed."

On the way back to Jodie's house Silk's van passed the bloody Indian, now lying next to a telephone pole, and the undersized Rat talking to a cop. The policeman was seriously shaking his head from side to side.

"I just don't understand it," said Silk trying to make conversation. "There hasn't been a fight in the Albatross in ages. It's a peaceful place!"

Silvia stared straight ahead seeing nothing and hearing nothing, her lips pressed tightly together in stern disapproval. Jodie sat silently beside her, not knowing what to say, and Silk was completely numb. His attempt to entertain Silvia had been destroyed in a fluke fight. Now what would she think of him, a man who took a woman to bars where there were brawls? She probably thought there were fights in there every day. He glanced at her perfectly composed face out of the corner of his eye. There was no doubt about it; she was really disgusted.

When Silk let the two women out of the van in Jodie's front

155

yard, Silvia walked directly to the front door and disappeared inside. She didn't say thank you, nor did she look back. Jodie lingered for just a moment.

"I'm awfully sorry, Silk," she said. "I shouldn't have invited us to go along with you. I had no idea this would happen."

"It couldn't be helped, Jodie," said Silk. "Apologize to Silvia for me, will you? I know she doesn't want to speak to me. God, what a mistake. And Jodie," he continued. "I have to get out there and try to catch some fish. When Cowboy gets back from Seattle, tell him to round up all the fishermen: Claude, the Locals, and anyone else who is interested in helping me do something about this new fishing regulation. He'll know what I mean. We have to have a meeting as soon as I get back from this trip. Can we use your house? We'll bring the food and drink."

"Sure," said Jodie. "I'll do anything I can to help. I know this new law is going to hit all the fishermen pretty hard. Now, don't worry about Silvia. I'm sure she won't blame you for what happened at the Albatross. She'll be alright. She'll have forgotten all about that fight by the time you return from this trip."

"I hope so," said Silk. "Thanks, Jodie, for everything."

CHAPTER

16

DEEP TROUBLE

John Silk cursed and slowed the CRACKANOON II to a standstill to make way for the large oncoming barge and the laboring tugboat. He was too late to sell his fish to the Port Bane Fish Company, and he was late for the meeting at Jodie's house. He watched as multitudes of white seagulls circled above and around the rectangular craft which held a deckful of fish tanks stocked full of salmon fry. This was one of the main problems their meeting was all about. Silk knew the tiny salmon in these tanks were coho raised on the Windover Ocean Ranch up Bane Bay. The biologists, who worked at the fish ranch learned slow, but they did learn.

When they had first released their baby coho from the ranch, down the concrete chute into Bane Bay, the seagulls got almost all of the tiny fish before they reached the open sea. Now they put tops on the tanks and barged them out. What biological device, he wondered, lay inside those tiny bodies which would tell them where to return when the time came to be slaughtered in the factory at the head of that chute?

"The fishermen's nightmare," Silk muttered. "Windover Enterprises." They, and companies like them, were behind the new coho regulation! They had paid off their lobbyists and politicians in Washington D.C., and got the bill passed. That law had effectively limited this fishing season. He wished the seagulls could swoop down, rip the lids off the tanks, and devour the fish. Like him, they knew the fish were in the tanks, but being only birds, they could do nothing about it. Well, he wasn't a seagull and he intended to do something about Windover Enterprises.

Silk glanced out the open cabin door and saw that several of the gulls had found a perch on the stern of the boat. They cocked their heads sideways and stared at him, looking very wise in the way of seagulls.

"Sorry boys," Silk spoke indulgently. "You just as well fly back to the fish plants to get your dinner. I don't have a thing for you."

The CRACKANOON II moved slowly by the burdened tug and on toward the dock. Silk felt the nerve strings in his stomach playing a discordant rhapsody. Coffee wasn't going to help.

He reached into the small refrigerator and got a beer. He was in deep trouble and he knew it. By offering to save the CRACKANOON II and to fish her for someone who didn't seem to care about a boat, he had put himself on the line. As if the fishing wasn't bad enough; as if his money problems weren't the worst; why in the hell did he have to have this additional suffering over a woman who didn't give a damn about him or the boat? The whole mess he had created was not only absurd, but illogical, idiotic, and asinine. He cursed himself for his ineptness in dealing with Silvia like he would deal with anyone else. But somehow with her the rules were different. Those clear, green eyes intrigued him. And her voice, so musical, so low. He couldn't even look at her without staring like an ass. Maybe the smart thing to do was to keep his eyes off her.

Cowboy stood on the inside finger of Port Dock 3 and watched Silk glide the CRACKANOON II smoothly up the narrow channel. She was a glorious craft to watch with her trolling poles stretching high on either side of her antennae, the curved white planks that flowed fore to aft, and the natural wood anchor shield that marked her bow.

For a moment Cowboy remembered his first troller, and a weak feeling of nostalgia swept over him as he watched the sleek boat glide easily into her slip and stop just at the moment her sharp prow touched the rubber bumper. He reached for the curled rope on the foredeck and snugged it fast to the rung on the dock.

"The way you docked this old boat made me think of Cap," Cowboy said. "Takes a real good skipper to handle a boat like that. Here, have a swig of scotch."

"Thanks, Cowboy." Silk accepted Cowboy's proffered bottle, tipped it up, and took a long swallow. The fiery liquid burned his throat, but it calmed his green stomach.

"You're two hours late," said Cowboy. "I've been waitin' for you to come in."

"I didn't realize I was so far south. Did you get your land with the stream and the house?"

"Nope, the deal fell through," said Cowboy patting his middle. "I still got the money right here in my money belt."

"You're crazy, Cowboy," said Silk. "Some bum will hit you over the head and steal all your life's savings from you. Take my advice; go to the bank and put it in a safe deposit box."

"Ah, you know I don't trust them bank bastards, Silk," said Cowboy. "They're dishonest as shit. I'd put it in a tin can and bury it if I had some land. No. I been a thinkin' things over, Silk. That lady give you 50% of the CRACKANOON II cause you saved the boat for her, and you do all the work fishin' for her. The way I'm figurin' right now is you ought to take the cash here, like I said, and buy the CRACKANOON II outright. You don't want to own no boat with no lady do you?"

Silk laughed and put his arm around Cowboy's shoulder.

"No, thank you. I already owe you too much money, and if I paid the boat off, I still couldn't catch enough fish to make my payments. No, Silvia isn't the problem. It's the Fish and Wildlife Department we're up against. Their new rule is real trouble. That's what this meeting tonight is all about. If we don't take action soon, there may not be any fishermen left to fish next season. I think I should let the boat stand as she is for awhile. Now let's get over to Jodie's house before everyone leaves."

But Cowboy had his nose in the air. He sniffed in one corner then down in the fo'c'sle like a woman following her nose to some particle of rotten food in her kitchen.

"What's that funny smell I smell?" he asked. "I never smelt nothin' like that on a fishin' boat before."

"Aftershave lotion," said Silk.

"Aftershave lotion!" exclaimed Cowboy. "On a boat? You never used that stuff before have you? Hell, you didn't even shave first. So that's why you never caught no fish. They smelt you! Level with me, Silk. What's goin' on with you?"

"I was just going to the meeting at Jodie's house, and I didn't want to smell like a troll on a fishing trip, so I showered, changed clothes, and put on some Old Spice."

"But ain't you supposed to shave first before you pour that stuff all over you?" laughed Cowboy. "Your face looks like a old porcupine. Hey, I get it! You're tryin' to shine up to that lady Silvia!

159

I seen her for the first time this afternoon. Boy, is she some looker. I'll be damned; John Silk is after a lady. No wonder you want to leave things like they are, and you can't fish worth a shit. Your mind ain't out there."

"Lay off, Cowboy," said Silk seriously. He switched off the cabin lights and walked out on deck. "What I think and do is my business. I'll drive."

By the time Silk and Cowboy reached the turn off to Jodie's house, they could hear the roar of voices all the way up the dirt road. As they entered the house, they could see the reason for all the noise. The frontroom table was cluttered with left-over food, and the guests seemed crammed into the room either sitting on the floor or camped on the available chairs. You could cut the cigarette smoke with a knife. All of the people present were verbalizing vociferously.

Silk's heart did a double take. There was Silvia sitting on the couch where she was smiling and listening to Captain Semyonov.

"I didn't know you brought Semyonov," he said to Cowboy.

"Hell yes," said Cowboy. "I thought he'd been stuck in that barracks long enough. Nobody's lookin' for him, so I thought he just as well come into town and get acquainted. He needs to see how us Americans do things. Looks like he's takin' advantage of the opportunity. I think I'll go pay my respects to the lady."

Cowboy's words stung Silk, but the last thing he intended to do was to show a public interest in Silvia. He moved to the table, scraped the remaining spaghetti from the bottom of the bowl onto a plate, and picked up a chunk of garlic bread.

As he stood and ate, he studied the people in the room. They were so interested in their own conversation that not one had even noticed when he and Cowboy entered the room. There were the Locals, Esteban, Claude, the Captain, Cowboy, and most of the prominent fishermen and their wives. Just the right men to talk to if the fishing season were to be saved. Jodie was in the kitchen scraping plates. Lillian was in a corner; and of course, there was Silvia on the couch between Semyonov and Cowboy. He could hear her low voice speaking in conversation with Semyonov. Cowboy was leaning over to listen. He couldn't understand the words she was saying, but the crystal tone was like music to his ears.

Silk shut his mind off, and he turned his attention to the men.

There were more important issues at stake than his preoccupation with Silvia, so he'd best get started.

"Can I have your attention?" he asked. The buzz in the room stopped and there was silence.

"We were waiting until you got here, Silk, before we got down to business," said Doug.

Silk knew that whatever he said, Doug and Al and Neil would support him and help him get his message across.

"I asked Jodie to call you people here to talk about our fishing problems this season," he continued. "Suddenly we've been shut down with only a twelve day coho season, maybe, to look forward to this year.

The Fish and Wildlife says that if the numbers of salmon increase in the areas where they're decimated, they're going to give us an additional twelve days for coho, but there is no promise this will ever happen. Twelve days will allow the dories to survive, but it won't affect our overall season. At the end of this year the politicians will have a strangle hold on the fishing industry. Next season they will only give us a two Chinook to one coho ratio and no coho season at all. With no coho season the dories will be gone, and that is the beginning of the end for the commercial fishing trollers.

What we've counted on this year is the coho, but they've prohibited our catching these salmon without any logical explanation. They claim that there are no cohos. We know that's an obvious lie. We know there are plenty of coho. Hundreds of them are floating dead behind our boats because we've had to throw them back after they've hooked our lines. It's a waste and a disgrace. The God awful truth is that the Chinook salmon is a thing of the past. They, in their wild state, try to go back to their spawning grounds up the rivers to lay their eggs. They find that their spawning grounds have been polluted to where no fish can live to spawn. Before they can lay their eggs, they swell up and die in all the shit. There is no possible way for the millions of Chinook eggs to survive. The Chinook salmon is a wild fish. Unlike the coho salmon the Chinook cannot be domesticated by man. Her nature is to go back to the same spawning grounds she came from to lay her eggs before she dies.

Because of modern man's imbecility, which we call progress, the Chinook's spawning grounds are filled with toxic chemicals from pulp and paper mills. Prolific algae, nutrient wastes from feed lot

161

and pasture drainage, and process wastes from canning and starch manufacturing plants pollute our salmon streams.

Packing house wastes, fertilizer drainage from farms and orchards, and domestic sewage from urban centers clog our waters. Roadbuilders', landslides, aftermaths of logging, forest fires, and parasites from plankton organisma, help to kill our fish.

I could go on with listing pollutants until hell freezes over. Forest sprays, defoliant sprays, dumping of wastes, domestic sewage, mine tailings and chemicals, and industrial effluents are just a few. Hell, I've forgotten the dams. The damn dams, and I'm not trying to be funny, cause a 5% to 70% mortality loss from turbines, fish ladders, and supersaturation of water which contains nitrogen, oxygen, PCB, and insecticides. Why, the poor salmon don't have a ghost chance in hell.

That some of these crazy Chinook escape to continue up the rivers to spawn, and that any at all survive to return to the ocean is a minor miracle. Therefore, it's obvious why there are so few Chinook in the ocean. The industries that are paying off the bureaucrats in Washington D.C. know this. They pass the law: two Chinook salmon to one coho ratio because they know the Chinook is a thing of the past. Just like they know that we, the commercial troll fishermen, are going to die out exactly like the Chinook salmon.

These men don't care if hundreds of coho have to be wasted to accomplish their purpose. They don't care if legends of fishermen, who have supplied the world with food for hundreds of years, become extinct. They've already set up their fish farms and ocean ranches which contain millions of coho. They can afford to wait a year, two years, three years, whatever it takes, because they know they will eventually have no contention from independent fishermen. In time they will have controlled the ocean's resources.

Doug, you and Al and Neil and Spence and Wilks remember ten years ago when there were fifteen hundred fishing boats in Port Bane. Now there are less than three hundred. We even lost Cap and Trevor because of the government's foreign fishing policies."

"Yeah, and we didn't even get to bury the fuckers," Graveyard said furiously. Silk paused at this outburst before he continued.

"Now, they're squeezing us out of the ports with their legal jargon. We're all here in Port Bane because of an honest profession called commercial fishing, but commercial fishing as a way of life will end here if we don't act at once."

"Just tell us what to do, Silk," piped up Neil. "Maybe we still have a chance."

"I'll do anything I can to them shitasses," said Wilks.

"Us Locals are fired up and ready," said Graveyard. Everybody talked at once. The whole group had suddenly become animated. Silk glanced around the room, and his eyes rested on Silvia. She was looking directly at him. Was it a look of contempt, or was she interested in what he had to say? He couldn't tell. He blushed, embarrased and quickly looked away.

"To stop this government control," he continued. "To thwart big business interests, we have to be heard. We've organized the PBFA into a union, but as yet, nothing we've tried has worked. Last winter the fishermen sent letters to our congressmen, our state legislature, and our governor, and no one paid any attention to us. The salmon prices are rock bottom this season. We can't seem to bargain with the fish companies; the public doesn't care; we're being eaten alive by the trumped-up laws passed by our legislature. The whole fishing industry is corrupt from top to bottom...."

"What the fuck are we goin' to do about them fuckers before it's fuckin' too late, Silk?" asked Spence vehemently.

"We must first make ourselves heard, Spence," Silk said. "The only action we have left is to call for a congressional investigation of the whole fishing industry. If that doesn't work, then we strike, whether we lose our boats or not."

"How do we go about gettin' one of them congressional investigations?" asked Doug.

"Our only chance is writing letters to the U.S. Justice Department in Washington D.C.," said Silk. "If every boat owner and every one on the bay front who is interested, writes a letter to the Justice Department, perhaps they'll take notice. What do you all think?"

"We can all write somethin'," said Graveyard. "And we can get ever' one on the bay front to write somethin'."

"Yeah, I can write," said Rat. "But what am I goin' to write?"

"I think the thing to say is just what we feel about this mess," said Jodie. "Don't you think that's the best thing to do, Silk? Just say what we feel? Or do you think some kind of form letter would be better?"

"I think you're right, Jodie," said Silk. "We all understand the current situation, and I believe a personal statement from each one of us would be the most effective. Now, if someone would be responsible for getting 300 envelopes addressed to the Justice Department, that would help expedite this process tremendously."

"I'll buy the envelopes," said Cowboy. "And I'll buy the stamps to mail them out, too."

"I and Silvia will address 300 envelopes and put the stamps on them," said Jodie.

"Good," said Silk. "You Locals see that all the letters are picked up from the fishermen at the docks and brought back to Jodie's house. You are welcome to use my van if you need it."

"Right," said Bent. "We'll do her."

"If you fishermen don't have time to write a letter because you're out fishing, have your wife write a letter for you. Alright?" said Silk.

"No problem," responded several of the fishermen. "Our wives feel the same way we do."

"Let's make it quick, my friends," Silk said finishing off his presentation. "The sooner we get the letters off, the sooner we get results."

Silk looked around the room and figured it was going to turn into a late night. He had said his piece, and he wasn't in the mood to talk to anyone or to get drunk. The best place for him was outside in the dark. He'd smoke a cigarette or two and wait until Cowboy was ready to leave.

The air was soft, and it was clear for a night in Port Bane. The stars were bright enough to see the constellations. Down the bay Silk could see the bridge lights, and further beyond, the blinking of the last buoy. Tomorrow would be a good day for fishing; he could feel it. There was no use seeing or thinking about Silvia anymore anyway. He had made a fool of himself three times, and three times was more than enough.

He was leaning on his van, finishing his second cigarette, when he heard the front door of Jodie's house open and close. Someone was already leaving. That was good. Cowboy and Semyonov would soon leave, and he could go home and get some sleep. He didn't bother to turn his head to see who was departing. He wanted to be alone in the dark.

164

He was thoroughly disgusted a moment later when he heard footsteps behind him. Somebody had seen him, now he'd have to talk.

"I saw you go outside," said a soft low voice behind him. "And I wanted to say something to you before you left."

Silk was so flustered he dropped his half-finished cigarette and ground it out with his foot. The voice belonged to Silvia, and she had once again caught him at a disadvantage. He had better pull himself together fast and think of something decent to say.

"I just came outside to think and smoke," said Silk defensively.

"I was impressed by what you said tonight," said Silvia. "I really don't understand the problems of the fishermen, but your presentation seemed solid."

"I guess I was kind of selfish," said Silk. "In reality, I was thinking about our boat."

"I'm perfectly happy with our arrangement," continued Silvia. "You have done exactly what you agreed to do."

"Let's not sing my praises yet," said Silk. "Let's change the subject."

"Very well," said Silvia. "It was kind of you to take Jodie and me to the Albatross the other day. It was the first time I had been out in Port Bane. I want you to know that I realize what happened in the tavern wasn't your fault. I was frightened and shocked at the blood and the fight, and I hated the ugliness. Perhaps I over-reacted."

"I asked Jodie to apologize to you for me," Silk said.

"She told me what you said," Silvia said.

There was silence for a moment, but Silvia showed no inclination of going back into the house. Silk couldn't think of anything to do except roll another cigarette.

"Mind if I smoke?" he asked.

"Certainly not," said Silvia. "I'd smoke with you, but I haven't any cigarettes. Would you teach me how to roll one like that?"

"It's too dark," said Silk. "But I thought while Cowboy and Semyonov are still visiting inside, I'd drive out to the lighthouse to have a look at this weather. If you'd like to come along, I could roll one for you in the car."

"I'd love it," said Silvia laughing a silvery laugh.

Silk parked the van facing the sea outside the wire barricade that surrounded the old Port Bane lighthouse. He turned off the engine and rolled a cigarette for Silvia then one for himself. They sat in silence for a considerable period of time while they smoked.

"She's some edifice," Silk said nodding toward the lighthouse. "Those black cliffs we drove across are composed of dark basaltic breccia, a conglomerate probably containing ores of iron which are highly magnetic. On occasion they deflect the magnetic needle on a boat. That's why we never pay any attention to our compass when we're trolling close-in toward shore. That conical-shaped building is ninty-three feet high, and it's eighty-one feet above the ocean. That makes one hundred and sixty-two feet all together. The structure itself is built out of brick and iron, and it's the second oldest active light tower on the Oregon coast."

"It is beautiful in the moonlight. I've always admired lighthouses a great deal," said Silvia. And then she added in a teasing voice. "Is this where you take all your women, John?"

"To tell you the truth, Silvia," said Silk, "you're the first woman I've brought here. In fact, until you came to Port Bane I'd never taken a woman anywhere. It's not that I'm adverse to them. It's because, well, I just haven't had the time. I've been too busy with fishing boats and fishing."

"I understand, John," said Silvia. "I'm glad we came."

The word 'John' kept ringing in Silk's ears. It had been so long since anyone called him by that name. Who was it that last called him John? Elizabeth Baird. But that was over ten years ago. She had meant everything to him then, but she wasn't anything at all compared to Silvia.

"Why do you call me John?" he asked.

"Because that's your name, isn't it?" said Silvia. "It seems funny to me that everyone calls you by your last name. Don't you like the name John? I'll call you Silk if you wish."

"No, John's fine," said Silk. "It's just that I'm not used to hearing it, that's all." He'd best change the subject now or he'd be talking about himself and a lot of personal bullshit he didn't ever want to think about again.

"After you left the house the other night, I started thinking. Here I am lucky to be alive, and I'm spending all my time feeling sorry for myself because I lost Matt and Jerome. Especially Jerome, he was all the family I had. Anyway, I decided to make a change, to look around me, to see the people, to start living again."

"Does that change include me?" asked Silk hopefully.

"If you want it to include you, John, then it does," she said looking directly into Silk's eyes.

At that instant Silk thought Silvia's words were the most important words he had ever heard. He wanted to take her in his arms and kiss her. Instead, he took her hand in both of his and said,

"I'd like that, Silvia. Let's take a short walk and look over the edge, then I'll drive you home."

The night was beautiful. The moon was a silvery circle on which you could see the shadows of its oceans. A gentle breeze wafted in from the northwest bringing the fresh ocean smell on its breath. Light from the moon reflected off the flat ocean until the sheen disappeared out of pure distance. At the base of the cliff the large, dark ocean swells died in gigantic bursts of foam as they spent their final energy against the uncompromising land mass. Silvia and Silk stood motionless on the edge of another world.

"John," Silvia said softly. "When you go out there, promise me you won't get killed."

"I promise," he whispered.

Silk never understood what happened next. One moment he was standing on the cliff holding Silvia's arm, and the next thing he knew, she was leaning against him with her head on his shoulder and he had his arms around her. He was trying to understand how this maneuvering had taken place when he felt her warm lips touch his.

167

PART III

BILLY KYLE
"BUCK" TANNER

CHAPTER

17

SEATTLE

The long, silver, humped 747 was cruising smoothly at 35,000 feet; at least that's what the captain said over the cabin intercom after the bell had rung and awakened Tanner. The cross-country flight from Dulles Airport in Washington D.C. to Seattle had been boring and uneventful. He had slept most of the way after the brief stop at O'Hare in Chicago, and all he had missed were the thousands of miles of dull gray clouds that covered most of the brain mass of inland America. Far below the soaring plane the clouds were beginning to break up into floating puffs, so Tanner could finally get brief glimpses of the dark green forests. The bell toned, and the captain spoke again.

"We are now crossing the Cascade Mountains. Over your right wing you can see magestic Mt. Rainier, and on your left is the now famous Mt. St. Helens. If you look closely, you may still see steam rising from her crater. Further down range to your left you can see Mt. Adams, and still beyond, Mt. Hood. In a short time we will begin our approach to Seattle. We hope you have enjoyed a pleasant flight."

Dr. B. K. Tanner yawned, stretched, and took a sip from his stale martini. He looked down the length of the long silver wing tip, and there in the distance, gleaming in the sunlight above the scattered layer of clouds, stood Mt. St. Helens. She was still magnificant after the tremendous explosion which had disintegrated her pointed top. Sure enough, he could see a puff of steam from the glowing crater inside. The other white-clad peaks in the distance were also stately; and underneath, through the holes in the cloud cover, he could see the dark green of forests. Silver-blue patches seemed to be scattered at random on the lush green background. These must be mountain lakes. The whole spectrum seemed to have come from an artist's brush.

After the concrete jungle of man-made monuments underneath the smog of Washington D.C., this country looked like a portrait of some undiscovered planet.

What was he doing in Washington D.C. with two boys to raise in the confines of an ugly eastern city? Mentally he made himself a note: if he could finish his business on the coast, he would take a one day hike in the mountains, maybe even buy a fishing rod and do some fishing.

Dr. Billy Kyle 'Buck' Tanner was an impressive man by anyone's standards: 6'2" tall, sandy-haired, broad-shouldered,large-boned, and solid as a stone wall. His 'MIKE HAMMAR'profile branded him "investigator" even before he cleared his throat to speak. But his detective eyes were the dead give away; you couldn't escape that uneasy feeling when his unblinking eyes focused on you.

Tanner was a self-made man. He had grown up in the small community of Tahoka, Texas, the son of a poor sharecropper. The eldest of thirteen children, he had been fourteen years old before he could buy his first pair of shoes. From the time he was six he was working in the cottonfields alongside his dad, but then his dad had died, and he had had to support the family alone. Big for his age at fourteen, he was doing a man's job in the town running a filling station.

His first break in life came when he was selected valedictorian of his senior class at Tahoka High School. This achievement earned him a scholarship to the University of Texas where he worked at least two jobs in order to send his family the money they needed to survive. He had to squeek by on his scholarship fund.

In four years Tanner graduated summa cum laude and entered law school at the same university. Three years later, still working two jobs, he received his law degree. Thank goodness, he had a good mind to go with his strong physical stature. The only activity he had really missed was high school athletics, but now he realized his sacrifices had paid off. From the University of Texas he had been hired by a prestigious law firm in Washington D.C. After working as a successful attorney for five years, during which he continued to support his mother, brothers, and sisters, he married a beautiful Washington D.C. debutante, Vivian Kramer, whose father was a senator. He soon found himself working mostly on various minutia assignments for the U.S. Senate, until one day Senator Kramer, in his devotion to his beautiful ambitious daughter, used his influence among his constituents to see that his son-in-law was elevated to the Privy Council of the U. S. Senate.

Celebration was in order; champagne was popped and gifts were opened. But the senator's gift, a solid gold nameplate with the letters B.K. 'BUCK' TANNER was by far the most spectacular. Of course, the senator had no way of knowing that Tanner thought the name Buck was silly, even though he usually introduced himself as Buck Tanner.

The nickname had come about over a debate in the law school at the University of Texas when he had bucked one of his professors over an issue about the Alamo and statehood. He had won the debate and at t᠁e same time the nomen Buck, and he was called Buck for the remainder of his law career. To him nicknames seemed absurd and belonged back in the college days, but if the name worked, well, use it, and a nickname seemed to be a prerequisite in Washington D.C. Most of the members of the senate and house of representatives, and even a president or two, had nicknames on their nameplates, and 'Buck' fit in perfectly with 'Slim', 'Stretch', 'Bull', 'Red', 'Stew', 'Dusty', and a host of others. The name had turned out to be a valuable asset when he was introduced to these stalwart gentlemen. The only problem lay in the fact that he hadn't earned the name in a war or on the football field.

Tanner had been fascinated by the senate investigations going on down the hall from his new office and found himself stopping by to listen to the committee investigative forces relentlessly badgering some poor soul. He soon caught the attention of Judge Elton 'Eel' Powers of the Justice Department. Judge Powers kept a keen eye out for quality people he could fit into his scheme of things. He looked for a man with superior knowledge, ambitious drives, and a penetrating personality. Someone like himself! He could always use another good troubleshooter with a ready smile and a sharp wit to handle some of the unrest going on in the field, so he had hired Tanner as an investigator for the U.S. Justice Department.

Tanner realized that he was still thought of as a young upstart by most of the department heads and that investigator, or the more mundane title, troubleshooter, was low man on the totum pole. But that position could lead to other nominations. With hard work, he might, just might, climb into the main stream of success as a politician or a judge. He was issued a badge and a permit to carry a weapon, and he was delegated full authority to deal with his investigations as he saw fit. However, he found himself away from Vivian and the boys most of the time, and Vivian was always bitching at him.

Up to date, Tanner's assignments had not been as exciting as he had imagined when he started working with Judge Powers. There was that business over the union session in Pittsburg, the women's lib upheaval over smoking in hotels in New York City, and a multitude of others he couldn't even remember. He had just returned from his investigation of the corn price scandal in Kansas, but before he could transcribe his salient facts into a report, he had beem summoned to Judge Power's office. Powers had a problem over a salmon controversy on the west coast.

While Tanner was in Kansas investigating corn prices, the Justice Department had suddenly been bombarded with over three hundred letters from Port Bane, Oregon. Boat owners and fishermen were complaining about salmon prices and shortened fishing seasons.

Their letters were filled with information and statistics about salmon which contradicted the lobbyists in Washington D.C., who, for the most part, were hired by large business concerns. They charged the government with fraud, neglect, and internal corruption. Worse, they demanded action immediately, and they even asked for a congressional investigation of the whole industry. The focus of the controversy appeared to be over the impact of ocean-ranching in Bane Bay where the Port Bane fishing fleet was anchored. On one side was the Oregon Department of Fish and Wildlife and Windover Enterprises Inc., who owned the ocean ranch; and on the other side were the anti-private aquaculture and pro-wild-fish forces spearheaded by the Port Bane Fishermen's Association. Tanner's mission was to quiet the upstart organization and work out a compromise before the next year's salmon season was set.

Tanner smiled to himself as he remembered the meeting in Judge Power's office.

"You have to leave for Port Bane, Oregon, immediately," the judge had said nervously.

"But judge, I need a break! I just got in from Kansas. I still haven't compiled my report from those findings, and I haven't seen Vivian and the boys for days," he had objected.

"Listen, Buck. You're my best man. I wouldn't consider asking you to take this trip unless I thought it absolutely necessary. When we receive one or two random letters from some non-descript place in the field, it's not that important, and we usually trash them. But when we get three hundred of the things from a single location, and all of them saying essentially the same thing... now, that could mean trouble! You've got to handle this problem for me, Buck."

"All of the letters say the same thing?" Tanner had asked.

"Essentially, yes," said Judge Powers. "Some of the letters were erudite, others simplistic and the rest called me every name in the book including: no good asshole, dumb sonofabitch, slow-witted numskull, punch drunk fool, confounded stooge, filthy pig sty, bloat gut, senile old fart, and one even called me a "sheet". You've got to get out there right now, Buck. Investigate and report your findings back to me so we can head off any trouble that might be developing. Lobbyists, environmentalists, and big business interests have got wind of the news, and they're interested in the controversy. They're bringing big pressure to bear on me. If you can work out a satisfactory compromise between these two forces, the dispute can be halted, and the storm that is brewing here in Washington can be quelled."

"But, Judge Powers," Tanner had protested. "I know absolutely nothing about the west coast fishing industry."

"That's good!" Judge Powers said. "That way you can be objective."

Now, here he was flying out to the west coast over some controversy about salmon with an assignment to solve some differences between commercial fishermen and private industry. He had ordered salmon in the restaurants of Washington D.C., and the fish seemed hardly a reason for such a controversy. Hell, fishermen caught fish, sold them, and then you ordered them and ate them.

To Buck the controversy seemed simple. The fishermen wanted long seasons; private industry wanted control of the fish they raised; and National Marine Fisheries Service wanted a say over the matter. An easy compromise among three interested participants was the answer, and that's what he would go after. After his success on this trip and a peaceful settlement, he, Buck Tanner, would demand a promotion and a new job.

"We're making our final approach to the SeaTac Airport," came the captain's voice over the intercom. "We hope you have enjoyed your flight with American."

Tanner had been watching from his window seat, and from the air the skyline and tall buildings of Seattle looked like every other

American city except that he counted seven islands in the midst of a large body of water. He wondered how long it would take him to get to Sandpoint from Seattle.

His schedule was tight so he wouldn't have time for sight-seeing. First he had a one hour session in Sandpoint with a Dr. Blackburn, the Director of National Marine Fisheries Service, who was supposed to brief him about another problem. Then he was to catch a second flight to Portland, Oregon, where he would lease an automobile at the airport and drive the hundred or so miles to Port Bane, Oregon. Once there, he would meet with the Port Bane Fishermen's Association.

The salmon problem sounded dull as hell, like the corn in Kansas. Why couldn't these people handle a minor controversy which was happening in their own back yard? Oh, well, that's what his job was all about, bringing the pieces together and making people happy. After his work with the farmers, this issue over a single species of fish should be easy to settle.

Tanner's adventures started almost before he stepped through the deplaning doors at the Seattle International Airport. A pretty, long-legged female took his briefcase and smiled.

"Fran Oliver here," she said. "Your driver and Dr. Blackburn's assistant."

"Buck Tanner," he held out his hand. "Washington, D.C."

"Shall we press on?" she said.

Before Tanner could pick out his suitcase from the revolving carosel, Fran was headed for the exit. He caught up with her at the N.M.F.S. sedan, where she was already sitting in the drivers seat with the motor running.

"Buckle up, love," she said. "I'll have you there in a jif."

The car lurched forward and around the traffic on the wrong side of the street before it ran a red light. Tanner reached for the front dash with both hands.

"You must be from England," he said gritting his teeth and watching the speedometer climb.

"Sussex. Bred and born," she said looking him over. "You must be a brave sort. Most American's buckle up."

"I can't find the damned buckle," Tanner said disparately. "What in blazes is the speed limit in this state anyhow?"

"Doesn't matter," said the woman. "We'll be there before you know it."

Tanner never did get his seat belt fastened. He hung on with both hands as the English lady wove in and out of the heavy traffic and swerved into a parking place in front of a three-story building. He followed the blue clad Fran through the maze of offices without a word. When he finally met Dr. Blackburn, he was still shaking.

Dr. Blackburn was a short, dumpy man who had gone to fat. With his bald head and little bright eyes behind his wire-rimmed glasses, he looked more like an H & R Block accountant than a fish expert. Tanner felt like a lumberjack in a suit compared to the rotund man.

He had been sitting here for fifteen minutes now, and every time the pudgy man began a sentence, a button on his phone lit up. At last, he shouted into the phone,

"Tell whoever calls that I'm out for the day. There's not a damned thing I can do about it anyway. I have a visitor from Washington D.C., and his time is limited."

"Problems?" asked Tanner finally relaxed.

"A war," said Dr. Blackburn. "A war which we can't possibly win. The N.M.F.S. is caught between a rock and a hard place, and I'm going to end up the fall guy."

"I didn't know there was a war going on in Seattle." Tanner smiled pleasantly.

"It's not a war, actually," answered Blackburn folding his stubby hands on his desk. "If it was a war, we'd kill the damned animals and be done with it. But no. We can't touch them."

"Which animal do you want to kill?" asked Tanner."

"Sea lions," said Blackburn. "The whole problem at Lake Washington started back in 1916 when the Lake Washington ship canal and it's locks were built," said Dr. Blackburn as if Tanner knew what he was talking about. "The change in the level of the lake shut off the steelhead's natural migration route to Puget Sound. However, the fish found new routes, and now we have a run of 4,500 steehhead, including 2,900 wild natives which swim up the eight mile-long canal from Puget Sound to Lake Union and Lake Washington, then into the streams that wind through the Seattle area.

Two years ago, instead of the sixteen hundred wild fish we need to perpetuate the run, the counters found only four hundred.

174

When we looked around for the causes of the decimation, we found the sea lions as the only probable cause. Last year the game department tried to get rid of these mammals by harassing them with seal bombs. When that didn't work, we played sounds of the orca whale, which is the only natural enemy of sea lions. That didn't work either.

This season, the sea lion non-breeding males from California showed up a month early. We tried taste aversion first, baiting weighted dead steelhead with lithium chloride to make the sea lions sick. That worked for about a week, then they learned they could eat the unattached fish.

When the wild run joined the hatchery run this month, we bombed them again. Yesterday we set off seventy-two bombs and still lost thirteen fish. That left us our only option. We had to catch the sea lions and return them to California.

We brought in a gill-netting boat and laid net across the two hundred foot channel. However, the net went only twenty-four feet deep. A ten foot tide made the water thirty feet deep, and the sea lions just swam under the net. We've spent all day sewing a twenty-four foot addition on to four hundred feet of fish net. That's what all the phone calls were about. If we don't catch them tomorrow, I don't know what we're going to do. Hell, there's a crowd of one thousand spectators watching the action, including sea lion protestors and everyone else that wants a cheap shot at N.M.F.S." Tanner was silent.

"So there you have it," Blackburn finished. "A battle between six marine mammals and the game department, the Washington Department of Fisheries, the U.S. Army Corps of Engineers, and us, the National Marine Fisheries Service. The six sea lions want the steelhead for dinner, and our agencies want to preserve the run."

"That seems like a lot of firepower against just six sea lions," said Tanner. "You mean to tell me that all those agencies and the people who run them can't handle six animals? It seems to me that the lives of six sea lions are not as important as the 4,500 steelhead. Why don't you just shoot them?"

"That's easier said than done," said Dr. Blackburn suddenly standing. "If I had my way, that's exactly what I'd do. But you see, the problem is so complex that a simple solution like shooting them will only bring the wrath of the people down on our heads." He began to pace the floor, his short legs moving quickly but not covering much territory.

175

"Steelhead are a prime sport fish. Sea lions are one of the endangered species listed under the Marine Mammal Protection Act of 1972. When one species threatens another and man steps in, the situation has social, political, economic, cultural, environmental, biological, and emotional implications. Our agency is caught right in the middle of a war between the worlds, the world of the sea creatures versus our world. The sports fishermen and the people who raise the fish want them protected, and the environmentalists and conservationists demand that the sea lions remain alive. We're trying to solve the problem in the only feasible way, yet we're getting a black eye for trying to save the lives of both species. Hell, every agency head of every involved organization is in Seattle. The whole business is costing the taxpayers a fortune. The stupid people who want to save the sea lions should think about their own pocketbooks."

Secretly, Tanner was already on the side of the sea lions. He wished he had time to drive down to the channel and see these creatures for himself, but he had a plane to catch.

"I wish I could help you," he said to Blackburn who had now sat down and was drumming his fingers rapidly on his desk. "But I'm from Washington D.C., and I know absolutely nothing about steelhead trout or sea lions. Judge Powers briefed me before I left, and he said you might have some information that would help me with my problem in Port Bane, Oregon."

"Oh, yes, that," said Dr. Blackburn. "I talked to Judge Powers by phone, and when he said he was sending out an investigator from Washington to survey the west coast fisheries, I asked to see you. I have prepared folders of information from N.M.F.S. on every fishing industry on the west coast: commercial, joint venture, ocean-ranching, net-pen-rearing, and even sports fishing. This information should be helpful in your investigation."

"I really appreciate your effort," Tanner said. "I have only the information I received in Washington, and those are mostly complaints that have been flooding our department. To be frank with you, I have never seen a Pacific coast salmon unless it was a fish that was accidentally slipped into one of our supermarkets back east. This material is definitely the kind of information I need. But mainly I have to visit each of these industries and objectively gather the facts about salmon. I understand that none of the people who catch these salmon are happy with the current regulations."

176

"That is essentially correct," said Blackburn. "It's mainly a case of not making all the people happy all the time, but that's not why I asked for you to stop by. Like I told Judge Powers, there is a chance your investigation might be able to help us with another problem. You scratch my back, and I'll scratch yours so to speak."

Tanner hated cliches. It seemed like lately he was always running into the kind of mind that used cliches as an excuse for intelligent conversation, and they were always irritatingly punctuated with a wink or a jab in the side. Now, here was another one, a wink this time. He was anxious to be on his way.

"And the problem?" he asked impatiently.

"To make a short story short," said Dr. Blackburn staring meaningfully into Tanner's eyes. "I'll put it in a nutshell. A few months ago we hired a foreign fisheries observer out of Port Bane. These young observers work for N.M.F.S. and help us keep track of the foreign fishing ships that are allowed to fish within two hundred miles of our coastline. We have one on every ship: Polish, East German, Japanese, Korean, and even Russian. Anyway, this man's name was John Silk, and he was, or seemed to be, one of our best observers. His reports were always accurate, on time, and informative. I checked them myself. For two months the activities on the MURMANSK, (That was the Russian ship he was assigned to), seemed perfectly normal. Then for some reason this John Silk jumped ship. To the best of our knowledge he left on an American joint venture boat called the OTTER." Dr. Blackburn took a resigned breath. "When we asked the Russians about the observer, they wouldn't admit to knowing anything about why John Silk left the ship, but they seemed to be angry about it. And then, when we tried to contact this John Silk, he was nowhere to be found."

"Are we talking about foul play here?" asked Tanner thinking of an obvious answer.

"No, no. Nothing like that." Blackburn waved the question aside. "No Russian would jeopardize his fishing rights for a man! In fact, there is no foreign ship out there that would take a chance on losing the rich fishing grounds off our west coast. They know a good thing when they get it! It's just that they won't communicate with us about anything except fish. The truth is that they probably don't know any more about what happened than we do.

We have never before had one of our observers leave a foreign ship without permission, and the only complaint we've ever received about any kind of mistreatment aboard one of these ships was a woman who radioed in one day that she was being molested by some of the Japanese deckhands. Frankly, we wondered about what kind of woman she was." Tanner figured a generalization without an investigation was typical of N.M.F.S.

"The point is this," continued Blackburn. "I'm just curious to know what the hell happened to our observer John Silk. At first we didn't pay much attention to the problem. We went ahead and sent a replacement for him. We figured he was probably ill or tired of the long, boring hours and somehow didn't have a chance to get in touch with us. Young people these days can't handle any real adversity you know; they're irresponsible. Hell, they don't care about anybody but themselves. They don't give a damn that we have to turn in our reports to the man upstairs!"

"And what is it you want me to do?" asked Tanner.

"I understand this John Silk is from Port Bane," said Blackburn. "He's a fisherman and is said to be one of the founders of the Port Bane Fisherman's Association. Since you are going to interview those people, I thought you might run into the man. We don't hold any grudges here at N.M.F.S., but for our records we would like to know what really happened aboard that ship. Our whole foreign fisheries observer program depends on the good will of our foreign fishing fleets, and to be frank with you, for some reason, since the incident, the Russian fleet is not being very cooperative."

"Why don't you send someone from your own organization down to Port Bane?" asked Tanner. "I think anyone from N.M.F.S. would be better qualified than I am to find this John Silk."

"Don't think we haven't already tried that," said Dr. Blackburn. "But you see, Port Bane is a very close-mouthed place. Everyone on the waterfront is hiding something in his past, and they won't talk to strangers. They all have aliases, and when our men tried to question them, no one there would claim that he had even heard of a John Silk. We don't even know whether his name is real or not."

"Well, if your people can't find him, I doubt very seriously if I can. But if I run into him, I'll certainly be very glad to cooperate with you," said Tanner.

At that moment Fran entered the door.

"Sorry to disturb you, Dr. Blackburn, but we must dash off to

178

SeaTac or Dr. Tanner will miss his connections."

"Thank you for your help," said Tanner rising. "I hope you catch your sea lions."

"Oh, we will, we will," said Blackburn extending a chubby hand which disappeared in Tanner's huge grip. "But remember, if you come across a John Silk, let us know immediately."

Tanner picked up his briefcase and walked out the door. Port Bane, Oregon and the Pacific Ocean would certainly be a refreshing change from another bureaucrat in Seattle.

When he reached the car, Fran was already in the driver's seat with the motor running. Tanner walked around to her side and spoke through the open window.

"Thank you, ma'am," he said cordially. "But I'll catch a cab since there's still plenty of time before my plane."

CHAPTER

18

SALMON STRIKE

Tanner pulled the drapery cords, and the purple-flowered outer curtain and misty-white inner curtain slid apart. The panorama of the Pacific Ocean unveiled herself.

As far as the eye could see there was an azure-blue sky and a deep lapis lazuli ocean punctuated by glistening white waves rolling in a rhythm of their own magic. This was Tanner's first perspective of the Pacific Ocean, and it almost overwhelmed him.

A second glance to the left told him there was some kind of barrier built of giant, basaltic, moss-covered rocks stretching out into the ocean. The waves on the outer edge were exploding into great bursts of scaling white foam. To the right was a protruding green penninsula with a tall white lighthouse sitting on top. Tanner took a deep breath. This must be like the place SIR THOMAS MORE described. "An imaginary and remote country, the epitome of perfection." He was beside himself. If old Powers had known he was sending him to UTOPIA, he would have come himself.

Strolling on the beach were a few people, couples mostly, arm in arm, walking casually, talking intimately. Where were the crowds and the beach chairs like at home on the east coast?

"Damn," muttered Tanner to himself, "these people don't know how good they have it. I'd better swim before the crowd hits the beach."

Last night, when Tanner had checked in, he hadn't realized there was an ocean outside his window, and the truth was he wouldn't have cared one way or another. By the time he had leased a Thunderbird at Portland airport, driven through the City of Roses, as Portland is known, found Highway 99, entered the Van Duzer Corridor with the damn blinding fog, a narrow road and the blackest black he had ever seen, he had finally come to Highway 101. After winding around the mountainside for another hour, and finally locating Port

Bane and the Port Bane Inn, Tanner was beat to hell. He hadn't even felt like lighting the fireplace or pulling off his clothes. He had barely managed to hit the bed. This morning, however, he felt like a new man. He had forgotten the wretched drive from Portland the moment he saw the blue Pacific Ocean.

He examined the suite of rooms he had rented. Excellent, he thought. They were large, clean, had pictures on the walls, plenty of light, and a colored television. A small refrigerator filled with set-ups stood next to a bar which held every kind of liquor he could possibly need for entertaining guests. There were logs in the fire-place, comfortable chairs to sit on, and a round table with a hanging light over it.

"Well, so much for that," Tanner said to himself. "Now for my before breakfast swim."

It had been years since he'd been to Galveston to swim in the ocean. Maybe he shouldn't call the Gulf of Mexico an ocean, but he'd remembered the warm water and the rolling surf.

He hummed GALVESTON, OH GALVESTON and stepped into the blue and white swimming trunks Vivian had packed for the trip. It was sweet of Vivian to remember to put them in he thought. He would have to call her tonight and tell her about the ocean.

He grabbed the big, white towel from the rack in the bath-room and ran barefooted down the hall taking the stairs at the end of the walkway two at a time. He raced down the beach access steps to the sand and across the sand to the edge of the surf. He pumped his legs up and down vigorously for a moment then took three giant strides like a broad jumper and made a flying leap into the surf.

Under he went, belly first, rolling over and over in the swirl-ing water. Everytime he managed to get a foothold on the slippery sand, he was tumbled again.

At last he felt his foot touch solid ground, and he gave one mighty thrust and burst from the water with a velocity that carried him three feet into the air.

"YEAOOOW!" he yelled to the people. "Don't come in. It's a solid block of ice out here."

Tanner hit the beach running, He grabbed his beach towel and bounded up the beach access steps three at a time. Now he felt like a congealed, giant wart. The worst part of it was, he thought he heard someone giggle as he ran past.

181

Powers had been right. The water here was too cold for swimming, and he'd damn near frozen his ass off.

"What I need is some red-eye gravy," he muttered to himself as he stood under the hot shower. "After that I'll get to my business, and neither Vivian nor Judge Powers will need to know about my little swim." He dried himself vigorously and dressed for the day; Harris tweed sports jacket, tan worsted trousers, gold silk shirt and brown loafers, and went down to breakfast.

Port Bane Inn didn't serve the gravy that he had imagined, but the hot coffee and American plate of eggs and bacon would do. He felt grateful as his still chattering teeth ground at the edge of the coffee cup. Swimming in the Pacific Ocean was definitely not his forte.

While he ate, he thumbed through some of his notes on the city. According to the map Port Bane was a small city of about 12,000 full-time residents, much smaller than the cities with which he was familiar. To locate the PBFA, he had only to find his way to the Pacific Coast Science Center which was only three miles away from the inn. They would definitely help an investigator from Washington D.C. to find his way about the community. They would surely know about a fishermen's organization if not John Silk. He finished his breakfast, stuffed the papers back into the tan leather briefcase, and strode to the parking lot. 'Thank God it was a small town,' he thought. He wouldn't be too enthusiastic about a place a hundred miles away.

He crossed the ancient Bane Bay Bridge and turned right toward South Beach. Shortly, he pulled into the parking lot of the science center. He locked the door of the rented Thunderbird and walked into the grey stone building in the same forthright manner he usually entered his own building. At least his teeth had quit chattering.

"Good morning," he said briskly to the prunefaced man at the reception desk. "My name is B. K. Tanner and I'm trying to locate the Port Bane Fishermen's Association."

"Never heard of it," said the receptionist not even glancing up to see who was speaking.

"Perhaps I didn't say the name right. I believe they go by the initials PBFA," Tanner said looking at a page of his notes.

"Nothing like that around here," the man said sleepily.

"Are you acquainted with a man by the named of John Silk?" Tanner pursued.

"That's a new one on me." The man shook his head unpeturbed.

"Perhaps one of the other people here have heard of the association or John Silk," asked Tanner nodding in the direction of three or four people sitting at their desks.

"Listen, Mister," said pruneface statically. "I've been here thirty years, and if I haven't heard of those people, nobody here has. As far as I know there's never been an organization by that name in Port Bane, and there's never been a man by the name of John Silk. What you might do is go to the Fish and Game Department over at South Beach Marina and ask them, if you're still not satisfied."

"Well, thank you," Tanner said. "I'll do that."

Tanner found every kind, shape and size of sports fishing boat at the South Beach Marina. He also found the sports fishermen cleaning up their boats and stocking them for the next salmon they intended to catch. But there was no one who had heard of the Port Bane Fishermen's Association or it's leader John Silk. They explained,

"We wouldn't have any reason to associate with that group, you understand. We're under different laws and have a different philosophy about fishing. Why don't you try the strip. Someone along the strip might have heard of them."

"Thank you," Tanner said. "Can you give me directions to the strip?"

"The strip is the same as Highway 101. There are businesses on both sides of the street."

For two hours Tanner walked from one end of the strip to the other. He had gone to the Oregon Fish and Game Department, the county courthouse, the chamber of commerce, the public utility building, and the telephone company, and he hadn't found a single clue to the PBFA or John Silk. The whole thing was beginning to puzzle him. Where did all those letters come from. And in a town this size, how was it possible that no one seemed to know about the Port Bane Fishermen's Association or the letters sent to Judge Powers? He had been at it since breakfast, and it was close to one o'clock. Maybe the thing to do was to go back to the inn, eat some lunch and talk to the innkeeper there.

"Where are the fishing boats that are supposed to be here," he asked."Where is your main port? Where are your fishermen? I've been trying to locate a fishing organization here for hours, and they seem to be non-existant. I know there have to be commercial fishermen somewhere in Port Bane."

"I hate to mention it to you,Mr. Tanner," said the easy-going manager. "But you're probably going to have to go down to the bay front area to find what you're looking for."

"What's wrong with the bay front area, and why shouldn't I go down there?" Tanner asked. "If that's where the boats are, that's damn well where I want to go."

"There are some places down there that are not too safe for strangers," the manager said hesitatingly. "Most of the characters down there are an unsavory lot. If you must go,watch your back."

"This is getting absurd," Tanner said to himself as he drove back to Highway 101 and turned left toward the bay front. "You'd think I was going to a Washington D.C. ghetto. This exercise is turning into serious private eye work."

He parked in the first vacent lot and walked past the shrimp plant where he watched a dozen women and men hovered over a conveyor belt and hastily sorting shrimp as they rolled by. He passed the Rip Tide Bar and Grill, Nick's Seafood on the Waterfront, the ice plants, the packing plants, and a pick-up unloading a huge pile of some kind of fish. He saw a vacant tavern, art galleries, souvenir shops, food bars, fishing gear stores, large nets being mended by bums, stray dogs, and a crowd of mean looking people sitting in every tavern. At last he decided the only way he was going to be able to talk with anyone along the bay front was to go into one of the many taverns. Before him stood a broken-down wooden structure with an unobtrusive sign on the window that said the Albatross.

'Well,' he thought, 'this one was as good as any.'

When Tanner placed his face next to the window and shaded his eyes so he could see into the dark interior, he found himself focused on a fierce bearded face which was staring right back at him. Shocked, Tanner took a deep breath and walked through the doorway. If he were ever going to get any information about the port, it was probably inside this bar.

As the door closed behind him, Tanner had the feeling that he was in the wrong place at the wrong time. Almost every stool at the bar was filled with bearded men in knitted hats of some kind, big boots or canvas shoes, plaid wool or sweat shirts, and overalls or

Levi's, all talking animatedly before large pitchers of beer. The nickel machine, slot machine, video game, and poker machine were going full blast, each with a gyrating man behind it. What the hell was going on here?

Tanner stood still for a moment feeling completely out of place in his silk tie and shined loafers.

"Ahem," he grunted. He had a habit of clearing his throat when he was unsure of a situation.

No one looked up or paid the least bit of attention to the stranger. To the bartender and fishermen he was just another misguided tourist.

"I'll have a glass of beer," he said more forcefully as he steppped up to the bar.

"You get a better deal if you order a pitcher," said the woman bartender with long straight hair.

"Give me a pitcher then," said Tanner amiably.

"You want a glass too?" she asked.

"Of course," said Tanner.

"Suit yourself," she said and grudgingly slid the two items in front of him.

Tanner was on his second glassful before he realized what the bartender was insinuating. Only he was using a glass. The other men were actually drinking from their pitchers. He quickly set the glass aside and took a big swig from the pitcher before he ordered a second one. He looked at the bearded men on either side of him, but neither of them showed any inclination for barroom conversation. They totally ignored him and stared blankly at the back wall.

"Folks don't say much around here," he said conversationally to the bartender.

Instead of answering him, she too turned her back and walked toward the far end of the bar. 'Damn,' he thought to himself. 'These people were not only tight-lipped, they were actually rude.' He noticed the big man at the end of the bar who had stared at him through the window. He was getting nowhere so he just as well start at the top. He picked up his pitcher, walked to the end of the bar and set it down beside the fierce looking fellow. The guy was actually larger than he was, but Tanner figured he could whip the big guy if it came down to it. He was getting pretty pissed about the whole situation. He had come all the way from Washington D.C. to help these people and here they were treating him like a leper. Well, if it took a fight to find out about the PBFA, then so be it. For a while the two men drank silently from their pitchers looking neither to the right nor to the left.

"I'm looking for someone," said Tanner at last.

The fisherman, whoever he was, didn't blink an eye. He continued to ignore Tanner and stare straight ahead.

"I'm looking for a man named John Silk who has something to do with the Port Bane Fisherman's Association," he pursued in a voice loud enough for everyone to hear. That got their attention. The electronic machines became silent and the conversations stopped. The big man next to him continued to stare straight ahead.

"Sheet," said a dark man two stools down. "El este no bueno pinarda narco."

"Where had he heard that word "sheet" before? Something was familiar so he must be getting close."

"I'm B. K. Tanner from Washington D.C., and I have business with the man called John Silk," he announced in an even louder voice.

"It's just another damned fed," said someone else. "What the fuck!" Then the conversations started up and the machines began to gong and rattle again.

'This is the damndest place I've ever been to in my life,' Tanner thought. He was really angry, but there didn't seem to be a thing he could do. He downed the last half of his pitcher, flung a dollar tip on the bar, and started toward the door.

"Mister," the bartender called after him as she scraped off his dollar. "If you want to find out about the fishermen, why don't you go down to the port docks instead of bothering us in here?"

"Which way are the docks?" Tanner asked grudgingly. Maybe he should have used bribery, but somehow he thought that wouldn't have worked either. These people just weren't giving out information.

"Up bay front a ways and it's on your right," she said.

Tanner parked the Thunderbird on Bay Front Road at the entrance of Port Dock 1 and walked across the wooden planked bridge to some of the biggest boats he'd ever seen docked in a port.

"Are these fishing boats?" he asked a fisherman who was winding a heavy rope around a steel bar on the walkway.

"Yep,"

"What kind are they?" he asked.

186

"Draggers, shrimpers and mid-water trawlers."

"Very impressive," Tanner said.

"She's a dragger, here," the young fisherman continued. "Fifty-footer. Takes three crewmen plus the skipper to man her."

"Are there any other kinds of boats in this area?"

"Yep," the fisherman said. "They're the small ones, the trollers."

Somehow, Tanner had gotten this fisherman's attention, but it was not long before he realized that he was on the wrong dock. The type of boat he was looking for must be the trollers, and the group that called themselves the PBFA, and the man called John Silk had to be at another dock.

"Don't know much about boats do you?" continued the fisherman getting into his subject. "You see this here net? All trawlers use a net. They're towed at slow speed on or above the ocean floor to catch fish or shrimp. The net is wide at the mouth and tapers back to the narrow cod-end which collects most of the fish. While the boat's fishing, the net is spread open by a big rectangular wooden or metal 'trawl door' attached with bridles and ground lines to each side or wing of the net opening. Doors can be flat, oval or slightly V-shaped, and a tow cable extends from each to a winch or pair of winches just aft of the pilot house."

"Are you a member of the PBFA here in Port Bane?" asked Tanner now that he had the man's attention.

"Nope," said the man. "I'm from Alaska."

"Have you heard of a man named John Silk?"

"Nope," said the man again. "Can't say as I have."

"That one over there looks pretty rusty," Tanner pointed down the dock trying to keep the man's attention.

"Fuck no!" the seaman said now back to his favorite topic. "She ain't rusty. She's a P-T Boat, made out of aluminum. She's a fucking hawk...hard charger...they had 20-mil cannons, 50 calibers on each side...well armoured. Run up against a Russian boat and fire on them with a S A M missile and they'd be gone in 'bout a minute. Get out about a mile, charge back and wave in Russian. It's all surplus property now, nobody can buy them. It's a damn shame. Them babies can travel at 60 knots, 75 miles an hour. She's got 2-18" exhaust, 2-18 cylinder napier delta engines. They develop 3100 horsepower per inch, and they like to run on JP-5 jet fuel.

Over there's the ocean going tugs. They're outdated now, 100' long, pulled ocean barges. When they got tanked up, it took 2,000 gallons of fuel..."

'Damn,' thought Tanner. 'I finally run into someone who will talk and it turns out to be a motor mouth.'

"I hate to cut this short," Tanner interrupted at last. "But my time is limited. Can you tell me where the trollers are located?"

"Right ahead," the seaman said. "Pull in at Dock 3 and you'll find 'em there."

Tanner parked the dusty Thunderbird in the Port Dock 3 parking lot. He took off his jacket, undid his tie, and rolled up his shirt sleeves. Now he wouldn't look so formal.

At the top of the ramp, which lead down to the dock and to the fishing trollers, a light-haired man in a uniform complete with a pistol and a club, came out of an old, shack-type building and stepped in front of Tanner.

"State your business please," he said in a stern voice.

"I'm looking for a particular fishermen," Tanner said.

"Do you own a boat docked here?" asked the policeman.

"Ahem, well, no," Tanner said. "But I'm from Washington D.C., and, like I said, I have business reasons to be here."

"No one is allowed on this dock unless they own a boat, mister." The steely blue eyes looked steadily at Tanner. "There are no exceptions."

Tanner realized the man would not hesitate to pull his gun if he made a move toward the ramp. He just as well give up today. What was he doing out here in the middle of nowhere anyhow where he was treated with suspicion, hatred, and possibly violence? Suddenly he could read the headlines in the Washington Post. "Justice Department Investigator shot in Port Bane." One thing was for sure, as soon as he got the information he came for, he would take the first plane back to his family and comfortable home. Trips like this could go to hell. He'd go back home, hire out to a law firm, and work on divorce cases.

By the time Tanner got back to the Port Bane Inn it was dark. 'Why wait any longer,' he thought. 'Why not leave on the first plane out of Portland. He could be here a month and never get anywhere.'

He poured a whole glass of scotch, sat down on the bed and took off his loafers. God he was tired. He should call Vivian, but she might pick up his foul mood and worry about him. There was no need to do that. He'd call her in the morning.

He stretched out on the bed, propped his head against the pillows, and just stared at the ceiling. Tomorrow he'd give it one more shot and then he'd go home.

Tanner was dozing when he heard a loud knock at his door. He sat up and listened. A moment later the knock repeated itself. Now who in the hell could that be? He didn't know a soul in Port Bane.

He opened the door a crack and stared at a tall red headed, wild-looking, unshaven man who was looking back at him with light gray, suspicious eyes.

"My name is John Silk," he said in an icy voice. "I hear you've been looking for me."

"Ahem," Tanner cleared his throat and opened the door wider. He was supposed to be the stalker, the investigator sleuth, who could track down anybody or anything, and he had gotten nowhere in this town. This man must have been in the shadows all along."I was beginning to think you didn't exist."

"Who the fuck are you?" Silk asked in a low voice. "And what do you want with me?"

"I'm B. K. Tanner from Washington D.C., and I was under the impression that you wanted to see me!" Tanner bristled. "I've been searching for you, by God, since I arrived in Port Bane. I'm here to investigate some of the complaints that your organization seems to be having. You can come in and sit down if you wish, and I'll mix you a drink."

"No time for that," Silk said. He quickly stepped inside the room and closed the door behind him. "You probably couldn't have come at a worse time. There's a strike going on here among the fishermen. It's a tense situation, and there's a possibility of some violence around the port until this thing settles down."

"What about the investigation you people called for?" Tanner asked. "Does this mean that I'll have to go back to Washington empty handed?"

"I have almost eight thousand pages of computerized material I'm willing to share with you, and I have a lot of sticky transgressions to talk over with you," said Silk, "but right now I have more serious problems. If you want this material and you want to hear the fisherman's side of the story, you'll have to come to my boat."

"I'll be glad to meet with you on your boat, of course," Tanner said. "When would you like to set the appointment?"

"Right now," Silk said. "I've only left the boat long enough to see you, but I have to get back and keep an eye on her. In a scenario like this anything could happen."

"That's fine with me," Tanner said. "It'll just take me a second to change clothes."

"You don't need to change," Tanner heard the order in Silk's voice. "The boat is warm enough, and we'll have you back in a couple of hours."

Tanner thought he must be crazy. Here it was, almost midnight, and he was sneaking out to some boat harbored in dangerous waters with a man he had never seen before. No telling what was waiting for them at the docks. He thought about telling this John Silk to go to hell. He'd get out of this place tonight and fly back to Washington D.C. But when he envisioned facing old eagle-eyed Powers with no report...either way he was caught, and he knew it. He picked up the keys to his room and put them in his pocket.

"Well, let's go then," he said, walking out the door ahead of Silk.

"We'll take my van," Silk pointed to the dark space just below Tanner's rooms.

"I could follow you in my car," Tanner offered.

"We'll take the van," Silk repeated.

"Suit yourself," Tanner said.

As they left the Port Bane Inn behind, Tanner was aware of the blackness of the night. It was just as dark here as it was when he was driving through the Van Duzer Corridor last night, except now there was no fog. He wondered if the people in this part of the world had eye strain, or perhaps a different optical structure.

This man, John Silk, was a silent sonofabitch. He had said nothing for the five minute drive to the iron chain-link fence he had seen earlier in the day. It was the same Dock 3 where he had met the uniformed guard with the scowl on his face and had refused him admission. This time, however, the guard was nowhere in sight as they walked down the ramp leading to the boats.

With it's tall trolling poles, pointed prow, and rounded stern,

190

the sparkling white CRACKANOON II was an object of beauty to Tanner who didn't know the first thing about fishing boats. As they approached her alongside the dock, a tall figure with something in his hands came out of the cabin and approached the rail. As they got closer, Tanner could see a man in a blue denim jacket with a lamb-skin collar holding a 30-30 Winchester. He was at least as tall as Tanner, and the drooping brown mustache which hung on either side of his mouth made him look like a bronco buster from Texas. Tanner felt his gut suck in when he realized the rifle was pointed at him.

"What the hell's going on, Cowboy?" Silk asked.

"Damned if I know," said Cowboy, spitting a wad of tobacco over the side into the water. "Who's the big dude with you?"

"Oh, him?" said Silk in a voice every bit as calm as the man on the deck. "That's Mr. Tanner from Washington D.C. It seems that our letters finally got some attention. He's out here to find out what's going on. Cowboy, I'd like you to meet Mr. Tanner."

"Howdy," said Cowboy. "We're glad you made it."

"Mr. Tanner," said Silk. "You go ahead and climb aboard there."

"Yeah, come aboard, Mr. Tanner," Cowboy said. "Hell, if you can do somethin' to help things, I might even go back to fishin' and become a American again. Silk's got a lot of bad stuff 'bout fishin' he's been needin' to give to somebody. The only problem is that nobody's been listenin' to him. Sometimes we all feel like we're bein' shit..."

Suddenly they heard a commotion at the top of the ramp, and somebody started running. They could see a dark figure charge past the fingers on the dock and turn toward the CRACKANOON II. Cowboy tightened his grip on the 30-30 and aimed it toward the running figure. His finger was squeezing the trigger.

'My God,' Tanner thought. 'He's the same guard that stopped me at the entrance to the dock this afternoon. He's after me!'

"Hold it, Cowboy," Silk said. "It's Pat McDougal, Hattie's husband. He took this guard job just to keep an eye on things for us. Wonder what he's so upset about. I'll go meet him."

Tanner saw the man whispering something in Silk's ear, and Silk came bounding back to the CRACKANOON II and started un-tying the ropes on the dock.

"We're taking off, Cowboy," he said. "Start the engine."

"Hey! Wait!" Tanner said. "What about me?"

"There's a change of plans, Mr. Tanner," Silk said. "It looks like you're going to have to make a choice. You can either leave the boat now, or you can stay here. If you stay on board, I'll be able to tell you everything about fishing you want to know, but we've got to get the CRACKANOON II out of here until this trouble blows over."

Certainly Tanner wasn't prepared for a boat ride. Everything he had brought with him was in his room at the inn. Yet, he couldn't afford to pass up an opportunity for on the spot information.

"I guess I should have brought a coat. Every stitch I have is at the room," Tanner said.

"Don't worry," said Silk. "We have plenty of gear on board. All you have to do is to decide. But do it quickly."

"What the hell," said Tanner. "Let's go!"

CHAPTER

19

SEA LEGS

Tanner sat bare-headed on the fish hatch of the CRACK-ANOON II and watched the lights on the bay front inch past.

Suddenly, he felt ecstatic and stood up. When he had arrived in Port Bane, he never realized that he would be riding a fishing boat out to sea in the middle of the night. He felt the breeze whip his hair, and for a moment he understood how a captain must feel as he watched his ship cut through the mighty waves on the ocean.

Halfway between the port dock and the bay bridge the CRACKANOON II stopped, and Silk and Cowboy came outside the cabin door and began to unfasten the ropes that secured the tall trolling poles on either side of the cabin.

"We always let down the trolling poles and attach the floppers before we hit rough water," explained Silk. "They give the boat stability and keep her from rocking as much as she might. Tonight we'll be running against the tide, and the wind's come up so we'll need every ounce of balance we can get. You'll see once we are in the jaws and hit the chops. They ought to be about six feet high from the middle of the jetty to the mouth."

"What are the jaws?" asked Tanner.

"They're high rock walls on either side of the channel where the Bane River hits the open sea," said Silk. "This stretch of water is called the bar. She'll be a little rough tonight so always make sure you're hanging on to the rail, or one of the overhead cables. We don't want you overboard before we get started."

Tanner grasped a steel cable that ran from the cabin to the stern and watched as they approached the long bay bridge high above them. Between the evenly-spaced rails on the bridge he could see the blinking lights of moving automobiles. Then they passed beneath the bridge, and except for a flashing buoy light, there was darkness ahead.

The black walls of rock seemed to close in on either side of the boat until Tanner thought they would crush the sturdy craft. The line of rocks he had seen from his room at the inn must have been the north side of this jetty.

Suddenly the first incoming waves struck the bow of the boat. Tanner lost his grip on the cable and was flung unceremoniously on his ass back down on the hatch cover. The waves became higher, and the CRACKANOON II began to rise and fall, and pitch from side to side, as she nosed through the thrashing water. Waves washed across her stern, and white foam swept over her bow. Tanner felt the shock of cold ocean water on his feet and looked down. His loafers were beneath the water. He was wet to his knees, and here came the swirling water again.

Tanner gritted his teeth and gripped the cable with both hands. He must be crazy. What was he doing out on a fishing boat with two men he scarcely knew, and the ocean threatening to wash him overboard any minute. What kind of investigation was this? His thoughts were interrupted by Cowboy who had been standing in the cabin doorway watching his discomfort.

"Pretty rough out there," said Cowboy spitting a wad of chewing tobacco over the rail. "You just as well come on inside the cabin where it's warm Mr. Tanner. There won't be nuthin' to see the rest of the night."

Tanner stood up and tried to make his body sway with the rise and fall of the boat. He sat down abruptly still clinging to the overhead cable with both hands. He was having some kind of trouble with his stomach. He suddenly felt a little queasy.

"No," he said. "I like it out here."

"Suit yourself," said Cowboy. "I just thought you might be a trifle uncomfortable in your shirt sleeves, sittin' with your feet in that there water."

Actually Tanner was in a hell of a shape. Not only was he freezing to death, but if he moved, he would throw up all over the deck.

"Sure is a nice boat," he said weakly.

"She's just a old sea tractor," said Cowboy. "She won't win any races, but she can catch fish. Well, you can sit out here as long as you want, Mr. Tanner. We'll be out of this rough stuff pretty soon, and then we can all relax. The way you're ridin' her, you'd think you'd done this before. Most flatlander's woulda' already been pukin' their guts out. Looks like you already got your sea legs."

At another time Tanner would have glowed at the compliment, but now all he could manage was to grit his teeth and hang on with both hands. He was sicker than a dog and afraid to move.

At some point in time—Tanner didn't know how much time had passed—the rhythm of the boat changed. She ceased to rise and fall and pitch from side to side. Now she was gliding, up one side of a widely-spaced swell, down the other side into the trough, then up the face of the next. As the motion of the boat changed, Tanner's stomach became calm enough for him to stand. At least he had toughed it out. He hadn't puked up his dinner. Now he could go inside the cabin with his pride intact and get warm.

When Tanner opened the cabin door, Silk and Cowboy were sitting at the table sipping black coffee. He staggered toward the table, his legs as stiff as a tin man.

"I wondered when you'd come inside, Mr. Tanner," said Silk with a straight face. "You're wetter than hell. What do you say if we go below in the fo'c'sle and get you some sea clothes. Cowboy, keep an eye out for anything that looks like a fishing boat."

Silk moved down the steps into the lower half of the boat as gracefully as a ballet dancer, and Tanner rocked right along behind him feeling like a bull in a China shop. He sat on the bunk and watched as Silk pulled out a chest from under one of the neatly made bunks. This portion of the boat was as well-kept as the galley.

As Tanner changed into long underwear, a pair of denim carpenter's coveralls, and a heavy wool sweater, Silk continued to rummage under the bunk.

"Mr. Tanner," Silk said. "To tell you the truth I'm perfectly happy you decided to come on this trip. Since you're from Washington D.C., and here to get the facts, this opportunity gives me a chance to talk with you. For years I have been studying the fishing industry and gathering any written material I could find on the subject. In this box are eight thousand pages of collected articles and computerized statements. You're welcome to look at anything in the box."

Tanner studied the large, cardboard container in amazement. The thing was filled to overflowing with paper and magazines. If the stuff were of any value, the information would make his investigation easier. Maybe he had hit upon a gold mine.

"Thank you, Silk," said Tanner. "Frankly, I don't know a thing about the fishing industry, and I would be glad to look at your material. I don't know about trawling, trolling, or foreign fishing. I heard the terms net-pen-rearing and ocean-ranching in Seattle, and I don't understand them either. I'll be glad to listen to anything you say about any subject."

"I'll just carry this box of stuff upstairs so we can refer to it if a question that I can't answer crops up in the conversation," Silk said. "Sit down here and let me get you a cup of hot coffee."

"I'd appreciate that," said Tanner.

He slid onto the wooden bench next to the wall, and took a sip of the hot liquid. Silence was the best policy until he could get ahold of his stomach. He took another sip, and this time he felt like he might survive.

"This strike is a bitch," said Silk. "I ought to be fishing, and here we are running out to sea in order to protect the boat."

"You can fish," said Cowboy. "It'll be light in a little bit, and there ain't nobody out here."

"No," said Silk. "I gave my word I wouldn't fish, and I'm going to honor it."

"I figured you'd say that," said Cowboy. "Well, we got us a nice little trip anyhow, and Mr. Tanner here can get acquainted with us."

"What about this strike?" Tanner asked. "Who's striking, and for what?"

"The fish companies are not paying a fair price for our salmon," said Silk. "The PBFA supported the strike. But some of our fishermen were so desparate for money that they couldn't strike or they would go broke and lose their boats, so they remained independent and continued fishing like the rest of the fishermen from most other Oregon ports. Now there are two sides, and fishermen are fighting fishermen. I kind of got caught in the middle. The night before last somebody sank two Port Bane fishing boats, and tonight there was a rumor that the CRACKANOON II might be the next target. I know most of the fishermen in Port Bane, and I don't think any of them would touch this boat, but I don't know the outsiders. I decided not to take a chance."

"I see," said Tanner.

Silk handed Tanner and Cowboy a beer and busied himself with the makings of ham sandwiches. Tanner felt like he could eat something now. Strange how, out here on the ocean, you could be sick one minute and hungry the next. The BUDWEISER in his hand tasted every bit as good as HEINEKIN which he drank in Washington.

"It's sportin' of you to come on out here with us, Mr. Tanner," Cowboy said, "'specially since we can't even catch a old salmon for you."

"I'd appreciate it if you didn't address me as 'Mr.'", he said to Cowboy."

"Yes sir," said Cowboy.

"And 'sir' is just as bad," continued Tanner.

"What're we 'sposed to call you then?" Cowboy asked. "Your bein' a important man from Washington."

"Well," said Tanner. "Back home they call me Buck, but I hate that name too. Silk is called by his last name, so why don't you just call me Tanner. I can live with that."

"I like it," Cowboy said extending a calloused hand that Tanner took in his large paw. "Tanner, by dern. Yep, I like it. You can go with me anytime."

"Do you own a fishing boat?" asked Tanner.

"Not any more," said Cowboy. "I just sold her."

"What kind of boat was it?" Tanner asked.

"She was a big one," said Cowboy, "A joint venture boat."

There was that term again. The same one that Blackburn in Seattle had used.

"Could you explain joint venture to me?" he asked. "I'm unfamiliar with the term or process or whatever joint venture is."

"You explain it to him, Silk," said Cowboy. "You've done the research."

"And you've had the hands-on experience, Cowboy," said Silk carrying three giant ham sandwiches to the table. "So you correct me if I get off base."

"Sound's like a plan to me," said Cowboy.

"Back in 1976, congress passed the Studds-Magnuson bill which was supposed to keep foreign fishing fleets from fishing our coastal waters."

"Let me tell the story," interrupted Cowboy. "I was one of the first joint venture fishermen. I finally got enough money together to buy my own fishin' boat. She was a sixty-foot stern trawler that cost me $275,000. I thought I was a settin' in tall cotton, only I couldn't get enough money fishin' to make the payments. Ever'time I got to catchin' fish, I had to come back into port and sell them to the local canneries so they'd be fresh. I couldn't stay out for more than five days at a time, and in them days salmon prices was always rock bottom. I was havin' a hell of a time and about to lose my boat when I heard of joint ventures.

Some shrewd businessman or politician used the Studds Magnuson bill as a excuse to invite foreign fishin' fleets to fish our coastal waters as long as they signed a contract with a American fishing company that they'd buy fish from American boats as well as catchin' their own. Then them foreign governments brought in their own factory ships to process the fish.

I studied the operation for a while, and I says to myself, 'hell Cowboy! You don't have a brain in your head. The answer to your problems is right under your nose. Why should a big boat like the OTTER come into port ever' three days and lose that there fishin' time? A boat like mine should fish six days a week, then take only one day off for maintenance. I figured right then that I ought to get involved with one of them joint operations and sell my fish to a foreign outfit so I wouldn't have to come into port. 'Course I'd get less for my fish, but them foreign fellers' didn't give a damn. They'd take anything that could be ground up. I wouldn't have to throw away them bottom fish no more.

Well, the OTTER came with a Alaskan fishin' license, so right away I talked to several fishermen in Port Bane who had big boats like me, and we banned together and started our own American fishin' company. Five of us took off to Alaska, and that's where I been ever' since until just lately when the Russians started fishin' for hake off the Oregon coast. That's when I come back home."

"I didn't know that there were any Russian ships within two-hundred miles of our shores," said Tanner. "I thought the Studds-Magnuson bill was a unilaterally declared 200-mile economic zone that was supposed to keep foreign interests off our shores and prevent other countries from depleting our ocean resources."

"That's what most people think," said Cowboy. "That's what our government wants them to believe. The truth is that them Ruskies can come to within three miles of our coast line as long as they're

198

doin' business with American companies or American joint venture boats. Right now Alaskan waters is just one big foreign fishin' ground."

"I don't believe it," said Tanner. "I haven't heard of our fishing boats selling directly to the Russians."

"That's what joint venture means," said Cowboy. "Not only them Russians, but them Poles, Koreans, Japanese, East Germans, Norwegians, anybody that wants to send in a factory ship and buy our fish. Right now there's a whole Russian fishin' fleet off the Oregon Coast. That was where I was sellin' my fish."

"What Cowboy says is true," said Silk. "He's just left out some of the more pertinent information. In exchange for allowing other countries to catch our fish, we are allowed to place our warships and submarines in any waters around the globe. Worse than that. Most fishermen think that joint ventures is simply catching fish in their nets and selling them to a, say, Russian factory ship for a profit. In reality the over-all use of the Magnusun bill for exploitation is devastating.

If the Americans don't get into the offshore factory business and run competition with other countries, the foreign fishing fleets will catch every salmon, halibut, black cod, rock fish, flat fish, hake, shrimp, king crab, dungeness crab, and tanner crab within two hundred miles of the west coast from Mexico to Alaska. I can see it coming within the decade. What's more, I can prove these statements by all the research in that box."

"All this is hard to believe," said Tanner. "And you say a Russian fishing fleet is right off the Oregon Coast. I'd like to see one of those big ships."

"Stand up and look out the windows," said Silk.

Tanner stood, and there were lights all over the horizon.

"I didn't know we'd headed back toward shore," he said.

"We didn't," said Silk. "Look more closely."

"Those lights are moving," exclaimed Tanner.

"Right," answered Silk. "What you're looking at is the Russian fishing fleet. We'll shut down for the night and take a closer look in the morning."

"I'll be damned," said Tanner.

While Silk shut down the engine, Tanner became aware that

he was again uncomfortable. He hadn't taken a leak since just before he left the inn. The pressure on his bladder was excruciating.

"Ahem. Do you have a restroom on board," he asked politely.

"You bet," said Cowboy, "and it's clean as a house 'cause a fisherman never uses it. Silk keeps it mostly for women, and he's never had one aboard. If you got to go, just step out on deck and hang her over the side. Make sure you turn opposite the wind, or she'll blow all over you. Fact is, I think I'll join you."

"Me too," said Silk.

As he stepped out on deck, Tanner became aware it was almost morning. In the semi-darkness he could see the lights in the distance as the CRACKANOON II, now drifting, topped out on a wave then slid to the bottom. He felt ill at ease and stared apprehensively over the side. If he stretched out his hand, he could almost touch the water, and the slapping of the waves against the wooden hull made a disconcerting sound. Silk and Cowboy, who had checked out the mast and running lights, were now standing with their backs to him facing the other rail.

"Be sure and hang on to that rail," said Cowboy. "There's been more than one fisherman fall over the side while he was a pissin'."

Tanner did as he was told. He grasped the rail with one hand and undid the buttons with another. He stood for one minute then another. Whether it was the close proximity of the water or the strange circumstances, he couldn't imagine. He knew Silk and Cowboy were watching him now, and that was almost as uncomfortable as his bursting bladder. Damn, he couldn't go despite himself. Embarrassed, he was almost to give up when it happened.

Without warning a huge black shape surged suddenly out of the water not ten feet in front of him. For a second the giant monster seemed to hang in mid-air towering over the cabin and the radar, then it sank slowly beneath the waves.

Several things seemed to happen at once. Startled out of his wits, Tanner let go of the rail and fell backward on the fish hatch flat on his back. At the same time he started pissing, and the stream went up the front of his coveralls and onto his face. He coughed, sputtered, managed to stand up, stagger to the rail, grasp a guy wire with one hand, and point himself over the side. Cowboy and Silk, who had witnessed the spectacle, were laughing so hard they could scarcely stand on the rocking deck. Tanner finished his business, and buttoned his pants before he turned around.

"What's so funny?" he said gruffly. "Haven't you ever seen a man take a leak before?"

"Not that way," said Cowboy, and he and Silk burst into laughter again.

"What was that thing, anyway?" asked Tanner.

"That was a killer whale," said Silk. "I've seen them come out of the water before, but never so close."

"First I almost freeze to death, then I piss all over myself," muttered Tanner.

"Just part of life on the ocean," said Cowboy trying to smother a new giggle. "You got to be ready for the unexpected. But we'll never tell if you won't."

"Cowboy, you take the wheel, and we'll find Tanner another change of clothes and a place to wash up," said Silk. "It's almost light, so we'll take a nap since we don't have to put out any lines today."

Tanner dressed quickly when he heard the engine start, and the CRACKANOON II began to plow through the swells. When he stepped into the cabin, Silk was standing at the wheel.

"Come over here a minute, Tanner," he said. "You wanted to see that Russian fishing fleet we talked about; well there they are. I've already counted eleven trawlers and the factory ship. We'll get closer to the big one so you can see it better."

Tanner looked across the grey waves toward the horizon, and sure enough there were several long shapes steaming in different directions. As the CRACKANOON II closed on the nearest vessel, he felt like a mouse in the now tiny troller coming upon a huge cat.

"Just how large is that ship?" he asked tentatively.

"She's about the size of three football fields," said Silk. "She's both a fishing boat and a floating factory. She processes all the fish those other vessels bring in, and in the interim between cod-end exchanges she fishes herself."

"I didn't realize the size of the operation that you and Cowboy were talking about," Tanner said. "How many fish are they taking?"

"All they can catch," said Silk. "And by rights those are our fish; those are United States fish. The Russians are vacumming our

201

ocean bottom with those huge trawl nets and sending all the fish they catch home to Russia; or, more likely, they're selling them to other countries and taking the profits."

"She's one hell of a big operation," agreed Cowboy.

"The end will come when our government legislates the American trollers like mine, and the sports fishermen, off the ocean," continued Silk, "and we let the foreign countries do our fishing for us. Go into any supermarket and you'll find most of the frozen fillets you buy are foreign caught."

In his imagination Tanner saw the entire American coast line filled with foreign fishing fleets. Then he saw beneath the water, and there were no more fish to be caught.

"Looks like I'm going to have a good report when I get back home," he said. "Nobody in Washington D.C. understands the problems we're dealing with out here."

As the CRACKANOON II came closer and closer to the giant factory ship Tanner could read the big, black letters MURMANSK written on the bow. Even with the cabin door closed he could smell the sickening stench of guts and cooking fish.

"Who authorized this fishing business with the Russians?" Tanner asked.

"Like I said last night, our government, the guys who employ you," growled Silk turning the CRACKANOON II hard to port. "The president, the congress, the state universities, and the N.M.F.S., and anybody else you want to name that's affiliated."

There was that name, N.M.F.S., and Silk had mentioned the organization himself. Now was as good a time as any to find out what Blackburn wanted to know about the observer who jumped ship.

"Ahem," he cleared his throat, "I was asked by my boss to stop in at Seattle at the N.M.F.S. to, perhaps, help them with a minor problem they were having. Dr. Blackburn mentioned something about an American observer having jumped ship. About the Russians acting strangely since that time. Have either of you heard anything about such an incident?"

The question had an impact on the two men alright. Silk looked knowlingly at Cowboy, and Cowboy at Silk. Some kind of message passed between them.

"I was wondering when I'd hear about that," said Silk quietly. "Are they looking for anyone?"

202

"Not that I know of," said Tanner. "Dr. Blackburn was just curious. There's been some kind of shake up in the Russian fleet since that time, and the relationship between the two organizations hasn't been friendly."

"I was the man who was working on that ship as a foreign fishery observer," he said slowly. "And I did jump ship with Cowboy's help. I had my reasons, but I'm not prepared to discuss my experience on the MURMANSK at this time. It's not that I don't trust you, but there are other people involved. Now, are you going to turn me in?"

"No," said Tanner. "And I won't see Dr. Blackburn again."

"Perhaps, I'll tell you the whole story later," said Silk. "But there is a man I have to consult first."

"That's fine with me," said Tanner.

"Good," said Silk. "Let's have breakfast. Do you want to cook Cowboy, or shall I?"

"I'll cook," said Cowboy going to the small refrigerator. "We'll eat, then I'm ready for a nap."

On the ocean, time had a way of passing without one's being aware of it. Tanner lay on one of the bunk's in the fo'c'sle listening to Cowboy's snoring. Silk was sleeping on the bunk in the cabin so he could keep an eye out for other boats. Full from the delicious breakfast of pork chops, scrambled eggs, hash browns, coffee and toast, Tanner felt warm and relaxed lying in the sleeping bag with his boots off. Sleeping on a drifting boat was romantic, with the peaceful slapping of the waves and the gentle rocking motion the only thing he lacked was Vivian beside him.

Tanner didn't know how long he had slept, but he was a-wakened by a couple of strange popping noises and the sound of something striking the wooden hull.

A second later he heard the heavy diesel come to life and felt the boat began to move. Silk's voice sounded over the speaker on the radio.

"CRACKANOON II to unidentified vessel. Your bullet just struck my boat, goddamn it."

His voice was interruped by three more popping sounds.

"Cowboy," he shouted. "Those bastards are shooting at us. Get up here and take this wheel."

Cowboy was out of bed and already up the stairs to the wheelhouse. Tanner, in his socks, followed him up. Cowboy took the wheel and Silk grabbed a rifle and ran on deck shoving shells into the firing chamber at the same time. A moment later he fired eight shots as fast as he could crank the lever.

In the distance Tanner could see a boat and two men busy on the back deck. It looked like they had knives and were cutting fishing lines. Silk fired eight more shots from the thirty-thirty, and then there was silence as the other troller faded into the distance.

After a considerable period of time, Silk came back into the cabin swearing and shoving more shells into the rifle's chamber.

"Those bastards," he said. "They were shooting at my boat."

"Did you hit them?" asked Cowboy calmly.

"Damn right," said Silk. "That son-of-a-bitch has more than one hole in his hull."

"Good," said Cowboy. "What do you figure made him want to shoot at us. We weren't doin' nuthin' that I could see."

"I don't know," said Silk hanging the rifle on its rack. "I didn't recognize the boat, so I know they weren't from Port Bane. Maybe they thought we might attack them because they were fishing."

"Well, you scared the hell out 'a 'em," said Cowboy. "They lost a expensive bunch of fishin' gear when they cut them lines. Serves the bastards right."

Tanner eyed the bullet hole in the front window. It couldn't have missed Silk by more than six inches, and the man was as calm as an Englishman who had just drunk his tea.

"Do these fishermen out here usually shoot at you?" he asked. "I'll admit that I was a little shaky."

"Not usually," said Silk. "In ten years fishing I've never been shot at before. But then again, I've never been involved in a strike before."

"We have to check out the boat, Cowboy, to see if there is any damage. Have you ever driven a boat Tanner?"

"Never," said Tanner.

"Nothing to it," said Silk. "Just hold her on this course, and if you see another boat, give us a yell."

Cowboy's voice came up from the fo'c'sle where he had been looking for bullet holes.

"Better get down here Silk. We got problems. One of them

bullets punctured a high-pressure fuel line, and there's already 'bout a inch of diesel down here. You got anything we can wrap this damn thing with?"

"I think I have some strips of inner-tube," Silk said. "If we can't fix it, we'll have to radio the coast guard, and get them to tow us in."

The sky was beginning to darken when Cowboy and Silk came up from the fo'c'sle. Tanner was still standing before the wheel, his feet spread apart, staring first to the right and to the left in case there was another boat. Both men were covered with grease and smelling of fuel oil.

"Thanks, Tanner. I'll take the wheel now," Silk said. "We've wrapped that hole as tight as we can with wire and inner-tube, but it's still seeping. We can't hold that pressure, and we'll be lucky to make it into port with the fuel in the tanks."

Silk cranked the wheel to starboard forty-five degrees. When the coordinates flashed properly on the loran, he set the automatic pilot, and shoved the throttle forward.

"Hello, CRACKANOON II! Do you read me? O.K."

Silk grasped the receiver and flicked the speaking switch.

"This is the CRACKANOON II, O.K."

"Howdy, Silk," said a gravelly man's voice. "This is Grave-yard. Thought I'd just give you a hollar and tell you the strike's over. Ever'body's happy now. They got a quarter raise per pound. You can come in now. O.K."

"Got you, Graveyard," said Silk. "we'll see you when we get in. O.K."

"How far do we have to go?" asked Tanner.

"'Bout thirty-five miles," answered Cowboy. "That white stuff you see ahead of us is fog, but don't you worry none. Silk can find his way home blindfolded."

Tanner stood up and watched, fascinated, as the CRACK-ANOON II enfolded herself in the swirling gray mass. Everything, including the waves seemed to disappear. Silk calmly switched on the radar, put his eyes to the screen, and twisted the dials."

"Better go topside and check the diodes," he said coolly to Cowboy. "Nothing is happening."

Tanner followed Cowboy out the cabin door and watched as the nimble man swung himself onto the roof of the cabin and began polking about the large white cannister with a screwdriver. He lifted off the outer shell and peered silently at the electronic equipment inside. The wind was blowing, the fog swirling, and the boat pitching. Tanner couldn't understand how Cowboy kept his feet, but the man somehow replaced the cover, swung down past Tanner, and stepped into the cabin.

"I sure hate to tell you this, Silk, but you're not goin' to have any radar this trip," he said. "Two of them there bullets blew that radar all to hell."

"Does that mean we're stranded out here?" asked Tanner.

"No," said Silk. "We'll go in blind."

"What does that mean?" asked Tanner.

"Not to worry, Tanner," said Cowboy. "Silk just means we'll go in without radar."

"What about the coast guard?" asked Tanner.

"Fuck the coast guard," said Silk. "I'm not about to let those bastards tow my boat and tear her up. This won't be the first time I've come in without radar."

Tanner, looking into the swirling mass, knew they couldn't see a ship even if it were right on them.

Hour after hour seemed to pass, then Cowboy said,

"There she is Silk, right in front. That's the outer buoy to the jaws."

Tanner stared. At first he could see nothing, then there it was, just to the right, not ten yards away, a blinking pinpoint in the white. How Silk had found that buoy in this gigantic ocean without a radar was more than he could understand.

"How did he find that light?" he asked Cowboy in a whisper. "That's impossible."

"Nuthin' to it if you know what you're a doin'," whispered Cowboy.

"O.K., you two," said Silk, giving orders. "Cowboy, you out on the bow, and you, Tanner, on the back deck. Look from side to side, and if you see a heavy shadow or anything that resembles a rock or a boat, shout at me. I'll only have a split second to react. We're going to thread the channel."

The next hour was the most anxious Tanner had ever experienced in his life. The waves from the outgoing tide washed over the back of the deck from both sides. He gritted his teeth and held his position. He strained his eyes into the dense fog; his eyes burned in his head; he didn't give a damn. He couldn't quit.

Then, Cowboy gave a war whoop.

"We made it! Look at the bridge, it's right above us!"

Tanner stared, and sure enough he thought he could see light through his watering eyes.

"I got to pull in them floppers and pull up them poles," Cowboy said. "It won't be long now, and we'll be dockside."

"Well, what did you think of your first trip on the ocean?" asked Cowboy as the three men walked up the ramp to Silk's old van.

"I thought it was a bit hairy," Tanner amazed himself with his understatement.

"It was that," agreed Cowboy. "Just a trifle hairy."

"How about a beer to celebrate the trip, Tanner?" Silk asked.

"I think I'll pass," said Tanner. His boots were still full of water and sloshed with each step. "I think I'll just go back to the inn and go to bed."

"Suit yourself," said Silk. "Cowboy and I are going to the Albatross and have a drink. I'll be by your room in the morning with all the material. If you'd like, I'll show you the ocean ranch and the rearing pens tomorrow. It'll take me some time to get that radar repaired, so I won't be fishing for a couple of days. Looks like I have a little time."

"That will be fine," said Tanner. The truth was, he didn't care whether he ever saw the two men again.

CHAPTER

20

CAT'S-PAW

Tanner must have fallen asleep the moment he hit the bed. He was still lying sprawled in the same position when he heard the faint rapping at his door become a loud pounding. He raised his head an inch or two off the pillow and looked around the white room with its purple flowered curtains. Where had he seen those curtains before? He sat up groaning and swung his stiff legs over the edge of the bed. Every bone in his body felt like it had been pulled apart and welded back together with a blow torch. He rubbed his sore ass gingerly and staggered toward the door through the piles of wet fishing clothes. Something in the room was beginning to stink like rotton fish.

"Ahem!" he said to the door. "Who is it?"

"It's Silk, damn it. Who do you think it is?"

"Go away!" Tanner roared.

"Open the fucking door!" Silk said. "It's time to get her going."

"Come back in a couple of hours," Tanner said.

"Like hell I will!" said Silk. "I'm not leaving until you let me in."

Tanner fumbled with the lock and opened the door a crack.

Silk grinned at Tanner and elbowed his way into the room. He was carrying a box of papers and a stack of clothes.

"You don't look too happy about seeing me this morning, Tanner. Didn't you get enough sleep?"

"No," Tanner growled.

"I brought some fresh clothes and the written material you wanted to take a look at. What we need now is a pot of hot coffee to get us started. Sound good to you?"

"Ahem," Tanner said. "No!"

"Are you this surly every morning?" Silk laughed. "How does your wife put up with a guy like you?"

Tanner wished Silk hadn't mentioned his wife. It reminded him of last night when he had called Vivian. In a daze he was trying to tell her about his hazardous trip on the ocean and how much he loved her. But the minute she heard his voice, she had shouted something like she thought he was dead, and something else that sounded like she hated his guts. Then she had hung up the phone. He couldn't figure that one out. Maybe she'd have been happier if one of those bullets had struck him!

"What makes you so chipper this morning?" Tanner asked. "Good night's sleep?"

"No," Silk said. "I never got to bed."

"Trouble?" Tanner asked.

"No," Silk said. "I stayed up all night mulling things over and drinking beer. The way I figure it though, now that the strike's over, my problems are solved temporarily. I can't fish until the radar is repaired on the boat, and that'll give me some time to work with you. So, if you'll tell me exactly what you need to take back to Washington D.C., I think we can get it ready in the next two days."

Tanner took the cup of black coffee Silk handed him and led the way to the round table in the next room.

"There's so much of this fishing business that I don't understand."

"I'll tell you what I know about raising salmon," Silk said. "I'll even take you on a tour of the two aquaculture facilities in Bane Bay."

Tanner took a long sip of his coffee before he spoke again.

"Explain this aquaculture to me in greater detail."

"Aquaculture here on the west coast is simply a method of farming salmon instead of letting the wild fish reproduce themselves naturally," Silk explained. "In 1971 congress passed an aquaculture bill that made it legal for private businesses to raise salmon as long as the fish remained in contained pens and were not released into the ocean. This bill opened the door for technologists and big business

209

to start a huge new industry that features salmon-in-a-pen as its main product. Right now in Washington and Oregon there is a net-pen rearing operation in every decent salt-water estuary, and applications are pending for a hundred more.

Net-pen-rearing is similar to raising chickens or pigs in closely packed barns. If you can, picture fifteen acres of net covered pens right in the middle of Bane Bay. There are thousands of coho salmon circling together in these nylon pens which are suspended in salt water. They are hatched from eggs from the bellies of salmon who would spawn naturally up our streams if they were given a chance. Once they are hatched, the smolts, or baby fish, are fattened for slaughter on dry cereal pellets which are full of antibiotics to protect them from diseases, and artificial carotenoids which give their white flesh a pink color. Wild salmon grow naturally, and get their color from eating shrimp or krill."

"I see," said Tanner. "What I don't understand is why this pen-rearing industry should affect the commercial fishermen and the wild salmon if they never leave their pens."

"Hell, SEABOOM has a floating field of one hundred and sixty pens up in Winchester Bay that covers more than twenty acres of water. Each pen holds as many as 20,000 to 30,000 fish. The company here harvests over two million pounds of domestic salmon annually. The majority of fish are sold fresh, whole-boned, or as butterfly fillets. Those bastards can sell fish three-hundred-and sixty-five days a year to supermarket chains. What's more, they can sell them any length, from pan-sized nine inch to as large as they come. One day companies like SEABOOM will control the protein sea life around the world. In short, they are forcing lower prices for inferior fish, and we can't compete because our seasons are limited by the Fish and Wildlife Department and the N.M.F.S.

But competition aside, there are other negative effects from these pen-rearing operations. Ecologically, fish shit and this waste plus uneaten feed forms up under the pens and kills all the surrounding marine life. The antibiotics they feed the salmon infect other sea animals and the entire marine ecosystem. Some experts say these artifical drugs pumped into the salmon will eventually harm humans.

What would happen if diseased salmon eggs or diseased salmon infected the wild fish? Disease, nutrients, and a stress condition from being penned are all symptoms of pen-rearing. A disease

210

can be devastating to a pen load of fish, like the new fungus that struck the pens of two year old coho salmon right here in Bane Bay. Only five hundred out of five thousand fish are left. Diseases never heard of before are showing up in pen-stressed fish.

If some action isn't taken every saltwater bay from Oregon to Alaska will be cluttered with salmon pens. I read somewhere that by 1996 one out of every seven salmon eaten will be pen-raised."

As Silk was talking, he was pacing around the room.

"Now, I'll tell you something about that perverted bullshit ocean-ranching, too." Silk slapped his hand with a fist.

"But how does ocean-ranching differ from pen-rearing?" Tanner questioned. "Isn't it the same?"

"Take AQUAPROTEIN right here in Bane Bay," said Silk. "It's a subsidiary of WINDOVER LUMBER COMPANY, called WINDOVER ENTERPRISES, INC., but they don't want anyone to know they've changed from the lumber business to fish. Their domestic fish have almost completely destroyed the wild salmon that used to swim up the Port Bane River.

They've built a manufacturing plant. They've put in a fish ladder or runway so the salmon can be turned out into the saltwater estuary and they have a large number of rectangular water containers called raceways; usually from twenty to one-hundred and fifty feet in length and filled with from one to ten feet of water. And they release their fish directly into the ocean and plan for their return after they have matured two to five years later. They know that the salmon will return to the very fish ramp from which they were released. And when they do return, the plant is prepared for them. They're either netted at the mouth of the fish ladder or they're allowed to climb the ladder or ramp directly into the factory. There, they're headed, gutted, filleted, or canned, and the number of eggs needed for the next season are ripped from the female salmons' guts and placed in incubators. Talk about a setup. If the commercial salmon fishermen didn't catch a few of them, they'd have a hundred percent return. Can you imagine a setup like that, Tanner? They don't even have to invest in feed beyond the smolt stage. The ocean furnishes the food, and AQUAPROTEIN reaps the profits.

To make matters worse for the fishermen, AQUAPROTEIN and the other large ocean-ranches release their fish so that they return during the off-season when the commercial fishermen are not

211

allowed to fish!"

"Are these ocean-ranched fish susceptible to the same kind of diseases as the pen-reared salmon?" Tanner asked curiously.

"You're damned right," said Silk, "but not as badly.These ocean-ranch people raise the temperature of the water to increase the rapidity of coho growth. This increase in water temperature acts as a stress agent that allows infections to run rampant among thousands of domestic fish that already have no normal resistance to disease because they're in a controlled environment. Every bacteria, fungus, and gill disease found in wild fish are abundantly found in those raceways and incubators. New diseases never heard of before continually crop up so new drugs and chemicals are fed to the fish as well as the existing drugs and chemicals. Hell, it's a drug and chemical whorehouse. And if one of those drugged smolt is released into the ocean still suffering from a disease, it can transfer the illness to wild fish."

"The whole process seems amazing," said Tanner. "But I would have to see it to believe it."

"I brought the documented material about everything I've said, and I'm turning it over to you. You're welcome to any information you need to pull out and take with you," said Silk.

"Wonderful," said Tanner. "I'd like to spend some time learning about these aquaculture ranches and farms. No one back east has ever heard of these industries."

"I'll do better than tell you about them," said Silk. "I've had Cowboy arrange to get a Zodiac from Claude, and we're going to take you out to the pens. There are about a hundred of them along the south side of the bay. PENWELL FARMS, which is owned primarily SOUTH AMERICAN ENTERPRISES, which is called SEABOOM locally, is the outfit that owns them. We're also going to take you on a guided tour through AQUAPROTEIN, which is an ocean- ranch right over by Claude's place. Once you've seen these projects with your own eyes, I think you'll better understand what I'm talking about."

"I hope so," said Tanner. He felt bewildered by the amount and complexity of Silk's diatribe. "I already have some questions that seem contradictory to me, even though I probably don't have the background to understand the answers."

"Go ahead and ask," said Silk.

212

"Well," Tanner carefully selected his words, "You keep talking about coho being raised by aquaculture processes. What about Chinook salmon? Aren't they the primary salmon you catch?"

"They used to be," said Silk. "The Chinook are immune to most diseases. These aquaculture bastards have never been able to raise them in pens or on ranches, just the coho. Chinook must go through the natural spawning process of laying their eggs and letting nature take its course. As a result of destroyed spawning grounds, there are fewer of them now, and they are smaller. But any idiot can raise coho in pens."

"One other question," Tanner said. "I believe in free enterprise. Aren't pen-rearing and ocean-ranching efforts made to insure salmon are saved in sufficient numbers to feed the American public?"

"That's true," said Silk. "At least on paper. But the problems created by the artificial raising of coho salmon are greater than their value. They are weaker than say, a natural-born fish. Their flesh is soft from lack of exercise. They're susceptible to every disease known to the fish kingdom. They're treated with every chemical and medicine man has ever invented. As a result the public eats these same chemicals and diseases, and no one knows yet what they're are going to do to the human body. We just know that the quality of salmon has declined by fifty per cent since the advent of aquaculture."

"I see," said Tanner. "I realize that I have a lot to learn. How did you become so knowledgeable about salmon, Silk?"

"I've made it my business to understand the business I'm in," said Silk. "What I've learned is not common knowledge. You'll understand more when you see the pens and the ocean-ranch. Right now I think we should have some breakfast before we meet Cowboy."

Cowboy was waiting at the dock in Claude's zodiac. Tanner followed Silk into the small craft, and the man with the drooping mustache cranked the fifty horse Mercury into action. A few minutes later they pulled up to the loading dock of something that looked to Tanner like an island in the middle of the bay. The island was made out of a hundred overgrown chicken pens halfway submerged in water. In both directions as far as he could see there were pens floating side by side. Chicken wire stretched over the top and down the sides to

213

somewhere deep into the water. Planked wooden walkways led alongside the wire pens and up to a corrugated wall, which was evidently the caretakers headquarters and the butchering sheds.

The man who came out of the main shed and tied them up to the docks was bald-headed and had a steel hook on the end of his right arm.

"Good morning," Silk introduced them. "This is Dr. B.K. Tanner from the east coast. He's heard a lot about pen-rearing operations here in Bane Bay, and he thinks the company he works for might be interested in the industry. I'm Silk, and that's Joel Scott, and we're trying to help Dr. Tanner around town."

"Name's Gus," said the sun-baked man. "I'll show you around since there ain't nobody here 'sept me. I'm the caretaker. The man you'll really want to see is Mr. MacWilliams. He's the boss of SEABOOM. I just finished feedin' them puppies so you missed the big show."

"I'd like to have seen that," Tanner replied agreeably. His interest set Gus to talking.

"See that little bar dangling from the plastic feed jug above each pen? When the bar is bumped, the jug releases a handful of them little brown pellets into the water. The young fish just touch the bar, but the older stock have learned to swim by and give it a hell of a swat with their tails. Some of them big fish is pretty tame. After several generations of domestication, they won't even eat the herrin' that swims inside them pens. Fact is, they won't even recognize the herrin' that wild fish feeds on. Herrin's what I used to use as bait when I fished."

"You were a commercial fisherman?" Cowboy asked.

"Sure was," Gus said. "I used to fish on a trawler named CRYSTAL, at least 'til I lost my arm to a flying cable. Then I couldn't fish no more. When SEABOOM come in here in 1985, I applied for the caretaker's job and got it. A man's got to live you know, and there ain't many jobs a busted fisherman can do. But I hate this job cause I know what's going on. I wouldn't eat one of these suckers on a bet."

As the men walked along the boardwalks, Tanner was thinking about what Gus said. He could see that every pen was filled with swarming salmon, from tiny ones the size of minnows to large ones that weighed up to seven or eight pounds. As far as he could tell, the fish looked alright to him.

"Why wouldn't you eat one of these fish," Tanner was interested. He told himself he would need this information for his report, but in reality he was curious about such a blatant statement from an old fisherman.

"And eat all them chemicals they pack into these babies?" Gus said. "Not a chance."

"I thought you said you fed them pellets," Tanner continued. "Isn't that food?"

"Yep, it's food alright," Gus said. "But hidden in them pellets is steriods to make 'em plump up fast like they do them athletes. Then they put in a red dye that makes 'em turn bright coral so they'll be appetizin' lookin', and on top of that they put in penicillin, streptomycin and every other kind of chemical I don't even know about."

"Why do they give them all those chemicals?" asked Tanner thinking about Silk's words. "Are they diseased?"

"It's supposed to keep them from getting diseases, but the funny thing is that by givin' 'em them chemicals, them fish lose their natural resistance to diseases, and they catch ever kind of disease that comes along. Then they give the disease to the rest of the fish in them pens. Some of 'em go ahead and die, and that's when the men in the white coats comes and take 'em over to the science center and do an autoposy to see if they can figure out what they died of. The rest of the stock that's sick, but don't die right away, are given more chemicals to make 'em well. Some last long enough to get into the supermarkets and are sold to the public, but I don't think any of 'em ever gets well."

"What kind of fish are these?" Tanner felt himself becoming alarmed because the story was beginning to touch him and his family. He knew damned well they had eaten bright coral salmon purchased from the supermarket. In fact, he, himself remembered selecting that color because it looked better than the other fish in the case.

"They're the cohos," said Gus. "They're the only ones we can raise in these here pens."

"Aren't they also caught in the ocean?" asked Tanner.

"Yeah, but them's wild cohos. They're O.K., cause they get their natural color from eatin' shrimp and krill. The ones I go for is the king salmon, the Chinook. He's the wild salmon, the only one man can't figure out how to tame. They've tried to domesticate him but they can't. He won't live no place 'sept where he 'sposed to. He don't get no diseases 'cause he has natural immunity. That's the only kind of salmon I'd eat."

"How can you tell the difference between coho and Chinook? Tanner asked. "Aren't all salmon the same color?"

"No," Silk said. He knew what was going on in Tanner's mind. He knew the investigator was getting damned concerned. "You see coho are bright pink, almost reddish coral. Chinook is closer to white."

"That's right," said Gus. "And in my opinion, I think the government should force the supermarkets to put signs on ever coho that's pen raised, like they put on cigarettes. They should print on them..."This fuckin' fish may be hazardous to your damn health!"

Gus's attention was suddenly diverted to three or four sea lions floating near the pens.

"My main job out here," he said, "is to scare off them critters swimmin' out there. We can't shoot them, so we have divers go down under the water and build tinfoil scarecrows, and they got speakers that make a noise like killer whales. I don't think anything does much good though. There's always five or six of the critters hanging around. Last week they bit through the nets and killed six big fish. The fish didn't cost much, but those nets are over a thousand dollars apiece. Boy, was Mr. MacWilliams pissed. He threatened to fire me, but I didn't give a damn. I don't know what I'm doin' working in a place like this anyways."

As the men toured the rest of the salmon pens, Tanner was aware that every word Silk had said about the operation was true. He didn't have a very good taste in his mouth as they said good-bye to the old fisherman, Gus.

"You know," he said to Silk. "I'll bet there's a lot of money in owning that number of salmon."

"Millions of dollars," said Silk. "What you've seen is big business, and money talks. In another ten years SEABOOM and companies like them will own the whole bay, and every saltwater estuary on the west coast. But wait until you see the ocean ranch."

When Tanner saw the AQUAPROTEIN complex, he was reminded of a prison he had once toured in upper state New York. The three men drove along a narrow dirt road across an isthmus onto a grassy, wild peninsula that stretched along the south edge of Bane

216

Bay. On the western tip of the peninsula sat AQUAPROTEIN, which consisted of acre upon acre of land surrounded by a twelve foot metal fence with three feet of barbed wire slanted inward across the top. At a distance Tanner could see a large group of buildings, and what looked to him like concrete cattle troughs. Along the fence at every fifty feet were posted no trespassing signs. At each corner of the fence there stood a high, platformed wooden tower.

"It looks like they don't want anybody in here," Tanner commented. "Those structures look like prison observation towers."

"Same thing," said Cowboy. "And you're right. The company doesn't want anybody in there. I told Silk that he was crazy for bringing you way out here. The word out is that the place is closed."

"It isn't closed," growled Silk. "They just don't want the public to see what they're doing. We'll get in if we have to climb over that fence."

"I'm not even going to attempt to climb that fence," said Tanner.

"You don't have to," said Silk. "There's a gate open. Somebody must have brought in some supplies."

Silk steered the van through the half opened gate, and soon they pulled in front of a large building labled 'office'.

"Let me do the talking," said Silk. "If they know who we are, they'll throw us out before we have a chance to see anything. I've been here before."

As they stepped into the AQUAPROTEIN office, a man who looked like a combination of Godzilla and a prison guard, stood up behind a desk filled with computer terminals. He had a large head of black unruly hair that partially covered a scarred brow. His huge hands were knotted, and the broken knuckles stood out like lumps on a potato.

"How'd you get in here?" he growled looking from one man to another. He had the cold stare of a boar hog, wicked but intelligent.

"We're interested in ocean-ranching, and we'd like to take a look at your operation here." Silk explained in his usual mild manner.

"Nobody's allowed on these grounds without written permission from headquarters," he said. "We have one open house a year for the public, and if you want to see the operation, you come back in

217

the spring."

"What if I said this man here was a government investigator from Washington D.C., and he demands to examine your plant?" Silk asked maliciously. "I'm just curious."

"I'd say it don't make no difference if he's the President of the United States," said the gorilla. "My orders are to keep everybody out. He wants to come in here, he goes through AQUAPROTEIN channels. You're out of here, or I call in security."

"Alright, we'll go." Silk motioned with his hands to keep cool.

As they walked out the door, Silk moved toward the van, then he disappeared around the edge of the concrete building. Cowboy and Tanner were right behind him. As they rounded the corner, Tanner could see perhaps a hundred rectangular concrete tanks filled with water. From the middle of the tanks a larger concrete canal ran about five hundred yards down the hill and directly into the choppy waves of the bay.

"Those are the raceways, Tanner." Silk pointed to the concrete tanks. "That's where they raise the salmon. What they do is increase the temperature of the water and feed the baby coho chemicals to make them grow faster. Then the coho has a false sense of maturity, and they come back early. The concrete runway that stretches down into the bay is what the salmon swim back up when they return. You see, salmon always go back to where they came from. So these bastards have their ready made fish ranch straight from the sea. They have it controlled. That fellow in there lied. This place is in full operation. We can't go inside the plant, but they probably won't mind if we look around outside."

The words were hardly out of Silk's mouth when an electronic siren began wailing, and several men in white coveralls and safety helmets ran out a corregated metal door at the back of the building.

"He wasn't kidding," muttered Silk. "Let's get out of here."

They reached the van before the first of the men came around the corner of the complex. Silk gunned the engine, pulled a full turn, and they sped toward the electronic gate which was now beginning to slide shut. There wasn't room for the van when they arrived at full bore, but Silk never slowed down. The left fender struck the edge and burst through the gate, ripping the structure off its hinges.

"There goes my left fender," said Silk.

"What about the gate?" asked Tanner. "Won't they report us

218

to the authorities?"

"I doubt it," said Silk. "That would mean an outsider was asking questions, and they don't want any outsiders near the place. They'll just replace their own gate and keep their mouths shut."

"Sounds like there's something mighty fishy going on in there," said Tanner. "No pun intended."

"There is," Silk agreed. "WINDOVER LUMBER COMPANY, alias AQUAPROTEIN, is after control of the salmon resources on the west coast. You might get a guided tour of AQUAPROTEIN if you pull some political strings or went to Windover's headquarters in Portland, but you can bet you'd only see what they wanted you to see."

"I've seen enough," said Tanner. "With the documentation in your material, I can make my report. But let's not have any more adventures until I get on that plane in Portland. My legs are still shaking."

"I've showed you every way that fishing is being destroyed out here on the west coast, and you have the statistics to back it up. Now, that I trust you as an honest man, Tanner, there's something else I want to do."

"I'm afraid to ask," Tanner said with a laugh. "But what is it?"

"Remember when you first arrived in Port Bane, and you were concerned about some things that Dr. Blackburn told you in Seattle? Things about me and the Russian ship I was assigned to as an observer?"

"Yes, I remember," said Tanner.

"Well, what I want to do is introduce you to one man who understands the destruction of sea life even better than I do, or Cowboy here. He's seen it happen all over the world," said Silk.

"And who might that be?" asked Tanner.

"Alexi Semyonov, captain of the Russian refrigerator/factory mothership you saw out there," said Silk.

"You mean you want to go out there again and visit some Russians?" Tanner couldn't believe it. Not another trip.

"Nope," said Silk. "Captain Semyonov lives in Port Bane now. He's no longer with the Russian fleet. As a matter of fact, he left the ship when I did."

"You mean you took him off his own ship?" Tanner was stunned.

"He took himself off," Silk said.

"I'd love to meet this captain," said Tanner excitedly. "I'm too tired to meet the man this evening, but tomorrow night you and Cowboy and the captain come to my room. We'll have drinks and some food and visit for awhile."

"Sounds good to me," said Silk. "What time?"

"Let's make it around seven thirty," Tanner said.

CHAPTER

21

SITKA

Silk staggered aboard the CRACKANOON II and threw himself fully clothed on the bunk in the fo'c'sle. He hadn't admitted it to Tanner, but he had run the gauntlet of time and energy he used only when he was fishing. Now, after being up and about for forty-eight hours, he was completely exhausted.

He thought of the time when he could drink every night and still do a full day's work. He had laughed and bragged for years that if he ever found a temperate bone in his body he would have it operated on and taken out. "Well, old buddy," he said to himself, "you can stop bragging and laughing now; you have just run into some old temperate bones." Last night with Cowboy had finished him off.

He took a deep breath and shut his eyes. Everything was happening at once. Tanner was here from Washington, D.C.; the strike was over; the fishermen had won concessions from the fish companies; he and Semyonov seemed to have nothing to fear from the feds.

Tired as he was, he had a feeling of well-being that he hadn't had in years. Perhaps there was some hope for the future after all.

A natural hologram of Silvia swam before his eyes as it usually did these days when he was trying to go to sleep. Tomorrow he had the whole day off, and he would go see her. Maybe he would invite her to meet Tanner.

Habitually, Silk was up at dawn scrubbing down his boat; washing the dishes he, Tanner, and Cowboy had used; carrying the cans and garbage up the ramp to the large trash barrels that served the docks; arranging every pan, dish, and tool in its correct cabinet; and in short, making every inch of the CRACKANOON II shipshape. When he could no longer find any tasks to do, he took a shower, shaved, and changed into clean clothes. Now was the hard part, the waiting until a proper time to show up at Jodie's house.

He had been to see Silvia at short intervals between every

fishing trip. She was pleasant enough to him, cordial, communicative about subjects that didn't matter, and polite, like she was talking to a co-worker or a stranger. Any other man would have called her a prima donna. But Silk couldn't even think of her in those terms.

If he had tried to explain his deep feelings to anyone, say Cowboy, it would have been impossible. His relationship with Silvia was like she had one arm in front of her holding him at bay. She always queried him about his fishing trips and listened closely to his account, then she inevitably asked when he was going out again. When he explained he was leaving the next morning, the conversation stopped, just stopped. He asked about her, about Jodie, about the bay, about everything he could imagine, and she answered in monosyllables. After the one-sided dialogue they sat side by side in silence until he said he had to be going. She walked him to the door, and that was that, until he came to visit her after the next fishing trip.

Silk, for his part, respected Silvia and catered to her every need. He hung like a slug on every syllable, then when he was alone, he analyzed the conversation, looking for any kind of hint that he was special.

He had done his best to get her attention, but she was as remote as a distant planet. He couldn't understand her. Why didn't she just tell him to go to hell and forget about it? Why did she seem glad to see him after every trip? Why did she seem to worry about him? What kind of relationship had he fallen into with this woman?

The truth was, and he had to admit the fact to himself, he was smitten. He had everything: the CRACKANOON II, his friends, the ocean, and a profession...everything except Silvia. To him she had sure as hell become a requirement. He wanted to possess her, to hold her in his arms, to receive her undivided attention. Why? Why didn't matter. That question was beside the point. The point was he wanted her. He couldn't anymore get her out of his mind than he could rationalize why he had this great need to be near her. And what did he receive in return for his efforts? An absolute zero. Neither plus nor minus. Damn, the woman was frustrating. She was totally unmovable.

Silk felt the blood coursing through his veins. The excitement he felt was even stronger than when he realized there was a large Chinook salmon down there somewhere on his line. Today was the day. He had a legitimate excuse not to go out again until the radar was repaired. He could have thanked the stupid bastards who shot holes in it.

This time he would level with Silvia. Tell her exactly the way he was thinking. Tell her he cared for her. But Jodie's house wasn't the place. He must get her away from Port Bane, away from everything, from the ocean that had nearly killed her, and from Jodie.

An idea slowly formulated in his mind. He knew the place; he knew the wine; and he knew the steps. He would ask her to Sitka. No one knew that place except him and a few loggers. He thought about her reaction. At least, if she refused, he would know the truth about how she felt.

By the time Silk arrived at Jodie's door, he had purchased two fifths of his favorite wine, French claret, 1968 Bordeaux, estate bottled. He had also purchased two sourdough French rolls, two New York cut steaks, and two crystal wine glasses. He would ask Silvia on a picnic, his selection of location, and his choice of fare. He wouldn't buy knives and forks and plates like a flatlander. He'd use both his fishing knives.

When Silvia opened the front door of Jodie's house, she appeared so alluring to him that for the moment he almost completely forgot about his resolutions and the real purpose for which he had come. Silvia's long dark hair framed her face, and she was dressed in Levi's, a white blouse—the sleeves turned back to show her slender arms—and new, beaded moccasins.

"Hello," said Silk. "The radar on the boat is being repaired, and I have the day off, so I stopped by to see how you folks were doing. If you feel like it, I thought you and Jodie might like to go for a drive."

Damn, there he had done it again. He had invited Jodie rather than admit he wanted to see Silvia alone.

"Jodie isn't here, John," Silvia said. "Lillian came by this morning and said that she had an important personal message which was only for Jodie's ears. She said that the message had just arrived via crystal ball. The two of them went off together, and I'm stuck here by myself. If you're asking for Jodie to drive with you, you'll have to come inside and wait."

"No," said Silk relieved. He thought he had heard new warmth in her voice. "Let's not wait for Jodie. Would you come with me and keep me company?"

"I'd love to come," said Silvia, that low, musical note in her voice. "Where are we going?"

"That's my surprise," said Silk, a bright grin spreading over his face. "Someplace you've never been. All you need is a light jacket or a sweater." This state of affairs was a real break. At the moment he could have kissed Lillian for taking Jodie out of the way.

"I'll get my things," said Silvia smiling.

Silvia thought that Silk would probably take her to one of the gourmet restaurants on the strip, or on a beach picnic, but she was unprepared for the long drive to nowhere.

They had left the highway about twenty miles north of Port Bane, and for the past hour they had been following dusty logging roads over mountain ranges, down into silent brush-filled valleys, and across rickety river bridges. The roads were narrow, rutted, and dry like they had never been maintained, and the view consisted of bare brown hills cluttered with brush and tree stumps, or still-smoldering giant slag piles. Why Silk had decided to take this route was a mystery to her, but he was a silent, strange man except when he had something to say. Then his sentences were fluent.

A thin layer of dust had covered Silvia's skin and lay in rivulets on her now-wrinkled white blouse. Both of them seemed to have exhausted the busy repartee with which most couples pass the time, and for the last dozen or so miles, Silvia had become uncomfortable as she was jostled about in the front seat of the old van. She felt tired, dirty, and thirsty.

"Are we perhaps in the vicinity of where we are going?" she asked, cheerfully trying to mask her discomfort. "We haven't seen a sign of life for over an hour."

"I didn't remember it being this far or that the roads were so rough and dirty," Silk said. His eyes were focused on the winding tracks. "These are just logging roads, and they've never had a blade on them. This ride must be a trial for you."

"Nothing of the kind, John," said Silvia determinedly. "I'm sure you know what you're doing."

They had just topped a steep rise when Silk suddenly stopped the van.

"What seems to be the matter?" asked Silvia looking at the

clouds of dust that floated in light levels throughout the van.

"Look," said Silk in excitement, pointing his finger toward a deep valley which stretched like a narrow green meadow toward the base of a bald mountain in the distance.

The sight was one of the most breathtaking Silvia had ever seen. The narrow area was completly forested in green lush foliage out of which ran a trickling stream. From the center of this isolated island rose a green, cylindrical disk with a pointed top, hundreds of feet above the surrounding foliage. The massive structure looked like a tall skyscraper which dwarfed the other massed buildings.

"What is it?" gasped Silvia in amazement.

"It's a giant Sitka spruce," said Silk proudly. "The only one I know of in these mountains. It's literally beautiful down there. That's where we're going to have our picnic."

"How romantic," Silvia exclaimed demurely. "It's quixotic!"

Silvia sat on one of the giant roots beneath the Sitka spruce and watched while Silk prepared their picnic. He dug a pit with a broken shovel, built a fire, and placed a wire grill over the pit. While the wood was burning to glowing coals, he opened the bottles of French wine so they could breathe properly and carefully unwrapped two beautiful crystal wine glasses. He slit French rolls in half, lay thick steaks on the grill, and handed Silvia a hunting knife which was razor sharp.

Silvia stared at the wicked looking blade.

"What do I need this for?" she asked, gingerly holding the knife in the air between the tips of her fingers.

"That's your eating utensil," said Silk. He was beginning to love this day. "In fact, except for the wine glasses, the knife is all you have, and if I get surly, or a big bear comes for you, you can protect yourself with it."

Silvia laughed. "I've been on picnics before, John Silk, but I've never been on one quite like this, a combination of fine wine, New York steak, crystal, and a fisherman's knife. Where are your plates? What are we going to eat on?"

"You'll see," said Silk smiling as he poured the first glass of wine. "A toast to us, the Sitka, and the CRACKANOON II."

Silvia held the glass without drinking to his toast and said, "Rather a toast to you, John, for taking me to this marvelous place, and all you've done to help make my life in Port Bane bearable."

They touched glasses then, and Silk drank. Except for the night at the lighthouse these were the first words Silvia had spoken that indicated she realized he existed.

"Wait until you taste your steak," he said trying to keep his voice from shaking. "Now you just relax and drink your wine while I get them going."

Silvia watched while John took salt and paper wrapped in foil from his Levi's pocket. He carefully seasoned the sizzling steak on one side. He expertly turned the meat once with the large knife before he salted and peppered the steaks again. When the steaks were done, he speared each one and placed it between the roasted rolls. He handed one to Silvia and watched seriously while she took a bite.

"What do you think?" he waited for her reaction.

"It's superb," she said. "I've never eaten anything so tasty."

Again Silk smiled at the complement. Who would have thought that something so simple would make her happy?

"I call this a French picnic," he said modestly. "If you like, we can do this type of thing often."

Silvia gazed at something in the distance, and for a long moment she said nothing. Then she spoke as if she were talking to herself.

"Today is the first day you've come to see me in over a week, John."

The somber tone of her voice stunned Silk. It had never occurred to him that she thought about him while he was away fishing, or that she counted the days he was gone. Maybe she liked him more than he figured. He'd have to think about that possibility on the next fishing trip. Right now, he'd better concentrate on Sitka and the food and this lovely, strange lady.

"Let me pour you some more wine." He could think of nothing else to say. He reached for her glass, and the coolness of her fingertips touched his hand. Somehow he felt like he had just been burned. "We could climb out on that big root and dangle our feet in the water," He said quickly.

"I'd like that," Silvia smiled mischievously. "I can't remember ever before sitting on a big root and dangling my feet in the water."

The afternoon passed swiftly for them. They analyzed the giant spruce together; they waded in the icy, clear stream; and they took a short hike up creek to a small waterfall.

Silvia's sombre mood had changed to that of a joyful wood-nymph. Her dark waist-length hair swayed in rhythm to her slender body dashing from waterfall to creek to hiking through wild flowers. But Silk's mood had become more pensive. He hadn't as yet found an opportunity to say what he had on his mind, and time was running out.

They were sitting on a blanket, halfway through the second bottle of wine, watching the small fire crackle in the pit, when he finally cleared his throat and started to talk.

"Silvia, there's something I've been meaning to talk to you about. I somehow haven't been able to find the right words, but I've given the matter much thought, about you and me, about us, I mean."

"What have you been thinking, John?" she asked pulling up little blades of grass and stacking them on top of one another.

For a moment he wanted to shake her and yell "look at me, damn it! I want you to look at me instead of some tree out there, or a waterfall, or blades of grass. I'm here! I want your eyes to see me! I'm alive!"

He had better get ahold of himself fast. That sort of behavior would mean the end of him forever. He took a deep breath and tried again in his softest voice,

"Silvia, I've thought a lot about you, about me, about us, especially while I was out on the boat, about the way we met, about your coming to Port Bane, and about us together. I just haven't been able to tell you all my thoughts. Today, well today, was preplanned, so I'd have the opportunity to see you alone. And I've almost let the time slip away without telling you what I have on my mind."

"I've thought about you too, John," said Silvia. Her longlashed green eyes studied the red claret for a moment before she put the glass to her lips and took two sips.

"When I first saw you lying out there on the jetty, I knew something special had happened to me," Silk continued. "Someone important had come into my life. Then that night at the lighthouse, I held you in my arms and I thanked God you were alive...I love you, Silvia."

Silvia reached toward him and touched his hand with her cool fingertips, and for the first time since the fishermen's meeting at Jodie's house, she looked directly into his eyes; her eyes were misty. Silk thought he would choke with emotion.

"I love you too, John," she said. "You are the dearest friend I have."

Silk felt shock run through ever fiber of his body as Silvia went on talking. Wait a minute! Friendship wasn't what he had in mind! Had she misunderstood him? He tried to pull himself together and concentrate on what she was saying.

"When I first came to Port Bane, John, I was shattered both physically and mentally. I had just lost my father, and I went fishing with Matt to gather my thoughts together. We ran into that horrible storm, and Matt was killed, and I almost lost my life. While I was lying on that bed recovering, I couldn't decide whether I should live or die. I had absolutely nothing left, no clothes, no home, no one I cared for. I was almost twenty-seven, and I had achieved nothing of value in my life.

Then I got to know Doc Ox and Jodie, and the Locals came to see me. They all seemed alive and warm and interested in me, although I couldn't see how they could care about a complete stranger."

Silk poured himself another glassful of wine. He stood up and began to pace back and forth outside the blanket edge. He felt sick. 'but what about me?' he thought.

Silvia showed no recognition that he had even moved. She looked steadily at the blade of grass in her hand and continued talking as if she were explaining her feelings to herself.

"After that, you came to see me, and I realized you helped save my life. You saved the wrecked boat for me, too. And you go out on that dreadful ocean to bring me money.

I hate the ocean. In the back of my mind I worry about you out there by yourself. I wonder whether or not you will be killed like Matt. Isn't it ironic that the one reason we are together today is because of violence and death?"

Silk sat down on the blanket next to her and refilled her glass with wine.

"The ocean isn't that dangerous," he said reassuringly. "I've been fishing out there every season for ten years, and nothing's happened to me."

"But the ocean is dangerous, John," Silvia contradicted him.

"I checked out a book from the library. The book said commercial fishing is the most dangerous profession in the world."

"The book doesn't mean a thing," said Silk. "That information is for young college students who've never been at sea. I am an expert. I know what I'm doing."

"I'm not worried about your ability as a fisherman, John," continued Silvia. "I know you're good, but the ocean is like an unfeeling beast. It can kill you without warning. And what about the last trip when those men shot at you? Don't you think you might have been killed?"

"I didn't know you had heard about that, Silvia?" exclaimed Silk. "That shooting was a once in a lifetime thing. People don't shoot at you everytime you go fishing."

"I understand that," said Silvia, "but look at the situation from my point of view, John. How would you like to worry twenty-four hours a day about whether I was alive or dead?"

"Come fishing with me," Silk implored. "Just once. One trip will help to erase the bad memories of the storm that you have stored in your memory."

"Never," said Silvia. "I'm never going out on a fishing boat again, or even a ship! You, yourself, recently lost two of your friends while they were fishing. You can't convince me fishing is safe. Personally, I couldn't stand going through the same agony as those other women in Port Bane. Sometimes I wish we had never met, John. I wish I'd never agreed to save that ghastly boat. I would prefer it to be sitting on the bottom of the sea. I like you, John, and if anything happened to you in that boat, I would always feel it was my doing."

Now, she had used that word "like". At first she said she loved him; then she called him her dear friend. Now it was down to like! Regardless of his efforts things seemed to have swiftly regressed. The day had not turned out at all the way he had planned. He had expected to tell Silvia how he felt about her; and, he had taken a considerable gamble that he might lose her, but there had also been the chance that she might respond to his overtures. Nothing had worked. He had to buy time so he could think about the problem.

"I shouldn't have mentioned taking you out on the CRACK-ANOON II," Silk said in a serious tone of voice. "I promise you that I'll never mention it again. But there is a place I would like to take you tonight, if you're not doing anything. I'd like you to come with me to Mr. Tanner's room in the Port Bane Inn. He's the Justice Department Investigator from Washington D.C. I suppose you've heard about him, too."

"Yes," said Silvia. "Jodie told me he was here."

"Well, a few of us are getting together this evening." Silk feared he was running his words together. "Cowboy, Semyonov, Tanner, and I. I'd enjoy having you come with me, if you will. I want to show Tanner that we have pretty women here in Port Bane,too."

"Are you flattering me, John Silk, by calling me pretty?" she smiled.

Silk stumbled over his own words. "I wasn't telling anything but the truth. It's just that compared to all those other women...you...well..."

"Of course I'll go with you," said Silvia. "I'd enjoy meeting Mr. Tanner."

Silvia seemed to have completely forgotten her emotional fit about the sea, and she was once again her calm, charming self.

Silk was beginning to relax again. He hadn't won any points today, but he hadn't been tossed in the pit either. He would just have to take it step by step until she had a chance to understand her own feelings. He knew one thing with a certainty, he didn't intend to give up this lady.

"Would you like me to invite Jodie so you'll have company?" he asked.

"I enjoy your company, John," Silvia said.

"Thank you for today, Silvia," said Silk rising and extending his hand, "and for talking to me. I understand your feelings, but I don't know what to do about them. Well, let's clean up this mess and get you back to Jodie's."

"Thank you too, John, for this charming day," Silvia took a deep breath. "This picnic has been a welcome break to my tedium. It's been a long time since I've been treated so...so splendidly."

Silk glowed at the complement. If she only understood how he really felt.

CHAPTER

22

RIPTIDE

The day before Tanner flew back to Washington D.C. he did not leave the confines of his room. Whether he liked it or not, his experiences on the CRACKANOON II and in the Port Bane fishing community had changed his viewpoint.

He stood before the window to view the spectacular Pacific Ocean one more time, and he got depressed. The scene on the beach was not nearly as romantic as he had imagined it on the first morning after his arrival. Several tourists, dressed in pretty outfits, strolled up and down the naked beach cluttering up his beautiful view. He switched his attention to the surf: the white-crested waves crashing over the great rocks, sweeping the sandy beach until they lost momentum up the steep grade, then rolling back, colliding with their brothers into a muddy roar of sand and foam.

Again he had a feeling of dissatisfaction. That water was too damned cold to go swimming. This ocean was for salmon and salmon fishermen. Disgusted, he picked up the phone, ordered a pot of coffee, and turned his attention to Silk's boxful of material about the salmon industry. He had his work cut out for him.

He thumbed briefly through the manilla folders reading the neatly printed topics on the catalogued tabs. There were professional articles written by marine biologists and scientists with impressive credentials after their names. These articles explained certain aspects of every kind of ocean fishing he could imagine. There were newspaper clippings about particular problems and pending legislation. There was a large section that held thousands of pages of computer readouts concerning salmon availablity, hatchery releases of smolt, and salmon egg distribution around the world.

Tanner opened a folder that was labeled TBT Contamination of Pen-reared Salmon. He didn't know what TBT meant, but he had visited the pens and seen the salmon. The opening sentence of the article caught his eye. It said,

"Pen-reared salmon contaminated with tributyltin (TBT) are entering U.S. seafood markets according to a recent report released by the National Marine Fisheries Service. According to the report coho salmon reared in sea pens treated with TBT, sold as aquaculture products and purchased in public markets are found to contain concentrations of 0.28-0.90 of TBT. TBT has been described as the most toxic compound ever deliberately introduced by society into natural waters. TBT and organotin are used as wood preservatives, additives to bottom paints, and netting treatment which are used in salt water pens for rearing salmon."

'My God,' thought Tanner. 'Silk and old Gus knew what they were talking about.' Fascinated now, he skimmed the rest of the article, and certain underlined phrases seemed to implant themselves in his memory.

"The most common pen-reared salmon products entering the U.S. market are 'pan-sized' or 'baby' coho harvested as juveniles in pens from places such as Puget Sound, Bane Bay, or Port Angeles. These farmed salmon have proved popular with some restaurants and markets seeking to promote 'fresh' fish the year round."

and,

"TBT concentrations in salmon that died during the bioassay were nearly constant for all doses. This suggests that TBT continues to accumulate until a threshold concentration is reached in critical tissues and causes death. Low doses of TBT can impair the immune system of rats, which suggests that salmon raised in TBT-treated marine net pens may be more susceptible to the poisoning. The effect of TBT on the human organism is as yet uncertain."

The article concluded with two underlined statements.

"The use of TBT as an additive to bottom paints has already been restricted in Norway, France, and Great Britain following the die-offs of shellfish beds attributed to close proximinity of TBT infected pens." And, "High levels of TBT have been found in most marinas along the California, Oregon, and Washington coasts. This indicates that action similar to that taken by Norway, Britain, and France should be taken by the United States."

"This is a matter for the Federal Drug Administration," Tanner muttered to himself. "Surely they know about TBT in coho salmon. These fish should be banned from American markets."

This information was from but a single article. What all must the other eight thousand pages of material contain?

Tanner selected a thick, hard-bound volume from the bottom of the box. This document was Silk's daily diary which spanned ten years of salmon fishing. Besides the usual material concerning number of fish caught, bait, weather conditions, and breakdowns, there were biting personal observations about governmental laws and regulations which caused the deterioration of fishing conditions. One statement particularly caught his attention. It read,

"The problem with this whole industry is the people who sit behind their big desks in Washington D.C. and make laws that control our lives. Most of these officials have never seen the Pacific Ocean, and if a salmon popped up on their dinner plate, they would have no idea whether it was a TBT infected coho, a Chinook, a hake, or a bottom fish."

By middle of the afternoon, Tanner was still sitting in his shorts at the round table. Scattered on the floor were page after page of Silk's research on salmon-trolling, pen-rearing, ocean-ranching, and joint venture. He had read at least two hundred pages of documentation, and he had hardly touched the contents. He had no idea how he would ever digest and organize this volume of material, but one thing was certain. This trip was a winner. He had investigated and discovered serious problems in an area that everyone in Washington either didn't know about or had completely ignored. He had seen the Russian fleet; he had toured SEABOOM and AQUAPROTEIN; and he had been out on a fishing boat for two days. He had made his own observations.

To the best of his ability he had to agree with Silk. Aquaculture and foreign fishing fleets would be the end of American commercial fishing on the west coast. Furthermore, he had more than enough statistics to document every point. When he returned to Washington D.C. after this trip, he would make a real impact. Judge Powers would commend him on his success, and Vivian would be proud of him. Changes would be made, and he might even receive a promotion. Yes, sir, this trip would pay dividends.

The bad part of the whole scenario was that he would have to leave Port Bane. He liked the place, and he had particularly enjoyed the people. They had finally accepted him, and now they trusted him

233

to help solve their problems. Big business in America was preparing to stamp out another way of life. It had already destroyed the independent farmer, the rancher, and the loggers. It had rendered ineffective the working man's unions. Now it was in the process of destroying the American commercial fishermen. Given what he had already discovered about foreign fishing, pen-rearing, and ocean-ranching, he felt the coming catastrophe had already hit. Like Silk said, "If action were not taken immediately, within five years there wouldn't be an independent salmon troller left on the west coast." Perhaps this time the destructive process of big government and big business might be prevented. Certainly, he was going to give it his best shot.

Tanner glanced at his watch, and for a moment he almost panicked. The whole day had passed away, and here it was almost time for his guests to arrive. He just had time to shower, dress, and order food and drinks for the evening. He was excited about meeting the captain of the MURMANSK. That Silk had told him about his helping Semyonov to defect proved the fisherman trusted him. Quickly he decided on a menu. Vodka and caviar for the Russian; and scotch, beer, wine, and roast chicken for the fishermen and himself. He picked up the phone and put in his order to the inn management.

Port Bane wasn't Washington D.C., and these people weren't politicians. One thing he didn't want was to appear to be pretentious. He dressed in Levi's and a dress shirt. No tie. The meeting would be short; he'd meet the captain; the guests would leave, and he'd call Vivian. He'd go to bed early so he could get another good night's sleep before flying home. Everything was under control.

He answered the door at precisely 7:00 P.M., the time Silk had said he would arrive.

"Come in, come in," said Tanner and stood back for his guests to enter.

An exquisite, dark-haired lady entered the room first and smiled with her green eyes and pretty mouth. Tanner sucked in his breath. She looked like a vision in white. Next to the lady was Cowboy who was neatly dressed. His hair was combed and his handlebar moustache was glistening with wax. He was followed by a tall, gaunt man with a heavy black and gray beard and cold blue eyes, wearing a gray turtle-necked sweater, dungarees, and hunting boots. Behind the tall man was Silk; his shock of red hair still unruly; his clean

234

sweat shirt still faded and worn; but he was freshly shaven and smelling good. Silk made the introductions.

"Silvia, this is Dr. Tanner from Washington D.C." Tanner shook hands with the elegant woman. "Silvia is half owner of the CRACKANOON II," Silk explained. "And this gentleman is Alexi Semyonov, former captain of the Soviet factory ship MURMANSK. Alexi, Dr. Tanner." The two men shook hands. "And, of course, you know me and Cowboy."

"Help yourselves to your poison," Tanner led the way toward the bar. "I hope I got everyone something he likes to drink—my pardon; Silvia is it? Silk didn't tell me you were coming. May I order you a mixed drink from the bar? There is vodka, wine, beer, scotch, caviar, vegetable plate, chips, pickels, etc."

"Thank you, Dr. Tanner," said Silvia in a voice that sounded like a cross between an organ and Swiss Christmas bells. "Scotch will be fine for me. I think John brought me along because I had heard about you and wanted to meet you. I hope you don't mind."

"Not at all," said Tanner. If Silvia were Silk's wife or girlfriend, then he had underestimated the red-haired fisherman. She had presence, a savoir-faire that would fit perfectly in Washington D.C. What was she doing in Port Bane?

"Sit down around the table," he said smiling. He needn't have bothered speaking. Everyone had already taken a chair, and Cowboy was pouring the drinks. Silk rearranged the dishes of food so there would be room for the large, elongated chart which he carefully unfolded. Tanner realized that the tall fisherman already had his presentation planned.

"Now," Silk said like a college professor. "This is a chart I have drawn which exposes the whole commercial fishing scam. I want you to take a copy of this chart back to Washington D.C. with you. You have met the fishermen, seen the pen-rearing, the ocean-ranch, and the Russian fleet. This chart puts all these areas into perspective."

Tanner leaned over Silk's shoulder and studied the chart. There were black, red, blue, purple, pink, and orange lines criss-crossing the page. Each line connected with rectangles which named the departments of the U.S. government, minor government organizations, west coast rivers, kinds of fishing, and names of big business corporations. There were also names of individuals, some of whom he recognized. At the top of the chart were insurance companies, banks and lending institutions, American fisheries society, President, U.S. House of Representatives, U.S. Senate, Department of Commerce, and Pacific Fisheries Management Council. Other lines connected the U.S. Coast Guard, U.S. State Department, U.S. Justice Department, Department of the Interior, and U.S. Park Service. At the bottom of the page were squares listing private ocean-ranching and private pen-rearing. There were companies he was familiar with, and companies he wasn't. There were names of universities, state and private. There were names of important men and political figures.

"I see," Tanner said studying the chart trying not to show he was overwhelmed. "You're right, Silk. It's going to take me some time to understand all of this information, but I already have a few questions. Why do you begin the chart with the insurance companies and the banks and lending institutions?"

"This is a money and resource situation," answered Silk. "And those big money people control even the President of the United States. To present a true picture, you have to start with them."

"I see," said Tanner. "I can see names like individual hatcheries, pen-rearing, and ocean-ranching, but where are the commercial fishermen like yourself?"

"That's what our whole struggle is about," said Silk. "We commercial fishermen are ignored. We are not even included on this chart and have never had our say."

"That's amazing," said Tanner. "I see some names of men I know in Washington D.C., and I see some people I don't know. I notice they are listed under universities, private businesses, and government organizations. There is even one man listed in all these places, and he advises the U.S. Justice Department. What does this mean?"

"Once you recognize who really controls the fishing industry," Silk said, "you begin to understand who is pulling the strings. I have done years of research on the people you mentioned, and these men are a source of real graft in the industry. They are teaching at universities, getting money from private enterprise, and advising departments in Washington D.C., all at the same time. They are, for the most part, rich and powerful. They should be; they're taking payoffs from three separate organizations. Right now the graft and corruption in the fishing industry is worse than it ever was with the unions and government.   The man you mentioned advises the Jus-

236

tice Department. He has more hats than a well-dressed hydra. He is advisor and agent for AQUAPROTEIN, which is a subsidiary of Windover Lumber Company; he is chairman of marine biology over at the Pacific Coast Science Center; he advises your department in matters of fishing controversy; and now he has gotten himself appointed Chairman of the World Conferences on Aquaculture Development for the United Nations. He has been an Oregon State Representative, co-chairman of the Natural Resource Committee for Oregon, Chairman of the Oregon Fisheries Society, Director of the Pacific Fisheries Management Council, team leader for Sea Grants to Underdeveloped Nations in Marine Resource development, and President of the American Aquaculture Society. If one were to investigate his past, they'd discover he's taken every pay-off he's ever been offered. He has all ten of his fingers in the fish pie, and if he had more fingers he'd stick them in too. He, and men like him, are the source of our problems."

"Those are some assertions," said Tanner. "Can they be proved?"

"Damn right," said Silk. "You have eight thousand pages of documentation in that box, and a chart that will blow the hell out of the whole fishing scam. Why do you think our letters asked for a congressional investigation of the whole industry?"

"This is heavy stuff," said Tanner. "Since it's impossible for me to understand this whole spectrum in a single sitting, perhaps you could summarize it for me."

"That's not easy," said Silk. "The probem is so huge, and it has taken so many years to develop. As you are most likely aware, we have not been a true democracy since the advent of modern technology, probably since the Viet Nam War. We are now an economic dictatorship with a police state to enforce the monied powers. Money buys; money controls; money is power. He who controls the money controls the world, and money is made from natural resources. For years, under the guise of democratic principles, The U.S. has kept petty dictators in power around the world in exchange for controlling their populations and getting more than our share of each country's resources. Now this government principle has spread to the oceans of the world. Not only oil, minerals, and dope are involved here, but protein. Whoever controls the protein of the world will eventually wield the greatest power. And since the ocean is po-

237

tentially the world's greatest source of protein, the future struggle will be over the oceans' resources. The big companies in the United States understood this fact years ago. Companies that you're familiar with got into the fish business because they had the money, companies like WINDOVER LUMBER COMPANY and JANZEN SOUPS. They bought up great tracts of ocean frontage and began raising their own fish on private ranches. Other companies have started pen-rearing operations in the salt-water estuaries. They have the money to pay for these operations, and they reap the profits. They have the money to lobby and control government policy concerning the fishing industry, and they also control the salmon egg distribution around the world. Therefore, they have a voice in salmon-raising operations in other countries. This is fine with our government because the United States is reaping the benefits of these profits.

The control of the fishing grounds around the United States is also part of the power struggle. By trading our coastal resources to foreign powers in exchange for permission to place our ships and missiles in the waters around the world, we can control the balance of nuclear power. These practices will eventually result in greater and greater control of other countries. The individual fisherman is not even recognized in this great struggle. He doesn't matter. Private enterprise is dying. The sooner the small businessman disappears from the scene the better."

Silk finished speaking and rolled a cigarette. There was silence around the table. Suddenly he grinned and looked like a little boy who has just been caught pillaging the cookie jar.

"I know I sound like a madman, Tanner," he said apologetically. "Or a revolutionist or a political activist, but really I'm a small-time fisherman who wants to make a living. It's just that I've done so damned much research on the problem, and I know that I understand what I'm talking about. We, the American fishermen, just need to have our voices heard before it's too late. That's why you're here to help us."

"Well I'm sure I might argue with you about one or two of your statements, Silk," said Tanner. "I still believe in our government and in a democracy. That's why I work for the Justice Department. But right now I'm just getting to know the the fishermen's problems. One thing I will guarantee you. I'll study this material,

238

and I'll take your chart back to Washington D.C. I can assure you your voice will be heard."

"That's all I ask," said Silk. "Justice for one; justice for all. Does anyone else want to get in on this conversation. I don't want to monopolize the whole evening."

"Yes," said Semyonov who had been listening intently to Silk. "I would like to ask Dr. Tanner some questions." His voice was gentle, but it carried the authority of a sea captain.

"I would enjoy hearing your point of view, Captain Semyonov," said Tanner. " Silk has told me something about your experience, and it seems you have taken some criticism about your country's fishing policies."

"As Silk has probably told you, for many years I have commanded Russian fishing vessels of every size, up to and including the largest ocean-going, trawler/factory ships. I have seen the Soviet fishing fleet expand from one hundred and eighty vessels in 1954 to nineteen thousand trawlers and fish-factory ships in 1988. I have fished around the world and seen whole fishing grounds become depleted from the Barents Sea to the fishing grounds off the American east coast. Much of the world's oceans are now almost sterile because of over-fishing, not only by the Russian fleets, but Japanese, Norwegian, East German, and Polish fleets as well. Now these fleets are in the process of doing the same thing to Alaska and the Pacific west coast. I have of my own free will been a part of this destruction, and I have witnessed its horror with my own eyes.

I must agree with most of the complaints that Silk has against American fishing problems, but I will be the first to tell you that Russian fishing policies are much worse than the Americans, mainly because we have expanded our fishing armada. We have destroyed not only our own fishing grounds, but we have obtained permission to fish the waters off every continent in the world. However, the Russian fisherman has always respected the Americans for their principles, even though we have often thought you are not smart. You willingly destroy your own resources for political prestige. That I cannot understand."

"The Captain here has a million stories," Cowboy chimed in. "He was the main reason I sold the OTTER and got out of joint venture. I figured if I was helping the Ruskies to destroy American fish, I wasn't worth a shit. I didn't want no part of that anymore. That's

239

why I'm retired now."

"I must add this point," continued Semyonov. "America with all its faults is still a kind nation. These people have taken me in and protected me, whereas I would already be in prison in Russia. True, there are political problems, but in Russia they would condemn all fishermen who protested and destroy them and their families. In America the fishermen wrote letters, and they were heard. They sent you, Dr. Tanner, to look into the problems. That is what America is all about."

After Semyonov had spoken, a silence seemed to settle over the room. Then Tanner said,

"Captain Semyonov, if there were time, I would like to sit down and listen to your life's story, but I have to leave in the morning. I'll just have to see you the next time I come out to Port Bane. I want to tell you people how much I've appreciated my visit here, and I can guarantee you that the American system, as so poignantly was stated, works. With my experience here, and fortified with this material, I'm sure that my boss will carefully read my report, and there will be action forthcoming. If no one has anything further to add, I think I'll turn in early and get a good night's sleep."

"Nonsense," said Silk. "The evening's young, Tanner. All you've done since you arrived here is work. Silvia hasn't had a chance to get to know you, and you need to sample the night life of Port Bane. The chicken, the trimmings, and the drinks were wonderful, but now you need some music and some entertainment before you go back to Washington D.C. You just as well come down to the Riptide with us and relax."

"Yes, do come," Silvia smiled warmly. "I would like to hear about Washington D.C."

"In Russia," said Semyonov. "We work very hard, but, then we drink and we dance very hard too."

"Yeah," said Cowboy. "Let's go have a hell of a time."

"Tanner never understood whether it was the large amount of scotch he had consumed while listening to Silk and Semyonov, or whether he had forgotten to eat, or whether it was Silvia's smile, but he put on his coat and joined the party. "In Rome always do as the Romans do," he rationalized to himself. He would call Vivian as soon as he got home.

"Alright, I'll go," he said. "But you have to promise to bring

240

me home early. I have a phone call to make, and a plane to catch."

"You bet," said Cowboy earnestly." We'll just stay a little while. We always get home early."

Tanner would have stayed in his room had he realized that Cowboy was thinking about early in the morning.

The Riptide looked to Tanner like a replica of an old western honky-tonk bar. A large, rambling wooden structure on the far corner of the bay road, it had outside speakers spieling out some kind of discordant cacophony played by electronic bass guitars, lead guitars, and drums. Tanner thought the music sounded like a hundred alley cats fighting. It must be some kind of hard rock music he hadn't heard before, but he couldn't be sure. He was amazed that these people were actually going inside until Silk explained.

"I'm sorry we have to take you here," he said, "but it's the only place in Port Bane that I know of where you can get a decent drink this time of night. We used to have good western music at some of the other bars until the tourists showed up. Now all the bars cater to the college kids from the universities who drive over to the coast for excitement."

"I understand," said Tanner. "This place will be fine for one drink."

The five people entered the Riptide and the deafening din deluged them. Silk shoved two tables together near the back of the room. He placed the chairs around the table, seated Silvia and ordered a round of drinks from the waitress. Cowboy sat down next to Silvia and leaned back in his chair, obviously prepared to enjoy the music.

Two women with guitars, wearing tight leather pants and leather vests and knee-length boots were jumping up and down on the stage to the rhythm of another woman pounding hell out of some kind of key board. At the same time another woman was beating on a drum set. Everytime the band reached a certain crescendo, the two guitar players on the stage kicked their feet high over their heads showing their protruding crotches to the greatest advantage. The small dance floor in front of the stage was crammed with young couples jumping up and down like monkeys or apes or African Ubangis on a string.

The drinks arrived, and Tanner swallowed half of his scotch in a gulp. Maybe the booze would get him in tune.

Another ten minutes of repetition and the song finally ended. Cowboy stood up.

"Sorry folks," he said. "I can't stand nothin' but country western songs. I'm goin' upstairs and try my luck at some blackjack."

Tanner's ears perked up.

"Gambling is legal in Port Bane?" he asked.

"Not all forms of gambling," said Silk, "but there are six or eight blackjack tables upstairs that run all night. It's fun once in a while to play with a few dollars."

"I'd enjoy watching Cowboy gamble," said Silvia. "Do you mind if I join you, Cowboy?"

"Tanner, if the music is too noisy here, you might like to take your chances with Cowboy," Silk said.

"Let's go," said Tanner rising enthusiastically. "I'm one hell of a blackjack player."

Actually, Tanner had only played the game a couple of times in Las Vegas, but he had won some money, and he had won fifty thousand dollars on the screen at home playing against his computer. Gambling was in his blood, and he'd memorized a set of rules that worked on his software. Winning a few hands would top off his trip.

They took the handsome curved staircase up to the second floor where the card room was located. The long room consisted of a standing bar at one end where drinks could be ordered, and eight round card tables that held six players each. Tanner took one look at the set up and decided it was a small operation compared to the casinos in Las Vegas.

The place was overflowing with gambling patrons, but there was one vacant chair at the first table. Silk gestured for Tanner to take the chair.

"What about you, Cowboy?" Tanner asked. "Don't you want to try your hand first?"

"Since you're our company, you can go first. We'll watch you and drink somethin' at the bar," said Cowboy.

"Thanks, Cowboy," said Tanner smiling broadly. "I've studied the game a bit, and I should come out ahead. If there's anything I like, it's a couple of hands at cards. This is a high point in my visit. If I don't have any luck, I'll give you a shout, and you can take my place."

Tanner sat down and took a fifty dollar bill out of his billfold.

242

"Cut me in for fifty to start with," he said to the dealer, an oily looking fellow with streaked gray hair and a gold front tooth. Tanner winked at Silk as the dealer handed him the ten red chips.

"Sure looks like you know your way around a blackjack table, Tanner," Silk grinned. "If you need us to help you tote your winnings out of here, we'll be at the bar over there."

Silk, Silvia, Semyonov, and Cowboy adjourned to the bar while Tanner anted up ten dollars on two hands and took his cards. He hit a twenty-one first pop and collected his money. Yep, he could tell, tonight was his night.

He upped his ante to twenty and tried to remember the rules he had memorized against his APPLE computer. Somehow, they came through a haze, and the cardinal rule kept reappearing. Don't gamble when you're drinking. Tanner shook the haze away. That rule was for Vegas, not a podunk blackjack room in Port Bane. He shook himself to clear his head and remembered,

Stand on 13 or more when dealer shows 2,3,4,5,6.

Draw on 12 only if dealer shows a 2 or 3, otherwise stand if dealer shows 4,5, or 6.

If dealer shows 7 or more, do not stand until you have at least 17.

There are 16 "10's" and 32 other cards in a deck. The ratio is 36/16 or 2.25 other cards to one "10" card.

Always double down on hard 11.

With a hard 10 double down except against an ace or 10.

Yes sir, he had the rules down pat now, and he was flowing. He hit a king and a jack again, and won another $20.00. He opted for a second hand, and bet ten on each. Again he won both. In short order he had doubled his original fifty dollars.

He ordered another drink from the bartender and settled down to play some serious blackjack. At first he won heavily, and he was soon five hundred dollars ahead. Then, after his third scotch, he had trouble remembering the rules he had memorized, and the numbers to hit and the numbers to stand. He remembered Silvia putting an arm around him and Silk cheering him on.

Someone said he was in the hole five hundred dollars. His back ached, and the room had become foggy. He looked at the dealer and tried to focus on the gold tooth smiling through the dealer's thin lips. He reached in his billfold and took out his last hundred dollar bill. Somebody said it was a ten.

"Nevermind, give me the chips," he ordered doing his best to keep from slurring his words. Damn, he had spent all his money, but that didn't matter. He was having fun, and he'd win this round. He laid down the chips, but he never knew he lost the hand. His final memory of Port Bane was holding on a black three and a red king.

There was a roaring in Tanner's ears and for a moment he thought he was riding to the inn in Silk's old van. That is, until he opened his eyes and noticed a shapely leg and gray dress walk by his chair. Damn, his head hurt. He was on a plane going somewhere, and the exquisite ankle belonged to a comely stewardess who was now standing over him.

"Welcome back to the real world, Mr. Tanner," she said smiling at the hurting man. "Your friends put you on the plane and said to buy you a drink when you awakened."

"Where am I?" he mumbled incoherently.

"You're on American Airlines heading into O'Hare airport."

"I mean, where am I going?"

"Washington D.C.," she replied.

"That's good," Tanner said relieved. "That's where I live. Could you bring me a tough bloody mary?"

The stewardness came back with a drink and a note signed by John Silk.

"Thanks for everything," the note said. "You are checked out of Port Bane Inn; your Thunderbird was returned to the car rental company in Portland; your baggage and box of material is checked on the plane. The four of us had a grand time driving you to Portland. We'll be fishin' while you are pitchin'. It's nice to meet another man who knows how to party."

"I'll be damned," said Tanner taking a sip of his drink. "I must have passed out."

244

CHAPTER

23

D.C.

When Buck Tanner arrived at his house in Georgetown, he was glad to be home. The trip had been long and tedious, He looked at his small clapboard house with the smooth lawn, dark boxwoods, evergreens, and azaleas; to him, it looked like a castle except for the looming mansions on either side. Yet, in between these two bastions of wealth, this modest house, reminiscent of New England, displayed it's own sedate arrogance.

Tanner entered the front room with the polished hardwood floors and dropped his bags on the blue antique rug. He expected to be met by his two sons and Vivian, but there seemed to be no one at home. Subconsciously, he ran over his mental calendar, but for the life of him he couldn't imagine where his family could be. Perhaps he should call his father-in-law and find out if they were there. However, there was no need to rush. They were probably shopping. He would shower first and then make the call.

Tanner had just stepped out of the shower when he heard the car door slam. Damn, Vivian was home, and he had forgotten to get a bottle of champagne to celebrate their being together again. He dried himself hurriedly and was pulling up his shorts when Vivian burst into the bedroom. There she was, dressed in a long, white silk evening gown. Her black hair was clipped to one side of her head and curled in ringlets to touch the tip of one tan shoulder. Her black eyes were flashing with excitement. God, she was beautiful.

Suddenly he felt self-conscious standing there in the middle of the room in his boxer shorts. He wanted to cover himself with his hands.

"Ahem," he said. "Hello, sweetheart. Where are the boys?"

"Why, they're at mother's," Vivian said. "Mother's maid,Maria, is keeping them overnight."

"Wonderful," said Buck walking across the room toward her.

"Then we have the evening to ourselves."

"Buck, have you forgotten what day this is?" she asked.

"What day is this?" Buck said kissing her.

"Careful," she said backing away from him. "Don't mess my make-up, Buck."

"O.K.," Buck said. "What's the big occasion I'm supposed to remember?"

"It's the Senator's birthday," Vivian said. "Remember, I mentioned it to you before you left. I'm so glad you came home in time to escort me to the party. It's black tie."

Buck always wondered why Vivian and her mother referred to his father-in-law as "the Senator". The last thing in the world he wanted to do tonight was to go to some party. The Senator be damned. He wanted to be with his wife.

"You must hurry, Buck," Vivian was saying. "We should have been there an hour ago. Did you have a nice trip?"

Buck steered the Cadillac into the driveway of the Statler-Hilton hotel and handed the keys to the valet. What a screwed-up homecoming. Here he was at another Washington social function. His starched shirt collar and black tie hurt his neck, and he hated going through the same old rigamarole he had been through hundreds of times before. However, Vivian was ecstatic; that was some relief.

This was the only social event that the Kramer family hosted each year, and Washington D.C. socialites looked upon it as one of the main events. They vied for an invitation, and the city reporters had a field day. They knew that anybody who was anybody would be there. Even though congress was recessed for the summer, the legitative members, who were honored with an invitation, stuck around town until after the Kramer's shindig was over. They were afraid if they weren't seen there, the word might get around town that they had been ostracized by the Kramers.

They entered the green ballroom, and the whole spectacular materialized before Tanner. The enormous crystal chandeliers were blazing with light; hundreds of people in their formal attire were milling about the ballroom in a spectrum of color; the big band was playing a variety of swing, boogy, rumba, twist, waltz, tango and hard rock hits; a hundred people or so were dancing; some were eating; many were drinking; and all were talking. Cameras were flashing at

246

random while the red-vested waiters shuffled among the guests carrying large trays of long stemmed glasses filled with bubbly.

In the far corner of the room Buck spotted the bar. This was his cue. To hell with the Washington champagne.

"I think I'll get a scotch," he said to Vivian.

"Are you serious, Buck?" Vivian said in a low voice. "We have to make our presence known. As usual, I see the Senator and Mother surrounded by the media. I'm sure they'll want a few pictures with us. Come on."

Buck saw the Senator standing tall and shockingly handsome above the crowd. His jet black hair had just the right amount of gray at the temples. His face was tan and creased with a smile. His voice was deep and perfectly modulated. All eyes were watching him, especially his thin, dark blonde wife who adored him.

"Damn glad you could get back in time for my party, Buck," the Senator said. "Good trip?"

"I think I have a hot one this time, Senator," Buck said. "Something that may even make a few heads roll."

"Sounds intriguing, Buck," the Senator said eyeing his son-in-law cautiously.

"I'd like to get your views on the possibility of a Senate Investigation," Buck said.

"That big?" asked the Senator.

"I have eight thousand pages of proof," Buck said.

"Sounds like you've done your homework, Buck. I'm impressed with your progress." The Senator slapped Buck on the back, and a dozen cameras flashed the event. The Senator continued talking while Buck tried to rub the blind spots out of his eyes. "Just remember," the Senator was saying, "it's a long, hard road to enter the congressional camp. Maybe I should get with you on this matter before we take off for Palm Springs. I can give you some inside pointers"

"I'd appreciate any support you can give," said Buck.

"I'm going to circulate for a while, Buck," Vivian said.

"I'm going to the bar for that double scotch," Buck was now determined. "Want anything?"

"No, thank you," the Senator said.

"Champagne is fine for me," answered Vivian. "I'll make the rounds, and we'll meet at the buffet table in an hour."

247

Something told Buck that the next hour was going to drag a bit. But this time he didn't mind so much. He had the Senator's ear, and he felt that life was looking up for him. Before it was over, he might even take a seat in the senate alongside his distinguished father-in-law. Yes indeed, the future looked bright.

Buck had just finished his first double scotch at the bar and was reaching for the second one when he heard a familiar voice at his shoulder say,

"I'll have a bitter lemon."

The voice belonged to his partner in crime at the office, Russ Hamilton, a distant nephew of Alexander. Russ and Buck had hit it off the first day they met. It wasn't because they had similiar backgrounds; they didn't. Russ came from a wealthy family in Germantown, Pennsylvania. He was educated, articulate, immaculate, and serenely confident. He was also an attractive man with a thin build, thick, premature gray hair and inquisitive blue eyes. More than one woman had lost her heart over him, but at thirty-five he was still unmarried.

"Too particular, I presume," Buck had teased him one day.

"No, old man," Russ had answered. "Women are strange. Now and again I think I've found the perfect one for me, and then after we've been together for a season or two, she changes on me."

"Mind if I join you?" Buck asked.

"It's good to see you back, Buck," Russ said shaking hands. "I was lost somewhere in the woods of boredom while you were away. Did you have a good trip?"

"Yeah, I had a hell of a trip," Tanner said. "The country on the west coast is beautiful. I've never seen anything like the Cascade Range. And even though I didn't get out in those mountains, I had a phenomenal two day trip on a salmon troller."

"Did you catch any fish?"

"We didn't fish. Just took a little ride," said Tanner. "However, I did get to drive the boat. If you want to do something exciting sometime, drive a commercial fishing troller."

"Sounds like it would warm the blood a bit," Russ smiled. "I suppose you're back in Washington on the same old treadmill."

"On the contrary," Tanner said. "This is the first investigation I've felt is worthwhile. I've uncovered real dirt, Russ. Evidently, the Senator thinks it's worth a look see. He wants to meet with me before he takes off for Palm Springs."

"Sounds like you're onto something hot, Buck," Russ was interested. "I've been here for ten years, and no one has paid the least attention to anything I've done. Let me give you a little piece of advice though. If I were you, I'd cover my back. I don't know what your project's about, and I don't understand the significance of west coast fishing, but play it smart, and don't try to carry the ball by yourself. That way, if anything happens, you won't have to take the fall."

"Yeah, you said that right," Tanner agreed. "That's the reason I've already approached the Senator."

"I thought so," said Russ confidentially. "I've seen things happen more than once in Washington. Not in our department, mind you, but in related departments in the same complex. Just remember what I told you. Watch your ass, my friend."

"We're getting morbid, Russ," Tanner sipped at his drink. "Let's change the subject. I can take care of myself. I wouldn't say a word if I didn't have the goods. But you're looking rather down in the mouth tonight; got women trouble?"

"Not women," Russ answered. "Woman. The same woman. I can't live with her and I can't live without her."

"If it's that bad," said Tanner, "why don't you try living apart for a while and see how you go?"

"Are you kidding? She's loaded!" Russ said. "Besides, she has a lot of clout on the hill. Without her, something tells me, the way I would go is shaggy."

"Is she here?" Buck asked.

"Do you think she'd miss one of the hottest socials of the year?" Russ exclaimed incredulously. "She's the tiny, pretty one in the red gown next to the birthday cake. See? She's talking a blue streak and flitting about like a butterfly from flower to flower. She's taking advantage of her juicy fortuity."

"Like what?" asked Tanner.

"Oh, like exchanging cards with important people she's missed in the past. She'll be calling up the names on the cards she's collected; setting up net-working brunches; recruiting new members for the Thursday Breakfast Club; scheduling dinners. She's a professional diner-outer, you know. And, of course, the high point is getting her picture in as many newpapers and magazines as possible. She hangs out at swank restaurants like Le Cirque and Maisons where she spends a hell of a lot of money to table-hop. She says everyone in Washington does it, and if they don't, they should, or they won't last long in Washington. That little woman is ambitious. It's a good thing she's rolling in the stuff. With my salary I couldn't buy her a pot to piss in. Besides, sure as hell, she'd demand a gold one."

"Listen, Russ," Tanner grinned. "We've known each other a long time, and we understand what makes each other tick. I'm going to tell it to you straight, buddy. I think you've got it pretty bad this time. You've said too many things that makes me think you're trying to convince yourself she's not the one for you. Well, I don't think you're pulling it off, even to yourself. Think about it. This one may be the one you can't live without."

Tanner quickly excused himself when he saw Vivian standing near the buffet table. As he strode across the ballroom floor toward her, he felt the same old familiar thrill he always felt when he was near her. She looked like a tall, black-haired ballet dancer in her white gown and coppertone tan; the kind of tan only tanning salons can produce. Her mouth was smiling with white even teeth as she talked to a be-jeweled matronly lady who handed her a card. She continued talking, but Tanner noticed that her black eyes were not looking at the lady. They were cooly scanning the room for other faces. For an instant Tanner felt like he had just seen something he didn't want to see.

"Ready to go, dear?" he asked hopefully.

"Not until the Senator cuts the cake," she said.

The party ended at 2:00 A.M. Buck had been ready to go since midnight. By the time he put the Cadillac into the garage and locked-up for the night, Vivian was already undressed and in the bed. Tanner undressed quickly and mixed a couple of dry martinis.

"The party was a huge success," she said propping herself up on the pillows. "Wouldn't you say so?"

"No doubt about it," Buck agreed. He sat down on the edge of the bed and handed her a martini.

"You know, I've been thinking, Buck," she continued. "It might be a good idea for you to join the Gridiron Club or the Capitol Hill Squash Club. I see you as a monolith now on a high level spectrum, and it's proper time for you to contact your sort."

"What in the hell are you talking about?" Buck asked. "Those wolves would tear me up."

"Don't be an idiot, Buck," she said. "It's time you realize that you can get anything you want, just by wanting it enough; being ambitious enough, and exuding charm."

'Ah. The smokestreams of charm can be very dense,' Buck thought.

"I've got some ideas for myself," Buck smiled. "You shouldn't worry your pretty head over things that have to do with business."

"I've never been so insulted in my life," Vivian's eyes narrowed. "You know perfectly well what I meant. I was only thinking of you and our boys. In all my life I've never really understood men."

"You know, sweetheart, you're an angel," Buck put his arms around her. "Don't worry, I'll be alright, just as long as I've got you. I'm so glad you belong to me."

"I'm not a possession, Buck," she said pushing him back. "And it gives me an emotional crisis to hear you say that."

"What I mean is that I love you very deeply," Buck tried to explain. "You're so damned beautiful: the way you look, the way you dress, the way you handle people. I want to look after you. I was going to tell Silvia about you when I was in Port Bane, but I didn't think she'd understand."

"Who is Silvia?" Vivian sat up suddenly. "You never told me about a woman being there, Buck."

Tanner realized he'd just made a big mistake. How was he going to fix this one? He'd better think of something fast. He got up quickly and took her empty glass and started mixing two more dry martinis.

"Is she beautiful?" Vivian studied his face. "Is she younger than I am?"

"She's just a young woman that reminded me of you," Tanner said.

"I don't believe it," Vivian said. "This is a joke in the worst possible taste. You're seeing other women when you're away from home."

"No such thing, Vivian," Tanner said defensively. "I love you. She just happened to be at the party they had on the night I left."

"What party?" Vivian was becoming alarmed now. "You have parties every night you're away, and you invite young girls? So that's the kind of work you do!"

"I was working with the fishermen, and she was the only woman with us all the time I was gone." Tanner was getting mixed up.

"Do you mean she's a fisherwoman, and you were with her all the time you were gone?" Vivian demanded.

"She was in my suite only one night," explained Tanner. He was beginning to feel that every time he opened his mouth, he was sinking a little deeper into a pit.

"Why did you even mention her unless she's important to you, and what do you mean she reminded you of me?" Vivian continued. "I don't think I'm flattered."

"You would be if you saw her," Tanner said. "She's very tall and slim. And she has long dark hair like yours, even longer, and her skin is olive colored like she has a natural tan. And she's got the greenest eyes I've ever seen."

Tanner watched Vivian's large dark eyes get watery, and he knew he had said the wrong thing again.

"Ah, Vivian," he murmured and tried to put his arms around her. Vivian turned her back on him and pulled the covers over her shoulders. "I always stick my foot in my mouth. What I'm trying to explain is that I'm not interested in any woman but you. This woman is a fisherman's lady, and I could give a damn about her. I just wanted to tell you...I love you, and I'm glad to be home."

Tanner listened for a reply, but there was only silence, so he started over.

"Viv," he began. "We've never been to the west coast. I want to take you and the boys to Port Bane on my next vacation. A real trip, something besides Disneyland. Silk will even take the boys out on his boat."

Tanner got a reaction this time. Vivian turned on her side and looked directly into his eyes.

"Buck," she said. "I don't understand what's gotten into you. We've been in Washington, D.C. for nine years, and you're still working at the Justice Department for a salary half your worth. Now you want to leave Washington and take our boys out to some godforsaken place and possibly drown them in the ocean. Well, I'll tell you one thing, Buck Tanner. If, and when, we ever go on a vacation again, we aren't going to Port Bane, and my boys will never associate with fishermen."

"I'm sorry, Vivian." Tanner felt like biting his tongue for mentioning the subject. "Port Bane was just part of my investigation, that's all. I don't want to take you and the boys there if you don't want to go. I just wanted to tell you—this time, my investigation paid off. The salmon thing is going to be the opportunity I've been looking for all these years. When I finish with Port Bane, I'm going to the top of the department. I'm going to make a lot of money, and I'm going to give you everything you've always wanted. I'll even become a senator if you want. We'll send the boys to Harvard. We'll get another house. I'll buy you diamonds and a new Cadillac and some gorgeous furs. We can take long trips together, to Europe or any part of the world you want."

Vivian kissed his neck and lips passionately.

"I'm glad you're home, Buck," she said. "I'm glad to see you. Tell me what happened on this trip that is supposed to bring about such a dramatic change in our lives."

"Well, I can't tell the whole story, of course," Tanner said. "I'll just tell you that what I uncovered out there is a real government scam, and I'm determined to follow this thing through to the end. I'm going to win it, Vivian, and do you know who is going to get the credit and the promotion? Buck Tanner. You're going to see it happen, Vivian."

"I'm so happy to hear you say that, Buck. We've been working hard enough over the years. It's time we've arrived. I realize, Buck, that you were tired and that you didn't want to go to the Senator's birthday party tonight, but it was important for us."

"Don't worry your beautiful head about the party," said Buck feeling more comfortable. "I was happy to go."

"I think it would be lovely to lunch together tomorrow," Vivian said smilingly, her warm imploring eyes filled with emotion. "Would you like that?"

"Damn right," Buck said. "I'll take you anywhere."

"Why don't you meet me at Le Cirque around oneish?"

"It's a date," Tanner said. Damn, it was so good to be home with his beautiful wife.

Vivian threw her arms around his neck and hugged him tightly.

"Oh, Buck," she whispered. "Isn't Washington, D.C. an exciting city to live in?"

"And to work in," said Tanner.

CHAPTER

24

THE ANT

Whereas, Tanner was usually critical and impatient with the Washington D.C. crowds, this morning neither the traffic nor the heat bothered him. He was oblivious to the swirling activity. His mind was on his long, hard work, and the reward he would gain for the job he had just completed.

He felt optimistic. What had seemed a mundane trip to Port Bane had turned into an adventure and the successful uncovering of a real scam. After last night he and Vivian were tighter than ever, and best of all he had the inside track in solving the fishermen's dilemma. He felt a lightness in spirit and an enthusiasm for his job which he hadn't felt since he first became a Justice Department Investigator.

Tanner carried his briefcase and the heavy box that contained Silk's statistics through the glass doors of the triangle complex, past the guards, up the elevator to the fourth floor, across and around thirty desks that held computers and typewriters—tools of the trade for the pool of investigative reporters—and into his tiny office which contained a desk, two office chairs, a file cabinet, and a small round conference table. The place felt like home.

Thinking back on it, he hadn't done badly as an investigator in Washington. He had started out in that zoo of cubicles and desks in the big room and worked his way up the government ladder so that now he had his own office. True, his name wasn't on the door, and he still had a number instead of a title, but he had set himself apart from the rest of the ants even if his salary didn't show it. And his office had a window that looked out on the grassy expanse of lawn and trees forming a kind of inner-courtyard, a sure sign in Washington D.C. that he was getting somewhere.

The wall behind his desk contained two small paintings, splotches of color, modern art he supposed, that Vivian had painted in her college art classes. Across from his desk was a framed papyrus scroll with blue silk tassles hanging from either side. This decorative piece was his pride and joy, a gift from his mother upon the auspicious occasion when he had received his law degree from the University of Texas.

As he stood for a moment and read DESIDERATA, he thought back to the kind, loving woman who was his mother. The scroll had probably only cost her ten dollars, but the message and spirit it contained were priceless.

"DESIDERATA is not much for a gift," she had said handing him the scroll. "The poem is not a new automobile nor a watch, but the words of this unknown author are wise. If the words are adhered to, they will prove more valuable to you than any material object."

Tanner had read the document every office day for years, and he had never grown tired of it. The words seemed to calm his nerves and put him in the mood for the day. He especially loved the line that read, "And whether or not it is clear to you, no doubt the universe is unfolding as it should." There were times he wondered about the logic of that statement, but now he could see it was true. He had worked hard for years, in blind faith that he was proceeding in the right direction. Now he could see his latest investigation was going to pay off in materialistic dividends, and he was going to be the recipient. Yes, the world, Washington D.C., and the universe was unfolding like it should.

Before he went into action, Tanner lit a cigarette and sat for a moment at his desk planning an order of events for the day. First, he typed up a complete report of his Port Bane investigation, purposely leaving out the names of Silk, Cowboy, Semyonov, and the part about his trip on the CRACKANOON II. Second, he wrote a brief thank-you letter to Dr. Blackburn in Seattle, Washington, omitting the fact that he had found the man in Port Bane named John Silk. Finally, he removed the manilla folders from Silk's box of statistics and placed them neatly in alphabetical order in the top two drawers of his file cabinet. Now he was ready to meet the judge and make his report. He picked up the phone and dialed the extension for Judge Elton "Eel" Powers, the current Assistant Attorney General for the Land and Natural Resourses Division of the U.S. Justice Department.

The Land and Natural Resources Division represented the United States in litigation involving public lands and natural resources, environmental quality, Indian lands and claims, and wildlife resources.

The division was also responsible for prosecuting and defending criminal and civil cases arising under the federal wildlife laws and laws concerning the conservation and management of marine fish and mammals.

Although the cases filed by the division in the area of hazardous chemical wastes were the most visible and complex, enforcement of the clean air and water laws was also a prominent part of the docket. Thus, the division actually brought civil and criminal enforcement cases to trial, primarily on behalf of the Environmental Protection Agency for the control and decrease of pollution of air and water resources, the regulation and control of toxic substances, and the environmental hazards posed by chemical wastes.

Government investigators like Tanner usually worked with federal regulating agencies to ensure the compliance with federal regulations. They examined records, interviewed individuals, wrote detailed reports, and testified before courts and administrative bodies.

Tanner figured that for once he had received an assignment which dealt with important problems. From his brief scrutiny of Silk's material and from his hands-on investigation in Port Bane, he already knew the problem bridged most of the areas in his job description. For once he was not dealing with corn in Kansas or pigs in Nebraska.

His thoughts were interrupted by Power's secretary who returned his call and set his appointment for 1:00 P.M. in conference room B. Damn that meant he had to call Vivian and tell her that he couldn't make it for lunch. For a moment he thought better of the idea. He'd call Power's office and cancel.

Then common sense took over. Vivian and the children depended upon his success in the department. The department was his work, and the department was Powers. To cancel his appointment was to postpone the progress on his project. He picked up the phone, dialed Vivian, and begged off. She was perfectly amicable and said she would go to lunch by herself since she had already planned to do some shopping. He was amazed at how understanding she seemed.

By the time Tanner returned from a lunch of orange drink with hot-dog and kraut from a corner stand, Powers was already in the conference room waiting for him. Calm and cool he sat now at the end chair of the round conference table drumming on a blank manilla pad with a Cross pen. He was dressed immaculately in a gray silk suit, muted shirt and 'rep' paisley tie. In his fifties, still slim, Judge Powers' only blemish was a partially bald head with iron gray hair strung rakishly over the top, covering as much area as possible. As Tanner entered, Powers stood up to shake hands.

"I'm glad you're back, Buck," he said with his wide smile that showed a mouthful of matching capped teeth. "I hope you had a successful trip. Must have a lot to say if you already have your report written."

Tanner handed Powers his report, sat down, and removed a stack of hand-written notes from his brief case. "I had a good trip," he said. "Highly informative and interesting, but I'm glad to be back home." He sat silently while Powers read through his report, waiting for the invariable questions that would follow.

"In the personal recommendation section," Powers began, "you say that the conditions which make commercial fishing untenable stem from mismanagement and misappropriation of funding by our federal agencies, that these institutions cause the majority of fishing problems, if I'm interrupting the drift of your prose correctly. Aren't those accusations a little strong coming from someone who has only been studying the situation for a few days?"

"They are strong," admitted Buck, "and I meant them to be. You see, I not only studied the fishing conditions first hand, but I met a fisherman who has been doing research on the problems for ten years. Right now I have eight thousand pages of written material in my file cabinet to support my statements. I have enough proof to blow this fishing scam wide open, a real investigation. That's why I made this appointment immediately. I would like you to read my report and give me permission to follow this investigation through. This time the Justice Department has the wherewithal to solve the problem."

"I see," said Powers stroking a firm chin. "You feel that you have the facts and figures to support your charges, and you'd like to pursue the matter."

"That is correct," continued Tanner. "This thing runs from the smallest ocean troller on the west coast right on up to the President of the United States. We're dealing with major corruption and major issues here. I was lucky enough to find the people in the know and get the facts. Now I'd like to continue the investigation. The people who sent you letters requested a congressional investigation. I know that step is a long way off, but I feel we have already uncovered sufficient cause to pursue the matter. I'd like to have a shot at it. I just need to obtain your O.K. for the job, and also a recommendation about where I should start. You know, along which channels?"

Again Powers read through the report.

"Well, Buck," he said. "You appear to have done your homework thoroughly, and you have stated sufficient cause. But if what you say is true, you have a long way to go to make any of this verbage stick. You've uncovered some big problems here that aren't easy to approach. And they involve a lot of important people in Washington, a lot of men in high-powered positions who don't want their policies questioned. Just reading over your information, I'd suggest our department sit on your report for awhile. Let's hear some more from the fishermen."

"I'll tell you, Eel," said Tanner, "I have all the information in my file cabinet to expose this scam and prove that a lot of these people are purposely misusing the fishing resources of the United States. That's supposed to be what this investigative department is all about, isn't it, uncovering these problems and seeing that justice is done? Well I've done my research, and there is a problem. I'd like to see it through to the end."

"I don't know, Buck," said Powers. "This is a mighty big mouthful. But I agree with you that we have an obligation to present the facts. I just think we might take a little more time. Aren't you about due for a vacation?"

"Bullshit!" said Tanner. "I'll take my vacation after this fishing controversy is solved. Now I just want your permission to go to work on this project, and I want you to guide and support me. Now where do I start?"

"Damn you, Buck," said Eel. "I knew when you came to work here you were an idealist, but I didn't know you were a hothead. I'll tell you what I'll do. I'll let you run with the ball and start this investigation, but if there's one slip-up on our part, or there's one word that can't be proved, then I'm going to suggest we drop the whole matter. Is that agreed?"

"Agreed," said Buck. "I knew you'd come through, judge. Now let me give it to you in a nutshell. The fishermen didn't know enough about their own problems to write them in their letters. Right now there is collusion between state and federal agencies as well as between private industry and government organizations. There are individuals right here in Washington D.C., one here in the Justice Department, who is being paid by private, state, and federal agencies all at the same time.

I have statistics to prove my accusations. There are only two real avenues we can take to clean up this mess: a grand jury investigation on a federal and state level, and a class action suit against big business companies like SEABOOM and AQUAPROTEIN. We're going to have to sit down with everybody who's involved, all in one room, take down everybody's point of view, then sit there until we get the whole mess thrashed out."

"Slow down a moment, Buck," said Powers. "I said I'd support you in this investigation, not start a revolution. You sound more like an uneducated fisherman than a government lawyer. You've been in Washington long enough to know the channels. We have to work through channels."

"I know, I know," said Tanner. "What I'm trying to say, Eel, is this: There aren't two sides to this debate. There is only one side, and there has only been one side for years.

The Department of Commerce, the Pacific Fisheries Management Council, the National Marine Fisheries Service, the State Departments of Fish and Game, the State Universities, and the big private industries are all in the fish farming business on the same side. The commercial fishermen, are so small they don't even appear in the overall picture. We have to make a case for the commercial fishermen and take it to the public. I'd suggest we take it to the Washington Post."

"We have to take proper steps, Buck," said Powers. "We must listen objectively to all information and seek out the truth before we make recommendations. Only then can we refer the pertinent knowledge to the proper departments. These are the steps for proper investigative procedure."

"I understand," said Tanner impatiently. "This democratic process has been going on for years, and very little has ever come from it. Injustice is being done and that is exactly why we got all those letters. You know what we're really talking about? Money, big money! Do you know who they are? I can tell you because I saw them. They are SEABOOM, AQUAPROTEIN, JANSEN SOUP, PETROLEUM

259

U.S.A, SEA FARMS of CHILI, and the NEW WORLD SALMON COMPANY. These are just off the top of my head. There are a hundred other major industries trying to get into the fishing business at the expense of our natural marine resources which should be owned by all the people. These big multiple industries have given up on exploiting forests and oil fields. They've used up these resources and made their profits. Now they want to get their mitts into our final resource, the protein in our oceans, and they want to control the profits from this exploitation. You know as well as I do that they furnish all the suck in congress with their high-pressure lobbying.

Let me ask you about the logic of lumber companies and oil industries switching their interest to fish. The destruction of the ocean bottom is secondary to their profits. They don't give a damn. They're really after money and control. They know that ultimately, if we don't blow ourselves to hell with nuclear missles, they'll control the ocean's protein, and he who controls that food resource will control the world. The whole damn fishing industry has become involved in this struggle over money and power. The thing is a crying damn shame." Buck hoped that he had paraphrased Silk correctly.

"Calm down, Buck," said Powers casually. "You're emotionally involved. Everything you've said is a generalization. Generalizations don't make the laws or clean up the injustices in our system. You have to state the facts specifically, then you have to gather support and push them through the system."

"I have the facts, Elton," interrupted Tanner. "More than enough facts. I just want you to help me present them in the right place to the right people."

"Alright, Buck," said Powers agreeably. "I don't necessarily agree with your opinions on this one, but I'm willing to support you up to a point. I know that the same weeding-out political process has already taken its toll on the farmers, the ranchers, and the unions, and that it's simply a matter of time before it happens to the fishermen. As you would say, "Those who have the resources get, and those who don't have them git". But that, in this country, is called progress.

In Washington D.C., especially in Washington D.C., the individual doesn't stand in the way of progress. To take a stand on an issue like the fishing industry is like committing political suicide. In this particular case I'd let sleeping dogs lie, let time take care of the

260

struggle between the individual and industry.

You have a family and a promising career in the Justice Department, Buck, and I know if you continue your good work here, you're going to have some fine offers from private law firms. To pursue this subject further is like opening up a whole can of worms, and there are powerful political forces here who are going to fight you all the way. My advice to you is to forget this personal crusade, forget the injustices, forget the problem. It's not yours in the first place. Take Vivian and the boys on a vacation. Your job will keep until you get back."

"I appreciate your advice, Eel," said Tanner. "But I can't forget what I've seen with my own eyes. Besides that, I like those fishermen, and I sympathize with their problems. I've promised them some action on the issue, and I can't break my word. I'll take my chances."

"Very well," said Powers. "Against my better judgement I'm going to let you run with this one, Buck. Right now, in the Department of Commerce, there is a committee being formed to discuss the forthcoming issue on commercial fishing seasons. There will be representatives from the National Marine Fisheries Service, Pacific Fisheries Management Council, U.S. Fish and Wildlife Service, and even a senator or two. The whole business will be under the guidance of the Secretary, Dr. Wayne Dumas. I think I can waggle you an appointment with Dumas, and he'll no doubt steer you to the heads of the other departments. But after reading your report, I can guarantee you that these men are not going to like what you have to say. Your report represents a point of view that is not at present popular in Washington."

"You mean that I'll be representing the truth," said Buck. "That there won't be a single commercial fisherman present to help decide on the fishing seasons which affect their lives."

"I don't know about that, Buck," said Powers. "But I do understand Dumas and the organizations he represents. Unless you can provide new material about the fishing industry that these men don't know, I doubt that he will listen to you. Dumas controls a well-oiled machine that functions along traditional lines. He won't take kindly to suggestions that we investigate certain men in the organizations under his control. Unless you can make a strong case for an investigation, I'm afraid it'll be just another opinion in a long tenuous process, something that goes on every year."

"I'll take that chance," said Buck. "And I think the research in my file cabinet will make an impact."

"Very well," agreed the Judge. "The first meeting, I think, will take place in about two weeks. I'm sure Dr. Dumas will see you before that time. Can you have a statement ready by then?"

"You bet," said Tanner. "I'll be ready."

"Fine," said Powers. "But remember I've warned you, and you're on your own until you can make a case for our department to begin an investigation. In this system one must prove a criminal act before one can investigate. These men you are talking about are trained professionals doing their jobs. That's what Dumas is going to contend. They have been working professionally with these problems for years. You'll be the new kid on the block."

"Roger," said Tanner laughing. "I'll be ready with a written statement which will forever strike terror in the heart of Dumas. I'll start with this year's salmon season and go from there."

"Good luck," said Powers rising. "You're a good man Buck, and I'd hate to lose you."

After his meeting with Judge Powers, Tanner went back to his office to have a cigarette and contemplate his presentation to Dr. Wayne Dumas. This was the first time since he had been in the department that Judge Powers had given permission for an investigator to pursue a case beyond the primary report. Whatever he had written or said must have struck the proper chord in Powers. His trip to Port Bane must have been valuable, or Powers would never have turned him loose. For a moment he felt like a student again, like he had just won a scholarship to the University of Texas. There was a lot of work ahead of him, but at least he was in on the ground floor. He was ebullient; he was ecstatic! He should call Vivian about his accomplishment; he must call Silk in Port Bane. Instead, he went to his file cabinet and began thumbing through the manilla folders looking for articles on commercial fishing seasons. He had to select the proper material, and he had to program it on his computer. It was too early in the game to get excited. Still, he'd call the Albatross Tavern tonight, and if Silk were in from a fishing trip, he'd give him the good news. Tanner turned his friend, Russ, down for an after-work cocktail at Mario's and with reckless abandon threw himself into the evening traffic on Pennsylvania Avenue. He could care less tonight

262

that the trip took an hour of stop and go before he turned onto the peaceful avenue which led to his house. After the party last night and the hectic pace at the office today, a quiet evening at home with Vivian was the medicine he needed.

Vivian met him at the door dressed in white cotton slacks with a subtle charcoal blouse set off by a red silk handkerchief tied rakishly around her neck. Her dark hair was done in long glossy braids that hung down on either side of her face. Damn, she was beautiful. Unlike most women, two children certainly hadn't marred her looks.

Vivian kissed him generously on the mouth, took his suit coat, and led him passively into the den where a beaker of cold Beefeater martinis was waiting. He stretched out comfortably on the overstuffed sofa and allowed himself to be served by the most elegant woman in the city. He took a generous sip of the chilled martini and told her about his meeting with Judge Powers. She listened politely with a smile on her face while he explained their conversation in detail and the steps he was already working on that would launch his investigation. Had Tanner looked more carefully, he would have seen that the smile on Vivian's face was fixed. She was bored, and she had heard similar stories before. Tanner finished his exposition and refilled their glasses.

"So you see," he said, "that meeting was the reason I had to cancel our luncheon date."

"That's quite alright, darling," Vivian answered sweetly. "I understand what being a lawyer's wife means. I simply rang up Mavis, and we luncheoned at Le Cirque and then shopped. I purchased an outrageously adorable new gown you will love."

"I should introduce you to Russ's lady, Anita," Tanner grinned. "Seems like you two have a lot in common. She loves Le Cirque and socio."

"I think not, Buck," said Vivian. "We already have our social set. We haven't the need to cultivate new acquaintances."

"Probably true," said Tanner kissing his wife lightly on the cheek. "I'll leave that to you, sweetheart. By the way, where are the boys? We haven't been interrupted once since we sat down."

"I took them over to mother's," said Vivian. "I got back late so I thought I'd let them stay over again tonight. We can spend a quiet evening alone together. I thought I'd surprise you and have our dinner catered tonight. We're having roast chicken and white wine."

"I appreciate it, Vivian," said Tanner. "But I wouldn't trade restaurant food for some home-cooked meals."

"'Cooking is for professionals and old people, Buck. Isn't this much better, darling?"

"Sure," Buck kissed her on the cheek. "A catered dinner is fine with me, if that's what you have your heart set on."

Vivian's catered dinner was very bland, and it wasn't too filling. After they had finished eating, they turned on the stereo and listened to some Glen Miller. They were firmly ensconced in one another's arms when Buck glanced at his watch and noticed it was 11:00 P.M., time to make his phone call to Silk.

"Sweetheart, excuse me for a minute," he whispered in Vivian's ear. "I have to make a phone call."

"Don't you think it rather late to be calling someone at this hour?" asked Vivian.

"I'm calling Silk in Port Bane. It's three hours earlier there," Tanner said oblivious to the irritated tone in his wife's voice. "By now he should be at the Albatross Tavern. I have to tell him the good news."

"Fine, Buck," said Vivian looking at him with her dark eyes that now seemed to have cooled.

"I'll just be a moment, dear," Buck explained. "You have to understand Silk to realize how important this news will be to him."

"I don't understand why he'd be hanging out in a tavern if he were a good man," Vivian said coldly. "But do what you must. Place your call."

"I could never explain the primitive conditions in a fishing port," said Buck apologetically. "I'm sorry. I just have to call him there."

Buck got Hattie at the Albatross on the tenth ring, and a short wait later Silk's voice came over the phone.

"Hello?" he said, and Buck could still hear the soft Boston accent.

"Hello, Silk. This is Tanner calling from Washington D.C. I'm just making a little progress report."

"Good to hear your voice," said Silk. "What's happening Buck?"

"Well, I just wanted to tell you. I had a meeting with my boss today, Assistant Attorney General Judge Elton Powers. Although he didn't say it, I think he was impressed with our work. He gave me the go ahead to make a presentation to the Secretary of Commerce, Wayne Dumas. I'm expecting the meeting to take place in about two weeks."

"Sounds good to me, Tanner," said Silk impassively over the phone. "Seems like you got ahold of some pretty important men."

"At least we didn't strike out," said Tanner. "Is there anything more I should put in my report?"

"Not that I know of," said Silk. "I've told you everything I know about the problems. There's nothing to add from this end of the scene since you left."

"Thanks for letting me take that box of material with me," replied Tanner. "Having written proof will make all the difference."

"Don't thank me," said Silk. "I've been collecting that stuff ever since I've been fishing, but I've never done anything with it. I'm just glad it's in good hands and going to be used for a change."

"Well, I'll do my best," said Tanner. "How are the people in Port Bane?"

"Well the people in Port Bane never change," said Silk. "Fishing's been a little better than last year at this time, but I guess the most important thing that's happened is that Cowboy's purchased himself a house. I just found out."

Tanner could hear Silk laughing. "What's so funny about that?" he asked.

"Nothing, I guess," said Silk. "It's just that he didn't buy a ranch and cattle; he bought a condenmed house on X Street inside the city limits of Port Bane as an investment. He and Graveyard are going to jack her up and pour footings under her. The crazy guy only paid seven hundred and fifty for it."

"Seven hundred and fifty thousand dollars?" asked Tanner increduously.

"No," said Silk. "Seven hundred and fifty dollars."

"That's insane," said Tanner. "There must be something terribly wrong with it."

"There is," said Silk. "That's why it's condemned. But Cowboy thinks he can make a profit on it."

"More power to him," said Tanner in a positive voice. "For that price he can't lose. Well, I just thought I'd tell you the good news. Let me know if you think of anything that will help."

"I'll do that," said Silk. "I'll think about you while I'm out on

265

the ocean. Good luck,Tanner."

"Thanks, Silk," Tanner finished on a positive note. "And my thanks to all of you for helping me get to Portland and on that plane. I must have made a regular ass of myself."

"No, you didn't," said Silk. "You just had a good time, and that is something we all understand."

"Well, I promise you all that I'll make one hell of a presentation, and I'll keep you in touch with our progress here. Thanks again, and goodby for now."

As soon as Tanner had hung up the phone, he went back into the bedroom. Vivian was already in bed curled on her side with her back to him. Tanner cursed the phone call to Silk even though he felt obligated to make it. It had interrupted a beautiful evening with Vivian. Then he thought of Cowboy and Graveyard, and he laughed out loud. Those people in Port Bane were different all right. He'd never heard of a house for sale at under a thousand dollars.

CHAPTER

25

THE ANTEATER

While Cowboy and the Locals were busy clearing away brush, hauling garbage, pouring concrete for the footings and jacking up Cowboy's house on X Street in Port Bane, Buck Tanner was busy in Washington D.C. making final preparations for his meeting with Dr. Wayne Dumas. Secretary of the Department of Commerce.

The preliminary meeting was to be limited to fifteen minutes, but fifteen minutes was enough time for Tanner to alert Dr. Dumas of the putrid, pork barrel antics going on in his own department. He was confident that once Dr. Dumas had seen the chart which contained the names of the key figures involved in the scandal, he would expedite a freeze on any future attempts to control the coho season on the west coast and call for a thorough investigation of the fishing industry.

Tanner felt elated. He checked and rechecked the chart and the condensed report before he placed them in his polished leather briefcase and zipped it shut. Maybe he should have stamped the papers FOR HIS EYES ONLY, or CONFIDENTIAL, or SECRET, or TOP SECRET. But labeling wasn't important. Dr. Dumas would treat the report with utmost priority; and he, Buck Tanner, would be given a glowing compliment for his efforts. Yes sir, this meeting was big time and he was ready. He was really starting to track in his department.

Dr. Wayne Dumas sat behind his large mahogony desk admiring the framed certificates of education that lined the grass-papered walls of his office. When he first moved into the large rectangular room, he had refused any other decoration other than his own accomplishments. After arguing with him for thirty minutes, the interior designer, a young woman out of Georgetown University, became angry and had stalked out in a huff.

Dr. Dumas smiled at the experience. Who needed office decorations when he had a list of framed credentials that stretched from one end of the room to the another. And they were impressive. There was a degree in Agriculture from the University of Iowa, a Masters degree in Fish and Wildlife from the University of Colorado, and a Phd. in Marine Biology from the University of California. After his graduation from the University of California, he had applied for a position at a small private college in Oregon as professor in the School of Fisheries, and he had gotten it. From this humble position his rise was rapid.

There were certificates arranged in the order of achievement right up to his current position as Secretary of the Department of Commerce and Advisor to the U.S. Justice Department. There it was, his climb to power: Chairman of Marine Biology at the Pacific Coast Science Center, Advisor and Agent to AQUAPROTEIN, subsidiary of Windover Lumber Company, Oregon State Representative, Co-chairman of the Natural Resources Committee for Oregon, Chairman of the Oregon Fisheries Society, Director of Pacific Fisheries Management Council, Team Leader for Sea Grants to underdeveloped nations in Marine Protein Resources Development, President of the American Aquaculture Society, and now Chairman of the World Conferences on Aquaculture for the United Nations.

Dr. Dumas stood up, lumbered his way across the room to the glass windows, and looked out over President's Square. At a distance he appeared to be a man of above medium height with fine, gray, slicked back hair. But upon closer study this observation was an optical illusion brought about by his small, pointed head, receding chin, long, slender nose, and neck that seemed to dissolve into sloping shoulders. Unfortunately, his weight was in his thick arms and legs and his wide hands and feet. One of his formidable adversaries, a Disney cartoonist, had acidly depicted Wayne Dumas in cartoon. He had titled it "W.D., A Sucking Pismire Eater".

Besides his repulsive shape, Dr. Dumas had two atrocious habits which made his associates shudder: He often flicked his wet tongue out and licked his lips, and he cracked his knuckles backward, forward, and sideways.

Wayne Dumas had let neither shape nor size stand in his way. As a boy in Iowa, he had been too physically unbalanced for sports, and with this prerequisite for popularity in school, he promptly became a bookworm. He haunted the Des Moines Public Library, reading books that had pictures of American fish and wildlife. His mother, a staunch Baptist, talked continuously about man's sacred control of his own universe. His father, an avid hunter and fisherman, collected horns, hides, mountable marlin, swordfish, and large trout.

By the time he was fifteen, Wayne had incorporated the two activities into his own personality. He listened to his mother's religious fervor, and when he wasn't listening to her, he was stalking animals in the woods with his father. He had supported his sojourn at the University of Iowa by working for the U.S. Forest Service. At this juncture in his life he was just another faceless student. Instead of attending football games and other athletic events, for which he had the utmost contempt, he built a pond and began experimenting with raising high bred rainbow trout. His reasoning was simple. If he could produce a better, stronger fish, superior to any the sportsfishermen had yet caught, he could influence the state hatcheries to buy his stock.

By the time he had attained his first professional administrative position at the Pacific Coast Science Center, he was considered an expert in raising trout and steelhead. Those fields of expertise had led directly to raising salmon in pens and on ranches, aquaculture as it were. Soon he was known as the foremost authority in the field, but his state salary wasn't commiserate with the services he performed. Then Windover Lumber Company had come along and offered him a position that really had clout and paid well. They didn't care that he continued his job with the state. From these two prestigious positions, one above board the other under the table, the jump into politics was natural. Looking back on his career now, he could see that he had been born, not with a silver spoon in his mouth as his mother might have said, but with an insight of how to control the resources of his society and turn them into power.

On the long trips for the United Nations to Manilla, Chile, New Zealand, Hawaii, Russia, and Japan, he had plotted his final goal. He had already been effective in introducing salmon eggs from the United States into these nations. Now they had thriving industries of their own, many of them controlled by private American companies. In this position he could blend these private industries into an international fishery organization of which he would be elected president. All that was left to complete this plan was a final agreement among the controlling major nations. Proper legislation would ban all private fishermen from the ocean shores. Protein from the oceans could then be divided equally among the companies that controlled them, and the markets for these fish would be stabelized. As president of this organization, he could control the protein resources of the world.

The agreement among the private industries was already waiting in bills before congress, and they had been placed on the agenda of United Nations representatives. However, legislation by congress to prevent private commercial fishermen from fishing was at a standstill. These cowboys were harder to control than wild salmon. There was a bill currently pending which would effectively shut down the salmon seasons, but opposition from private individuals was rising. Letters to congress from fisherman's organizations on the east and west coasts had stopped the fishing control bill and inspired an investigation by the United States Justice Department. All lobbyists from the pen-rearing and ocean-ranching organizations owned by giant corporations had been unable to prevent this investigation. These commercial fishing bastards had always been a thorn in his side.

But this morning Buck Tanner from the Justice Department was coming to see him. He had made the appointment with the investigator merely to get on the inside of the investigation. The presented report would be channeled to his department, the Department of Commerce, where it would die a natural death. Without this stupid investigation the control bill would pass both houses of congress, and the results of this legislation would eliminate the commercial salmon fisherman from coast to coast.

Dr. Dumas looked at his watch. It was twenty minutes before his meeting with Buck Tanner. He sat down and smiled to himself. The pending bill was simple in its concepts. It would give the commercial fishermen a limited twelve day season in exchange for legalized estuary ownership by monied corporations of twelve pen-rearing operations and seven ocean-ranches in the states of Washington and Oregon. The twelve days selected by the committee were the days most likely to have stormy weather at sea. Before next year he would see that the twelve day coho season was eliminated by new legislation, and at the same time new aquaculture locations would be accepted by congress. With this continued plan it would not take long to starve out the fishermen once and for all.

At precisely ll:30 A.M. Buck Tanner was ushered into Dr.

270

Dumas' office. Dumas stood up and looked Tanner over for a moment and then motioned him toward a straight-backed chair across from his desk.

"It was good of you to see me, Mr. Secretary," Tanner said shaking hands.

"Not at all," Dr. Dumas said with an affected accent. "Sit down, Mr. Tanner. Since our time is limited, let us get right to the point. What do we have here that makes this meeting so imperative?"

Tanner unzipped his leather case and placed two stacks of documents on Dr. Dumas' desk. The first set was old hat to Dr. Dumas, the fishermen's plight, the shortened seasons, no representation, and so on and so forth.

Needs, needs, needs! Dumas got up from his desk and started pacing the floor. For hell's sake he thought as he gazed out the window, will the bastard never stop talking? I've heard this garbage a thousand years. He licked his lips and nervously cracked his knuckles. Why does he have to keep waving the papers under my nose to prove his point. Who gives a fuck?

As Tanner finished his report, he gingerly placed the set of documents back into the case and turned his attention to the second document still lying untouched on the desk.

Tanner unfolded a large white chart with rectangular blocks outlined in bold black ink. Words and names were printed inside each block. And the blocks ran all the way across the width of the chart and all the way down the length of the chart. It was set up into divisions of all the branches of the federal government. Net-working of the worst kind. Blue lines, red lines, green lines, and yellow lines crossed and criss-crossed the chart, from the President to the smallest block in the far corner which said 'salmon eggs'. And at the top of the chart, and above the President's block was inked in, in handwriting, the words: Banking, Lending Instutions, Corporations, and Insurance Companies.

As Tanner began to explain the chart, Dr. Dumas stopped pacing around the office. He stopped gazing out the window. He stopped cracking his knuckles and licking his lips. He stared at the chart for several minutes and, for once, Tanner knew he had hit a vein.
Dumas' head was whirling. How in the hell did this 'pissant' come up with a map like this! He knew many of the prominent capitol hill lawmakers who's names were listed here in black and white, and

271

every colored line on the chart lead to them and through them from hell and back! In fact, he saw his own name, and he was in shock. He thought 'that' had been taken care of a long time ago.

"Where did you get the material for your report, son," he asked Tanner cautiously.

"Actually, it's a compilation of events and names that range over a number of years and was just recently brought to my attention through an investigation I was directed to spearhead," explained the investigator.

"Well, my boy," Dumas continued. "That was quite a presentation."

"Thank you, Mr. Secretary." Tanner smiled confidently. "That is a real compliment coming from you."

"You presented a view of off-shore fishing on the west coast that hasn't, as yet, been heard in Washington. I'm sure you've researched the subject thoroughly before you wrote your report." Dumas said. "How long have you been working on this project?"

"Two to three weeks," said Tanner.

"Have you the material to back up the report?" Dumas asked.

"Every scrap of it, Mr. Secretary," said Tanner.

"That's good," said Dumas. "My suggestion is to turn your material over to the proper agencies that set policy and deal with the fishing industry, and let them handle the problem."

"I was counting on your giving me full reign to see this project through, Mr. Secretary," Tanner argued. "I feel I'm qualified to carry the ball since I've had first hand experience with the fishermen and their situation. You see, the fishermen aren't represented in Washington. They don't have a spokesman, and they never have. Well, they have one now, me. They've been at the mercy of every kind of bureaucracy on both a state and federal level for the last twenty years.

That bureaucracy is corrupt from top to bottom and has practiced every kind of lie and collusion ever invented. I'm seriously driving for a congressional investigation of the whole industry. There is a power struggle over our marine resources which will affect the environment as well as every individual in the world. I'll tell you, Dr. Dumas, now the fishermen have a representative in Washington, D.C."

"You say you have statistics that can support these accusations?"

"You damn right I do," said Tanner firmly. "This little report today is a tip of the iceberg. Off the top of my head I can give you three examples that will be made known to the public. The plot is simple. Private industry has become government, and government has become big industry. I'm talking about corporate takeover of the west coast fisheries. I'm talking about monopolies of ocean farming, fish processing, marketing, sales to foreign countries, and national food chains. The same giant corporation controls and reaps the profits.

Take Windover Lumber Company, for instance. This lumber company has subsidiaries in Chile, in England, and in Japan, as well as AQUAPROTEIN in the U. S. This company was responsible for the sale of American salmon eggs to Chile. These eggs came from private state-owned institutions who got the eggs from the government-controlled hatcheries for nothing. Talk about graft. The American people owned those eggs, but Windover reaped the profits.

I have this chart that traces certain individuals from Windover Lumber Company to state universities and to government agencies. A few men are getting paid by all three institutions at the same time. What do we call that kind of operation in the law courts, Mr. Secetary?"

"But I've been Secretary of Commerce for several years, and I'm familiar with my department, the Pacific Fisheries Management Council, the U.S. Justice Department, the National Marine Fisheries Service, and the west coast universities," said Dumas, "and I don't know of a single individual who is receiving a salary for more than one position. I think you're barking up the wrong tree, Mr. Tanner."

"No, Mr. Secretary. I am not." Tanner was adamant. "In the files which I gathered in Port Bane, I have the names of specific individuals who have violated every law in the book. Right now, there are at least three men who are being paid thousands of dollars by private companies, like Windover, and serving at the same time in political offices in Washington, D.C. I even have some materials on your research in pen-rearing before you became Secretary of Commerce, but I haven't had time to sift through all the statistics yet. I intend to follow up my report today. If it becomes necessary, I'll take proof of this investigation to the Washington Post, then heads will really roll. Mr. Secretary, I'm here to ask you to support me and the west coast fishermen in this investigation."

273

"Mr. Tanner, you sound like you're pursuing an interesting quest." Dr. Dumas paused and licked his lips. "I'll tell you what let's do. Since today is nearing the Fourth of July and everyone will be taking the long week-end off, why don't we continue our discussion after the holidays. Don't you think it would be more expedient to wait until everyone is back on The Hill. Is that agreeable with you?"

"That's very agreeable with me," Tanner said happily.

"You have a good time with your family, and we'll get together at the beginning of next week. My secretary will give you an appointment." He cracked his knuckles to indicate the meeting was at an end.

When Dr. Dumas closed his office door behind Tanner, he ran his long hand over the perspiration on his face. What was he to do now? This unexpected development compounded his problems and presented imminent danger to him, his associates, and his beloved department. Why, he could even be exterminated if it came down to a formal investigation. He had to act fast. There was only one thing he could think of to do short of ordering the extreme unction for Buck Tanner.

Dr. Dumas had known Judge Elton 'Eel' Powers of the Justice Department for many years. He was Tanner's immediate superior, and he knew him to be a man of precise habits. Since it was noon, he figured he would find him eating his lunch at Mario's restaurant. Powers, in fact, was never a minute late for his three daily meals, or a meeting, or an appointment. If he could reason with Powers, and Powers could pull Tanner off the case, then he would be saved a great deal of time and trouble. But if Tanner were allowed to continue his investigation, he might not only stir up enough opposition to stop the regulatory fishing bill already pending, he might also have a few scalps in congress. How had Powers allowed a hothead like Tanner to investigate the fishing industry in the first place?

Dumas tried to think back to the date he was still taking payoffs from Windover. That financial venture was just before he became Secretary of Commerce. Tanner would sure as hell uncover that. He had worked too many years to lose his position as Secretary of the Department of Commerce and his appointment to the United Nations. If Powers refused to coorperate, then he would simply make a phone call to a particular senator he knew. From there the avenues for Tanner's investigation would close one by one. Certainly, he,

Dumas, was not without resources.

He was correct. Powers was sitting alone at his favorite table in front of the corner window overlooking Pennsylvania Avenue, devouring a large plateful of steak and potatoes.

"May I join you for lunch, Eel?" he asked.

"Certainly, Mr. Secretary," mumbled Powers through a mouthful of steak. He chewed for a moment and finally swallowed. "I thought you might be in for lunch. Let's see. It's about Buck Tanner, isn't it?"

Dumas smiled, sat down in an overstuffed chair, and ordered a martini, very dry, and a spinach salad.

"You're right, Eel," he said turning his attention back to Powers. "That was an outstanding report on west coast fishing this morning, and from a side of the spectrum I never expected."

"Buck's quite a boy," said Powers sawing off another chunk of sirloin. "He's always been an idealist, but I never thought he'd get so personally involved with this fishing matter. Had I known, I wouldn't have given him the assignment in the first place. And I never thought he would come up with the kind of material he presented to you this morning. When he first discussed the matter with me, I thought it was some strong stuff from a pretty good attorney. He told me he brought back eight thousand pages of data, and he's working through all of it. Yes, sir, pretty strong stuff."

"You know what's going to happen if he is allowed to continue this crusade, don't you?" Dumas sipped nervously from his martini glass.

"I have a pretty good idea," said Powers. "Buck is stepping on the toes of some pretty powerful men. If I understood him right, he intends to pursue the matter to the top, then he's going to ask for a Grand Jury Investigation of government involvement in the fishing industry."

"Well, what are you going to do about him?" asked Dumas licking his lips.

"I'm going to do nothing," said Powers. "I talked to him this morning about getting too involved with marine resource problems. And I warned him about the consequences if he ended up on the wrong side of the money interests in Washington. That's all I intend to say except that I will wait around and pick up the pieces."

'Sure,' Dumas thought, 'and I may be one of those pieces.'

275

"Buck's his own man, and a bulldog," Powers continued. "If he really has proof of paybacks and conflict of interest in the departments he mentioned, he will raise quite a stink, but I doubt that he can get as far as a Grand Jury Investigation. I think it's best if we just wait this one out, Wayne. What do you think?"

"Shit! I'm not going to wait," Dumas said softly through stiff lips. "The fucker is questioning the integrity of the Commerce Department, and every agency from the U.S. Department of Fish and Wildlife to the National Marine Fishery Service. We have enough trouble holding such a wide spread body of public servants together as it is. We've all worked long and hard to get these damn, impending, ocean relegatory bills before congress, and I'm not about to let some fly-by-night, idealistic, investigatory lawyer postpone their passage. Is there any way you can transfer him to another department or send him on a vacation to China, for hell's sake, until the bills pass?"

"Not a chance, Mr. Secretary," said Powers. "Just cool down and think about it for a minute. There have been investigations by the Justice Department before, and by people who have more clout than Tanner, yet nothing ever came of them. Besides, I like Tanner, and I have no justifiable grounds to change his assignment. He's doing a job he believes in; he's just doing it too thoroughly."

"I see," said Dumas. Yes, he could see alright. He could see there was no way he was going to change Powers' mind. Powers was almost as bullheaded as Tanner. He never did like the man anyway. His whole investigative department was a crock.

"I'll monitor Buck's progress and keep you out of the rumor mill," Powers said.

"Well, Eel, I have to be going," Dumas rose to his feet. "I hope you are able to temper this potentially explosive situation. Maybe you're right. We'll just let the matter rest until Tanner makes his final report to the Department of Commerce, then we'll see. Can I pick up your tab?"

"No, thank you, Mr. Secretary," said Powers. "I always pay for my own lunches. Don't even put them on an expense account. Keeps the wolves away. What about your salad?"

"I wasn't hungry," said Dumas. "I don't usually eat lunch anyway."

Dumas went directly back to his office and dialed a number.

276

That Powers wasn't going to do anything to head off trouble was obvious to him; so, as usual, he'd deal with the crisis himself. And he knew just the man to call.

CHAPTER

26

PARADE

Pennsylvania Avenue in Washington, D.C. was once a blighted region of saloons, gambling dens, lodging houses, quick lunchrooms, cheap-jack shops, and cat-penny amusement places. From these establishments came the great regeneration in the first quarter of the century. Most of the old buildings on the south side of the avenue between the capitol and the treasury buildings, and for several blocks on the lower north side, were demolished; park areas were laid out, and the ambitious Federal Triangle development begun.

The avenue extends about seven miles in a northwesterly direction from the district line, across the Anacostia River on the John Phillip Sousa Memorial Bridge, and through its famous central section just west of Rock Creek. The Anacostia section between the district line and the river has developed into an attractive residential area, especially upon the Anacostia heights.

The steel-arch bridge, which was dedicated to the "March King," was constructed by the district and completed in 1940. In harmony with the upper stretches of the avenue, the six-lane structure is 1,590 feet long and has an overall width of 72 feet. Its broad sweep with 6-foot walks on either side is illuminated at night by a perfect line of lights. Between the White House and the Capitol, Pennsylvania Avenue connects the executive with the judicial and legislative branches of the government. This broad avenue is rich in historic association, traditions, and sentiments. No other mile of American roadway has provided the setting over so long a period for such pageants of national pomp and cermony. The inaugural parade of every president since Jefferson and the funeral processions of all the presidents who died in office have followed this course. Lafayette and Kossuth, the King and Queen of England, and many other foreign dignitaries have been acclaimed here. The homecoming armies of three wars have received tumultuous welcome. And since the grand review of the victorious Union Army, which marked the end of the Civil War, the 4th of July Independance Day Parade in Washington D. C., has been celebrated along this avenue.

Buck Tanner, Vivian, and their two young sons joined the fabulous number of patriots to witness this year's great march of victory celebration. The day was rosy with the first flush of dawn as people trooped to grandstands and lined sidewalks in order to extend a passionate tribute to the soldiers who would be marching. Hundreds of flags, marching bands, and floats would round out the pomp and circumstance. The whole day would be filled with patriotic sentiment.

In front of the White House was a pavilion all covered with flags, roses, and evergreens. Here, the President, First Lady and the military staff would review the parade. Grandstands for officials, dignitaries, and disabled veterans had been built on the opposite side of the avenue near the capitol where the parade would start. School children stood beside their parents in wide-eyed anticipation.

At nine o'clock sharp the cannon roared signaling the beginning of the parade. From down the avenue emerged faint martial sounds, and then the flags and their bearers, the marching bands, the soldiers, and the floats came into view.

The grand marshals, bearing the flags of freedom atop their white steeds, led the parade. Following the horses were the dignitaries of the city: the mayor, the president of the Chamber of Commerce, and other prominent citizens.

There were floats of Indians, covered wagons, a steam engine, a liberty bell, historical heros, a Columbus ship, and the invincible American Eagle. Some of these vehicles were pulled by horses, others by cars. One, in the form of a large drum, was even built over a bicycle.

The parade of floats was interspersed with some band or marching organization: high school bands, college bands, military bands, ethnic bands, murmers groups, the 300 piece "Jackie" band, the drum and bugle corps, the Legionaries, and the rifle regiment. The U.S. Cavalry Review—carbines, sabers, and redscarfs dangling around their necks—strutted proudly along. Elegantly costumed young equestrians on high stepping horses showed their stuff. Eight other white horses pulled the black caisson in memorial of the honored dead. This macabre spectacle seemed hardly to move because it

279

was accompanied by two bands playing SOUSA dirges in dirge time. The cadences of muffled drums proved to the listeners that all hadn't been roses and wine in this country's development. To offset this serious note, in an otherwise jovial parade, each marching band performed a two minute prepared drill before the grandstanders sitting in front of the capitol building.

Buck and Vivian had arrived early in order to get the good seats at the top of the grandstand across the avenue from his office. This was the one time each year that their family spent the whole day together. Vivian had graciously packed a picnic lunch of fried chicken, rolls, fruit and nuts, and she had laid out their riding habits for horseback riding in the park. Tanner had filled the family car with various rackets, balls, horseshoes, blankets, and games they wanted to play after the parade and the picnic lunch. Later in the evening they intended to watch the fireworks at the Washington Monument.

"I love a parade," Buck said, a relaxed smile on his face, and he began to sing. "I love a parade...the prancing of feet, I love every beat I hear of the drum. Oh, I love a parade...."

"Buck," said Vivian. "You're embarrassing me. Every time you try to sing that song you get the words mixed up with six other compositions."

Tanner had tried to sing or "harch" to JOHN PHILLIP SOUSA'S marches ever since he was a young boy in Texas. He had grown up on SEMPER FIDELIS, THE WASHINGTON POST, THE THUNDERER, THE LOYAL LEGION, THE PICCADORE, THE GLADIATORS, THE U.S. FIELD ARTILLERY, THE FLAG OF FREEDOM, THE HIGH SCHOOL CADETS, THE ARMY OF THE POTOMAC, and the greatest march of all time, THE STARS AND STRIPES FOREVER. Poor SOUSA. The guy never knew that a few days after he died in 1932, Congress introduced a bill designating his musical composition of THE STARS AND STRIPES FOREVER to be the national march of the U.S.A.

"I don't recall seeing this figure on a float before," Vivian whispered in his left ear.

Standing on a replica of the first horse car was the statue of a twelve foot tall, stocky militaristic figure. A fringe of white hair lay below a bandmaster's cap; the squared shoulders were impeccably straight; and the hands moved up and down with graceful restraint.

"It's the old maestro himself. This is the first year they've done SOUSA," said Buck. "It's about time!"

"Who?" said James, his youngest son.

"JOHN PHILLIP SOUSA," explained Tanner. "He composed a lot of the marches you'll hear today."

"I like this one," said Billy Kyle, Jr. "Look! Uncle Sam wants you!" He pointed at the imposing sixteen foot figure dressed in red, white and blue who represented the U.S. Government and the American people.

"I, of course, prefer the equestrians on their handsome mounts." said Vivian.

While Buck Tanner, his family, and the crowd of onlookers applauded the "Uncle Sam Wants You" float, two men dressed in dark business suits and carrying black leather cases edged their way through the crowd toward the tenth street entrance of the Justice Department building. The door was quickly unlocked with a key, and the two men slipped inside. The time was 11:00 A.M.

"It's on the fourth floor," the young one whispered. "We'll take the stairs. Don't forget to put on your gloves."

"This one should be relatively simple," said the dark man as he pulled on a pair of surgical gloves. "It's not often we're furnished the keys to do a job in broad daylight."

"Yeah," said the younger man, who looked like he hadn't seen the sun in a decade. "That's what's bothering me. There's something strange about this heist. I don't know how you feel, but going in the daytime is bugging me. Let's get it over with. I want to clear out of this place and quick. The deal smells fishy to me."

"Yeah," agreed the older man with deep creases in his forehead. "Maybe we're being set-up for some reason."

"Whoever's masterminding this one has got to be a biggy," mumbled the thin one fumbling with the lock. "I don't like it."

The men obviously knew what they were doing and had had years of experience in their trade. They walked directly to Buck Tanner's office, and the thin one unlocked the door with a second key which hung on a key-ring attached to his belt. While one of the men expertly worked the dial to the simple safe that stored the floppy disks for the computer, the other one emptied the file cabinet of its printed material which he stashed into leather cases. In less than ten minutes the office was completely ransacked. Tanner's evidence of the fishing industry scam had completely disappeared. These ex-CIA

281

men, thugs Tanner would have called them, were professionals. After break-ins in Lebanon, Cuba, and South America, a tiny office in Washington, D.C. was duck soup. In two hours the papers would be shredded and the disks burned, then they could collect the remaining two thousand dollars for the job.

On Monday morning Tanner marched into his office in high spirits. Life was good. He, Vivian, and the boys had spent a relaxing 4th of July weekend together. They had picnicked, played horseshoes, quoits and badminton. They had ridden horseback through Rock Creek Park. In the evening they had watched the July 4th fireworks at the Washington Monument and ate hot dogs and sauerkraut. All in all this outing was one of the most rewarding the family had ever shared together. Vivian had been unusually affectionate, and Tanner felt rested, fulfilled, and happy. He sat down at his desk and lit a cigarette.

At first glance his office looked exactly as he had left it. He switched on the computer and reached for the floppy disk in the carton. The disk wasn't in the carton. Maybe he had absentmindedly stored it in the safe. But why would he do that?

The moment Buck opened the safe he realized he'd been had. The receptacle was empty; cleaned out completely. He jerked open the file drawers. There were no file folders, and there were no files. Silk's material had vanished.

Tanner tore out of his office, tie askew, coat flapping, down the hall in between the multitude of desks and computers, cursing and knocking into chairs of busy workers. He didn't bother with the blonde secretary, but went right past her into Powers' office, slamming the door behind him.

Powers looked up at the unwelcome intruder with a cold quizzical stare, but Tanner was oblivious.

"My fucking disks and my files are gone, and so is my report," the distraught man exclaimed. "They're gone!"

"What do you mean gone?" asked Powers calmly.

"You know, gone, disappeared, kaput!" said Tanner.

"Nonsense, Buck. You've just misplaced them. They couldn't just walk off. Now, think for a moment. Could you have taken them home for the holidays?"

"Hell no!" said Tanner. "Give me some credit for my brains.

282

I'd know if I took a case full of files home. I tell you, Eel, some sonofabitch has been in my office and taken my material on the fishing industry. That's all that's missing."

"Well, we'll just have to see about this!" said Powers. "No one violates any department in this building. That would be a federal offense. Let's go down to your office and take a look."

Once again for Powers' benefit Tanner went through his file cabinet and his safe. The search ended with both men staring into each other's eyes.

"Are you sure it was here when you left Friday?" asked Eel.

"Hell yes," Tanner said. "I've been working on that stuff for two weeks, and I almost had the works transferred to computer disks. But the floppy disks are gone too, and the internal memory of the machine has been blanked out. Someone sabotaged my project over the weekend."

"In this case there's nothing to do but call in the police," said Powers rationally. "While you're making the call, I'll canvass the department. It doesn't seem logical that your office would be the only one ransacked. Other people must have material missing too."

Six hours later Tanner was the most frustrated man in Washington, D.C. He had called the city police. The city police had called the F.B.I. Powers had checked with every person in the department. Unless he was crazy, and he wasn't, his office had been the only one broken into, or unlocked as it were, and the only thing missing was John Silk's research.

"Mr. Tanner," the chief F.B.I. agent said. "If your office was broken into, it was a professional job done by professional people. There wasn't a single fingerprint or a single item out of place except what you say is missing. We have a scientific department which will analyze an item of dander or a single hair if it's there, but in this case we have uncovered nothing. Our investigation will continue until we are satisfied no evidence is going to turn up. At that time we'll discontinue the investigation. Of course, that decision will have to be made when we come to it, and approved by both your department and our department. However, you might as well know right now. We don't have a single clue to go on, and it's not likely one will turn up. My experience in these matters tells me the material has already been shredded by now, and the thugs have been paid off and shuttled out of town. A crime like this is a damn shame, and I hate to say it,

283

but I think you may as well kiss this one goodby. Wish I could be more encouraging."

Tanner was left sitting at his desk with both hands holding his head. Powers sat across from him.

"I'm sorry as hell, Buck," he said. "I don't know what to say. I realize how important this project was to you. Now, don't worry, I have another assignment you'll enjoy."

"I don't want another assignment," said Tanner. "I'm going to continue this one."

"Buck," Powers used his most fatherly voice. "I want to give you some advice. There are ups and downs in your line of work, and you've just hit a downer. You have to roll with the punches. You heard what the agent said. Face the facts, Buck, that material is gone. If you're worried about those people in Port Bane, forget them. They were there a long time before you ever went on this investigation, and they'll manage after you're gone. You let them handle their problems, and we'll get on with some more important things."

"I'm sorry, Eel," Tanner said. "I can't do that. I gave those people my word, and I'm going to stand behind it. Where I come from, if you give your word, you honor it. I'm going to tell Vivian. I know she'll feel the same way I do. And then I'm going to call Silk in Port Bane and see if he has another copy. If he doesn't, I swear to God I'm going to work from memory. Is that O.K. with you?"

"Suit yourself, Buck," Powers said. "You have my O.K. But remember, from now on you're on your own. I can't push the thing any more than I've already pushed it. Getting the appointment for you with Dumas was a damn sight more than I would have done for anyone else in the department."

"Speaking of Dumas," Tanner mused. "I want to say just one thing. There are only four people who knew about this project. I knew it, and I sure as hell didn't steal my own work. Vivian was with me all weekend, so she couldn't have broken into my office. I know you, and I trust that you believe in me so you didn't violate your own department. Do you know who that leaves? That leaves Wayne Dumas! He gave the orders. He hired the thugs, and he violated the U. S. Justice Department. I just don't understand how you can see the obvious and still keep silent. I'm going to confront the bastard."

"Now Buck," Powers reasoned. "What you accuse Dumas of doing may very well be true, but you can't prove it. Can you? Why

284

hell No! So don't make an ass out of yourself. This issue isn't worth it, and I'd advise you not to confront Dumas until you have some concrete evidence. As for me, I have a department to run, and no matter what I think, I have to work within the confines. You heard the F.B.I agent say there wasn't going to be any proof. Give it up, Buck, and for hell sake don't attack Dumas. We have to look ahead and work with what we have. We have other important projects pending."

"Bullshit!" Tanner said. "All I want is one more chance with Dumas. Even if Silk didn't make copies of the material, I think I can keep this investigation alive. I have the issues pretty well set in my mind."

That evening when Tanner told Vivian, he heard the same line Powers had used. In fact, if she weren't his wife, he would have thought the woman was Powers speaking. In his mind he carried the problem a few more steps, and he realized that what he had to say and do really didn't matter to Powers. It didn't matter to Vivian, to her father, to any senator or representative, or even to the President of the United States. According to Silk they were all guilty of surreptitious knowledge.

Frustrated, Tanner went into the den to make the worst phone call of his life, the call to John Silk. He had taken the fishermen's struggle into his hands and failed to take precautions because he hadn't understood the impact this information might have on the people in Washington, D.C. Now the worst had happened. If Silk didn't have another copy, he didn't know what the hell he was going to do.

As luck would have it, he got Silk on the first ring.

"Ahem.. Silk? Hello," he said. "It's Buck Tanner in Washington, D.C. I have a little problem back this way, and I need to ask you just one question. Did you happen to make a copy of the fishing material in that box you gave me?"

"Sorry, Tanner," Silk sounded steady over the line. "I never thought to make copies because I doubted the stuff would ever be used. Anything wrong?"

"The damn stuff has disappeared," Tanner said.

"That doesn't surprise me," Silk spoke as calmly as if he were steering the CRACKANOON II through the fog without radar. "It's about the way I figured it. Once they saw you had the goods on them,

285

there was no other alternative. They couldn't let you expose their political swindle, or there would have been a scandal from Oregon to Washington D.C. Don't blame yourself, Tanner. I guess that about finishes your investigation, doesn't it?"

"Losing your material does put a cramp in my plans," Tanner admitted. "But Powers will fix it for me to see Dumas again, and I think I can swing the project from there. Don't worry, I'm going to continue the battle. Because someone stole my work won't stop me. I may slow down, but by God I gave you my word as a man, and I believe in this investigation. I'm not finished yet, Silk."

When Tanner returned to the living room, Vivian was waiting for him. As he looked at her, he was surprised at the vehemence in her expression. The softness in those dark eyes, which had always expressed her love for him, had been replaced by a bitter, granite stare akin to hatred. Tanner was so upset, he felt disoriented.

"I never dreamed this investigation might cause a break-in," he said. "Someone didn't want me to tell the truth, so they broke into my office and took my work."

"You brought it on yourself, Buck, and you know it," said Vivian. "This time you've not only hurt yourself; you've endangered your family. Can you imagine what could have happened to me and the boys if those men had come to the house while we were home alone?"

"I don't think there was any danger to you and the boys at all." Tanner tried to think rationally. "I told you what the detective said. Those men were professionals, and they were only after one thing, my disks and the files which could expose some important people."

"You don't know that for sure, Buck. If the fishing scam you were working on was so important, what makes you think they wouldn't have killed for it?" Vivian continued to stare directly into Tanner's eyes. "You simply must think of your home and our security, and that does include the Senator and mother. I will not have them disgraced."

"Your father is a lawmaker," Buck said. "He of all people is very interested in just laws for this country. I intend to do exactly like I know your father would do. Now that I've started this fight, I have to see it through."

"I'm not supporting you on this fishing thing, Buck. You're

putting yourself and your family in a very sensitive position, and I will not have our reputation and integrity jeopardized over something that does not concern anyone in Washington, D.C." Vivian said threateningly.

"Don't worry, dear," Tanner said. "I'll take care of my problem. Nothing else will happen, I promise."

Tanner's next meeting with Dr. Wayne Dumas was arranged for Monday morning at ll:00 A.M., exactly two weeks after the break-in.

"This is one report that is going to get Dumas right in the neck," Tanner told Powers at the office that morning. "I'm positive he's the main cog in this whole fucked-up political mess."

"Remember, Buck," Powers warned him. "He has a lot of friends in powerful places."

"I know who his friends are," Tanner said. "I have them listed on the chart. Everyone of them is bought and paid for by those big companies."

"Think, Buck. You don't have a chart anymore," Powers continued. "My advice is to make your presentation the best way you can under the circumstances. Take the consequences, and if they're negative, shelve the project. Remember the byline in Washington. 'He who refuses to compromise is he who soon meets his demise'."

"That's why I like working for you, Eel," Tanner grinned. "You're always so full of good advice."

As it turned out, Tanner's next meeting with Dr. Dumas was worse than the first one. He was in the middle of joint venture as Cowboy had explained it when Dr. Dumas suddenly stood up and ended the discussion.

"Mr. Tanner, you have done an excellent job of stating your views concerning certain problems the west coast commercial fishermen may be having." He walked around his desk and cracked his knuckles for emphasis. "You seem to have touched on every phase of the industry, and you've given excellent recommendations. However, your views and recommendations do not appear to be founded on facts. I do not see a shred of evidence to support your contentions. Do you have statistics to support your claims about the salmon?"

"No, Mr. Secretary, I do not," Tanner said. "My office was robbed over the holiday weekend, and my material is gone. However, I would like to suggest a second investigation by a team of qualified experts who would gather statistics, and I would like to spearhead such an investigation myself."

"Mr. Tanner, I'm a busy man." Dumas licked his lips and rubbed them together. Tanner stared at the strangely shaped man and waited. Here it was, the moment he had been dreading. "I have an important committee meeting with the Department of Fish and Wildlife concerning the coho season, and I don't intend to keep them waiting. Good-by, Mr. Tanner."

"I would like to sit in on that meeting if I may, Mr. Secretary." Tanner was grasping at straws. "Perhaps I could speak to the committee."

"Absolutely not!" Dumas said coldly. "Good-by Mr. Tanner."

When Tanner got back to his office, he could tell that something was bothering Powers.

"What happened?" Tanner felt the new pressure. "Did Dumas call you and tell you to put a thumb on me?"

"Something like that, Buck." Powers shifted uncomfortably in his chair. "He said he didn't want to see your face around his office again. I think it would be best to cool things for awhile. I've decided to take you out of the investigative field and assign you to some home office details, Buck. It's a good thing to do at this time, at least until this controversy settles down a bit."

"That's fine with me," Tanner said. "I need to be close to the departments I want to contact. You know damn well I'm not going to drop this project without a last ditch effort, Eel, and I'm not going to let Dumas scare me off. I know now that everything Silk said was true. Dumas is as guilty as sin, or he wouldn't be pulling these shenanigans. Don't give up on me now, sir. Let me see this thing through."

"I'm not reassigning you as yet, Buck," said Powers. "I'm giving you some extra time, but I expect you to adjust. Stop being so damned sacrificial. Somehow you've put us both on the spot. You're my best man, and I'd hate to lose you. Damn, I'm going to have some sleepless nights over this fish business."

Over the next few weeks Tanner ran the gauntlet of Washington bureaucracy. He took the west coast fishing problem to the U.S. State Department, the Pacific Fisheries Management Council, the National Marine Fisheries Service, the U.S. Fish and Wild Life

Service, and finally, to the U.S. House and Senate. Not one individual in any department would even discuss the problem with him. He decided they were all in it together. A network headed by Wayne Dumas controlled the fishing industry. Tanner found that instead of making progress on the problem, he was losing ground. He began to stalk the committee meetings that had any jurisdiction over the west coast fishing industry. He hung around in the halls of the different departments in the house and senate. And then, one day, as he was standing near the senate chambers, he heard one of the committee members say that the decision had been made to cut the next coho season to less than half the tonnage they had had this season. He had better warn the fishermen in Port Bane.

"Are you sure about that?" John Silk asked over the phone.

"I'm sure," Tanner said. "You are only going to be allowed to catch one half as many coho, then the season will end. You'll probably have half the number of days you expected."

"Thanks for calling me, Tanner," Silk said. "We'll just have to gear up for the worst."

"How's Cowboy's house coming along? Got it finished?" Tanner asked.

"Well," Silk said. "They jacked it up and it fell down the hill. Cowboy had to destroy it. He's fishing with me now. Not many fish."

"I'm still pursuing the project. I want you to know that, Silk," Tanner didn't know what else to say.

CHAPTER

27

SENATOR KRAMER

Tanner had one more person he could go to; his own father-in-law, Senator Kramer. If anyone had influence around Washington, D.C., Senator Kramer did. Tanner had been putting this meeting off, hoping he wouldn't have to go to this length, but now he had been driven to his last excruciating step.

By the time Tanner's meeting with the Senator was concluded, and he had locked up his office and driven to Georgetown, the time was close to 7:30 P.M.

The house seemed unusually quiet for this time of night. The boys did not meet him at the door, nor did Vivian. Instead, she was sitting on the white sofa in the living room dressed in red hostess pajamas. Her black hair was piled in curls on top of her head and clipped at the crown with a ruby barrette. Her face was expertly facialed.

Vivian watched Tanner enter the room, but she didn't come to greet him. Instead, she sat unmoving, staring more like a mannequin with a mask than his usual bubbly wife.

"Hello, darling," he said and kissed her forehead. "You're looking unusually beautiful tonight. Where are the boys?"

"I'm having them sleep over at mother's tonight," she said stonily.

"Any special reason?" Tanner asked. "Have I forgotten another important occasion?"

"The Senator called me this afternoon after you left his office, Buck," Vivian said.

"He did? Good." Tanner took off his overcoat and sat down in the overstuffed chair across from Vivian. "Anything special?"

"He told me you have become a very unpopular personality on The Hill. He said that you have lost credibility. That you are becoming a nuisance, a bore and an embarrassment."

"Now, honey, don't get upset." Tanner moved to the sofa beside her.

"How can I have any self-esteem when I hear these things about the person I'm married to?" Vivian went on.

"Don't say things like that, sweetheart," Buck said.

"Face it, Buck. Would you say you have achieved what you are working for?" Vivian asked.

"No," Tanner admitted after considerable thought. "But I'm trying."

"For a long time I've tried to make you into the certain kind of person you should be," Vivian said. "I just wonder if you have the ability to be that person. I have been unable to see you in that light."

Tanner quickly analyzed her words before he answered.

"I'm not a loser, if that's what you mean."

"We've been married nine years, Buck," she went on. "You seemed to have had great promise the first five years we were married. But, then after that, there has not been much progress. I feel that I have grown and you have not."

"Viv, please." Tanner tried to hold her hand. "I love you and the boys. Why don't we leave this city and find a new beginning someplace else?"

"But don't you see, Buck? I don't want a new beginning with you any place, and I'm certain the boys wouldn't either! Do you want the boys to have to pay for your failure? You're insensitive to your sons' needs. I'm the boys' mother, and I must think of their future. My problem has to do with constancy. The boys have a home here; they have a future; and they have a prestigious school and important friends. They have a place in society already set up for them. I simply cannot tolerate anything that will jeopardize their standing in the community."

"I just don't know what to say anymore," Tanner said humbly.

"I have been thinking over what the Senator told me this afternoon, and I now realize that it has come down to choices, Buck," continued Vivian determinedly. "I have made my decision. I want us to separate. I will not be degraded. I will not have the Senator and mother humiliated, and I will not have the future of my sons jeopardized."

"Honey, don't do this to me." Tanner realized this conversation had become very serious. "There are a lot of things I'd do

291

differently if I could. A lot of things I'll change if I can."

"I consider this separation to be less my failure than yours, Buck," Vivian said. "I have been here. You have not. I have to believe that your leaving is the only step remaining. We simply cannot live together. I don't respect you anymore."

"What can I do to change your mind, sweetheart?" Tanner asked. "I'll quit the department! I'll go into private practice! I'll run for the senate! I've told you all this before."

"Tomorrow," Vivian continued. "I shall be at the club from ten until three. Please have your things out of the house by the time I get home. And please find yourself another lodging for tonight."

An hour later Buck found himself checking into a cheap room in central Washington, D.C. The Bagdad was an old run down, sleep-easy hotel just off the downtown section. There was a bed, a dresser and a toilet. Below Tanner's room he could hear a jukebox playing "Shades of the Night", and above the music he heard the clanging of dishes and the voices of chattering people in the small cafe where he would be soon eating most of his meals. Nearby, he heard the noisy motor of the local buses that roared up and down the streets all night. Likely one of them would be going by his office at the Justice Department, and he thought about how he could save gasoline money by hopping one of them.

This experience was like the worst possible nightmare out of a middle-class soap opera except that he couldn't turn off the T.V. He felt like he had lost control of his life. Was this possible? Was this going to be the extent of his life, the way it would end for him? What in the fuck was going on anyway? Tanner had to admit he didn't know.

He walked into the bathroom and looked into the mirror over the tiny chipped sink. He saw an image in the mirror which reminded him of a second rate detective who had just learned that someone had shot his partner. His tie was askew, a stubble of beard was beginning to cover his face, and his eyes were blurry from overwork, strain, and shock. He looked like death warmed over, and he felt miserable.

He was sick of Washington, D.C.; he was sick of his job; and he was sick of himself most of all. He thought of all the events that had happened to him in the last few hours, and he wanted to throw up. His brain was not able to encompass his dilemma. What in the shit was he going to do now? He had tried to do the right thing. He had tried to be kind to everyone. He had tried to be a good husband and a good father, and where had it gotten him?

Suddenly he felt furious. He knew he had done the best he could. He had fought for the underdog all his life. In fact, he was the underdog! Now, who in hell was fighting for him!

This bullshit is what he got for trying to keep his principles. Yeah, kicked in the guts. Kicked in the guts right at the time when he was the most vulnerable. Well, he'd show them. He knew he was right in what he had started out to do, and he was damned sure not going to give up in the middle of a fight. There was one group of people, the Port Bane fishermen who had been kicked and stomped just like he was, and he was going to win this struggle for them or die trying.

Tomorrow he was going to march right back into his office, and from there he would go to every office on The Hill. He was going to keep this project going. As he saw it, the only way he had a chance to win back Vivian's respect and love was to stick with the project, push it through until he got the senate investigation, and then stick it up their asses. He would prove to her, and to all of the sonsofbitches who were trying to kill him, that he was a fighter and could win against the odds.

Today was just a bad day for Vivian. He understood that. She didn't mean what she said tonight. But he would play along with her mood. She would feel differently after she had thought it over for a few days, and she'd see that he was right. He was a man with a force. He would win her back.

The first thing Tanner did the next morning was to call his house. He wanted to tell Vivian that he loved her and that he understood her frustration. But after twelve rings, he finally realized no one was going to answer the phone. Vivian probably had an early meeting at one of her clubs.

He thought back over last night, and he knew he was right on. He was going to get those bastards, and then he was going to laugh his head off. They deserved it, all of them, and that included the Senator. The scum had no right to interfere with his family like that.

God, the towel smelled like sour milk. Everything in the room smelled stale and chewed on. This was one morning he was going to be glad to get to the office.

Over the next week Tanner called his house at least once a day, but no one answered. He knew if he kept trying, Vivian would soon relent and talk to him. All their tiffs in the past had followed that pattern. He'd play it cool and give her some space. By now she would be missing him as much as he was missing her.

Then one day an operator came on the line with a canned recording which said the number he was calling was no longer in service. This situation was getting to be ridiculous. Vivian was carrying things a little too far this time. He dialed the information operator and asked if there was a new listing for the Tanner residence in Georgetown. He was promptly informed that there was a new listing but the number was unpublished, and the operator could not give it out under any circumstances.

Tanner had had enough of fun and games. He was going to get to the bottom of this. He dialed the Senator's residence number, and he heard the bozo butler's voice say,

"Senator Kramer's residence."

"Listen, Waylan," Tanner said. "This is Buck Tanner. I want to talk to my wife."

"Your wife is not here." Waylan's voice sounded like a zombie.

"Well, if she's not there, where is she?" Tanner roared.

"I'm not at liberty to say," said the Butler in the same emotionless tone.

"You fucking asshole, put Mrs. Kramer on the phone," Tanner shouted into the receiver.

"Mrs. Kramer is indisposed at the moment. Good-by sir."

"You imbecile, don't you hang up on ..."

Two days later an envelope was handed to Tanner while he was sitting at his desk in the office.

"What's this?" he asked.

"A restraining order, Mr. Tanner," the sheriff said.

Tanner took the letter and stared at it. He felt like he had been kicked in the teeth. A restraining order, of all things, prohibiting him from contacting his own family or the Senator's family, and he wasn't to set foot on either property. He was furious. He wanted to throw the sheriff through the window.

Tanner began to work harder and later at night. He didn't want to think, and he didn't want to go back to the empty, sleezebag hotel room. He couldn't sleep anyway; he was too overwrought. He felt an inexplicable eerie feeling gnawing at his gut which seemed to have something to do with timing. He felt that time was running out for him. The devil himself had a hand on his shoulder.

Powers wasn't helping matters either. For weeks now Powers had been hounding him every day to take another out-of-town assignment. He had tried to explain to Powers that he wanted to stay in pocket in case his family wanted to see him, and he was not going to leave Washington D.C. Powers was turning into a cold fish, and it seemed that everyone in the office had picked up on his mood. There were no more "good morning, Mr. Tanner" or "Buck, let's have lunch". No one stopped by his office door to chat for a moment, and not one person, in all the weeks he had been living at the Bagdad hotel, had visited him. It was like he had become a leper.

Then the day came when Powers stripped him of his office. Without warning, without so much as a word, he was moved to a desk in the room where the pool of investigators worked. He was assigned to do the same trivial paperwork shit that he had done when he first came to work for the Justice Department. His salary was cut one-third and the benefits the same amount. He was already short of money, for hell sake. If he kept going at this rate, he would be begging on the street corner.

Tanner decided to call Russ. He had to talk to somebody or go nuts. Maybe if he could bounce off Russ, he could get some of this weird dollop off his stomach and straighten out his mind. He would do the same for his friend if Russ needed him. Russ said "yes" when Tanner called, and they met that same day at a hole-in-the-wall cafe on Eighth Street.

"Do I look the same to you?" Tanner asked Russ as soon as their coffee had been served.

"You look the same to me, old buddy, except, maybe a little washed out," Russ said.

"Well, I must have changed somehow," Tanner complained. "I'm sure as hell being treated like a piece of shit lately. You tell me—what the hell has made me so bad all of a sudden? My wife kicks me out of the house; my in-laws won't talk to me; my boss demotes me; and everybody at work treats me like I'm invisible."

"If you're asking me what I think, Buck," Russ said quietly. "I think the word is out to get you."

"But what the fuck for?" Tanner said. "What have I done except try to do my job?"

"You must have stepped on somebody's toes," Russ said, "and I do mean somebody big."

"Then if I'm scaring somebody bad enough to put the word out on me, I must be close to a rat," Tanner said. "Why is it so damned hard to kill a rat?"

"You have to catch him first," Russ explained. "And don't forget, rats run in packs. There's always more than one. They have teeth too, big ones. Thay can chew through anything."

"Yeah," Tanner admitted. "And right now they've got me by the nuts."

"Well, I have to get back," Russ ended the conversation abruptly. "Big meeting this afternoon."

"I don't think you should be associating with me for a while, Russ," Tanner said quietly. "Some of those teeth may start chewing on you if they think you're too close to their meat."

"You can't scare me off like that, my friend," Russ said. "I intend to stay in touch."

Tanner's mind suddenly switched from Russ and focused on the rat. He knew exactly who the 'king rat' was. The beast was Wayne Dumas, Secretary to the Department of Commerce. Dumas had started the war against him. Dumas had thwarted his efforts. Dumas had masterminded the break-in into his office. Dumas had caused the rift with Vivian. Dumas had gotten to Powers and turned the whole world in the department sour on him. Who the fuck did he think he was, God?

Tanner looked at his watch. It was about the time that Wayne Dumas took his early dinner at Mario's restaurant. If he were lucky, he could catch the man and try to reason with him one more time. Surely he had one ounce of decency in him and wouldn't want to ruin him forever. But if Dumas refused to cooperate, he would resign tomorrow and leave Washington, D.C. His mind was made up.

By the time Tanner arrived at Mario's restaurant the place was buzzing. The restaurant was filled with government employees, senators, secretaries, and lobbyists. Tanner spotted Wayne Dumas sitting at his regular table in the corner next to the window. A strange man was sitting across from Dumas, and they were eating something that looked like chicken.

When Tanner was a few feet from their table, he saw Dumas glance up and look at him before he quickly turned his attention back to the chicken and the strange man. Tanner felt like a fool, standing there, being ignored by a rat like Dumas. He wanted to jerk the clumsy bastard up and shake the shit out of him. This was one time Dumas wouldn't sneak out. He had to walk past Tanner first, and Tanner decided he was going to stand there if it took the rest of the night. Sooner or later the bastard was going to have to face him. That time might as well be now.

"Mr. Secretary," Tanner sidled up to the table. "Excuse me for interrupting you, but I need to talk to you about something important."

"Why, hello, Tanner," Dumas smiled his wet smile at Tanner like he hadn't seen him standing there. "Sit down and join us. What are you drinking?"

"Nothing," answered Tanner. "It's a private matter Mr. Secretary."

"Sit down and relax, Tanner," Dumas said amiably. "It's probably about fish farming and your lost records. You can talk in front of Carlton Forbes here. He's originally from Norway, and he's from the part of the country where you did your last investigation. He's the general manager of the pen-rearing operation in Puget Sound out in Washington State. He's here from N.M.F.S. to lobby for an extension of their pens. You're planning to add about a hundred, aren't you, Carlton? Quite a contribution to the local economy I would imagine. You should talk to him, Tanner. He might change your mind about commercial fishing."

"He probably wouldn't care to hear what I have to say," growled Tanner. "If he's in the pen-rearing business, we're on opposite sides of the fence."

"Glad to meet you," said Forbes and extended a paw that was bigger than Tanner's. "If you want to know about pen-rearing, I'm the man to ask. I was in the business in Norway before I got the job for Noraqua in Puget Sound. Norwegian money is behind the major part of our operation so we're a very inexpensive addition to the local economy."

"I toured a pen-rearing operation," Tanner said. "And it was the ugliest business I've ever seen perpetrated on the ocean." He turned his attention back to Dumas. "I just want to know one thing, Mr. Secretary. Did you have anything to do with the break in at my office?"

"I'm not a crook, Buck," Dumas answered smiling. "I just disagree with your analysis of the fishing industry. I warned you at the outset about the opposition you would face if you took an unpopular position, and now you've found out the hard way."

Tanner took a deep breath and stood up to leave. The Norwegian stood up with him and put in a parting jab. "You are the man Dr. Dumas was telling me about. The man who likes fishermen. Ha! We are far ahead in Norway. There is no longer such a thing as commercial fishing. In America the small fisherman is stupid and uneducated. If you will pardon me, I think they are shit."

As Tanner stared at the large Norwegian, an image of Silk, standing at the wheel of the CRACKANOON II, swam before his eyes. For some reason he could never explain, he doubled up his fist and struck Forbes squarely on the jaw. The big man went down like someone had hit him with an axe handle. When he fell, he landed on top of the table loaded with food and drink and slid into the lap of Wayne Dumas whose chair also went over. Two of Mario's bouncers grabbed Tanner at the door and held him there until the Washington, D.C. cops arrived. They handcuffed him before they carted him out to the patrol car. On the drive to the police station, Tanner realized he had lost Vivian.

He was fingerprinted and booked and placed in a jail cell by himself. Eventually, he was told he could make one phone call. Well, who was he going to call? He couldn't call Vivian because he didn't know the new phone number. He couldn't call the Senator, Waylan had been given the word to hang up if he called the house. Russ was at Cathy's, and he didn't know her last name. Dumas wouldn't give a hoot if he rotted in jail. That left Powers.

After Powers paid the bail and dropped Tanner off at the Bagdad hotel six hours later, Tanner thanked him and invited him up for a drink. Powers said no, but he told Tanner to come by his office the first thing in the morning, and they would have a little chat.

When Tanner walked into Powers' office bright and early, he felt better about the world. But for the life of him he couldn't feel any guilt about hitting the big Norwegian oaf who was mouthing off

298

in Mario's restaurant last night. He sat down in the chair Powers indicated and took the cup of coffee handed to him.

"Thanks again for bailing me out last night," he said. "It was damn decent of you."

"It's too bad you're such a hothead, Buck." Powers leaned back in his chair. "This is our fourth meeting resulting from this crazy obsession you've developed over those fishermen in Port Bane, and it's our last. I'm having to terminate your employment as of now. You've become the laughing stock of the people on The Hill, and you've discredited the Justice Department. You're to receive your final check in the mail."

"Just like that?" Buck shook his head in disbelief. "Just because I hit some moron last night?"

"Hitting a visitor to the city was the topper, of course," Powers said. "But I want you to know the truth. This termination is not my doing. The word came down from the attorney general himself and in his own handwriting. By the time I got to the office this morning the notification was already on my desk. I'd hang in there longer, Buck, believe me, but if I did that, I'd be putting my own neck in a noose."

"Is there any alternative?" Tanner asked hopelessly.

"Not an ink spot," Powers said. "I think you're finished in Washington, D.C., at least for a couple of years. You know how cliquish these people in government are. Now, if you want me to, I'll be glad to write you a letter of recommendation commenting on the excellent quality of your work, but I'm afraid it won't be worth much salt in this city. Maybe the best thing for you to do is to move to a new location where nobody knows you. Try to get a new start."

Tanner went to his desk in a daze and began to pack the accumulation of items that had been with him since his first day in the department. When he picked up the framed motto his mother had given him of DESIDERATA, he stared at it for a long time. Suddenly he saw words in it he had never noticed before. New meanings began to take shape. Words like: "As far as possible be on good terms with all persons; Speak your truth; Enjoy your achievements as well as your plans; Keep interested in your own career; Be yourself; Take kindly the counsel of the years; You are a child of the universe, and whether or not it is clear to you, no doubt the universe is unfolding as it should."

When Tanner read the last line about how the universe was unfolding as it should, he felt his anger begin to rise. His universe wasn't unfolding the way it should! He had no universe left to unfold! He'd been tarred and feathered, boiled in oil, ridiculed, stomped on, laughed at, persecuted, and strung up to dry. He jerked the motto out of the frame and tore it into pieces. He threw it in the trash can and the frame on top of it. He never wanted to see that damned poem again as long as he lived, and if anybody mentioned DESIDERATA, he was going to knock the hell out of them. Justice and the pursuit of happiness! Justice. Justice. Justice. He wished he could throw the whole Justice Department in the trash can and everyone in Washington D.C. with it. Good riddance.

Tanner kept his hate for the city going strong. But it was not until the third week after his demise at the office that a knock at the door gave him the final punch. This was the first time anyone had called on him since he'd been in that fleabitten dump. Maybe the caller was Vivian. Maybe it was Powers wanting him to come back to the office. He opened the door with great anticipation and was handed a letter-sized envelope stuffed to the brim with some kind of papers from Lindley, Beeson, and Jacobs, Attorney's-at-Law. Tanner ignored the woman messenger while he opened the envelope. A moment later he felt himself go blank. He thought he was going to pass out.

Preliminary Divorce Decree legal jargon. Vivian was filing for a divorce. He sat down on the bed too weak to yell or cry.

Tanner was in shock for days. He never left the hotel room, rarely had a mouthful of food, and he hadn't seen or heard from a soul he knew. After a time, however, he began to revive enough to think about his situation. He was barely maintaining. No law firm was going to hire him. Vivian was gone. Powers was through with him. And the truth of the matter was, he was tired. On top of the whole mess, he was running out of money. He looked at his bank account and his balance was so low he didn't have money enough to buy gasoline to escape. There was nothing left to do but to sell the old Plymouth, pay off the Bagdad, pack his clothes, and get out of town before it was too late.

Two days later Tanner sat in the Washington, D.C. bus station with two bags of luggage, a bus ticket, and $59.00 cash in his pocket. 4:00 A.M. Not many people in the bus terminal this time of the morning. Janitors mopped the floors. The loud speaker announced buses. Heels clicked. The drafty doors swung open and shut.

Tanner glanced across the huge room and saw a billboard of Uncle Sam with his top-hat, his striped pants, the stars on his tails, and the devil in his eyes. 'Uncle Sam Wants You' he said in bold letters, pointing his finger right at Tanner. The sonofabitch, Tanner thought. He had always admired this figure representing the United States of America and respected what it stood for. Now, when he looked at the billboard, he knew the truth about Uncle Sam. Instead of standing up proud and straight, he should be bending over kissing somebody's ass.

CHAPTER

28

THE FINGER

Tanner's telephone call on July 5th did not come as a surprise to Silk. He had simply told Tanner to forget about the missing box of material and had gone about his business. Tanner's last phone call, however, was unexpected. He had miscalculated Tanner's tenacity. The man was still trying to expose the plight of the Oregon salmon trollers, and he didn't have a leg to stand on. He was also policing the committees for any rumors of cut-backs, ratios, and regulations that could adversely affect the Port Bane fishermen. Silk decided he would buy the guy a drink and compliment him on his guts should their paths ever cross again.

August 1st. The Department of Fish and Wildlife in Oregon opened up a thirty day coho season, and the fishermen's prospects for the season brightened considerably. Silk and Cowboy had begun catching three hundred coho a day, and like the dory fishermen, they came in every night instead of staying out on five day extended trips. The price per pound was $2.35, and they were making a profit.

August 7th. After the first week of the new coho season the fish processing companies lowered the price of the salmon to $1.00 a pound. They claimed that the law of supply and demand had come into effect because of the salmon glut on the market.

On August 12th. The Department of Fish and Wildlife closed the coho season again. That meant that for the remainder of this year's fishing season the two Chinook salmon for every coho ratio was again put into effect. Buck Tanner's phone call to Silk had been true; rumor had become fact. Hope for a successful fishing season by the Port Bane fishermen had been extinguished like a pinched candle.

A survey by N.M.F.S. studies by marine biologists from coastal universities, a report by the Oregon Department of Fish and Wildlife, and strong lobbying interests by powerful organizations had made their impact upon Secretary Wayne Dumas and the United States

Department of Commerce. On paper they proved that a full thirty day coho season by commercial salmon trollers would seriously harm the fishing industry on the west coast.

Three hundred salmon trollers in Port Bane knew this was a lie. Each boat had been hooking a hundred coho salmon a day. They couldn't keep the damned things off their lines. Now they were back in the same old mold. Boat after boat crept into port with a half dozen Chinook lying in holds of melting ice. There were still too many coho in the ocean and not enough Chinook.

The tragedy of the whole futile exercise was the throwing of coho and shakers back into the ocean where long strings of dead or dying fish floated white behind each boat.

The dory fishermen, who had no cabins on their boats and could not stay out overnight, were finished.

The commercial trollers, one size up, were not much better off. One fisherman became so incensed at the terrible waste of beautiful salmon, he sold his boat and retired. Another fisherman, instead of throwing the coho back in the sea, hid twelve in a gunny sack to feed his family. When he sold his fish to the cannery, his wife accidentally flung one of the illegal fish up on the cannery dock to be weighed. A sharp-eyed worker spotted the coho and reported the fisherman to the authorities. The state police searched his boat and busted him for twelve illegal coho. There was no way he could borrow the money to pay his fine, so the dock authorities confiscated his troller.

Fisherman after fisherman could not afford his dock fees, and every week the port dock grabbed another salmon troller or two. These beautiful vessels were scavenged for parts before they were burned at the Port Bane Salvage Company yard. The fires burned day and night.

The fishermen were by nature a superstitious lot. In the black recesses of his mind, each remembered some past sin that he had committed, and each wondered whether or not he had himself to blame for the evil time that was upon them. One man blamed the vicinity of the moon and Jupiter. Another laws of probability. More than one fisherman blamed his bad luck on Lillian's messing with the stars, but they didn't dare mention this fact out loud because they were afraid she would hear about it and conjure up a worse curse.

Sometime during the second week after the coho season was closed, Silk and Cowboy were fishing several miles northwest of the rockpile when Silk felt the hackles on his neck rise. He glanced up from his work to see just what the trouble might be. The misty sun was beginning to sink toward the horizon in the west, and around them were several trollers in the Port Bane fleet. In the distance Silk could see the fishermen working the decks, probably going through the same process as he and Cowboy.

"What's up?" asked Cowboy flinging a twenty pound, flopping salmon aboard.

"I don't know," said Silk trying to get his fingers in the fish gill. "I just had a funny feeling, like something's not right."

"You're just imaginin' things," said Cowboy eying the salmon. "Now that's one nice Chinook, the third one today. Maybe we're into 'em."

Silk turned the fish over and looked carefully at the spots. "I hate to disappoint you Cowboy, but this mother's a coho. We'll have to throw her back."

"Bullshit!" said Cowboy. "Hit her in the head. If that's a coho, she's the biggest one we seen in days. I'm goin' to keep her. She's already good as dead anyways. Some shark'll eat her, or the birds, if we don't."

Silk raised his eyes involuntarily and stared into the distance. "Well, this is one coho you're not going to keep, Cowboy, look!"

Cowboy looked in the direction Silk was pointing, and there, bearing down on them at a good twenty knots, was a sharp-prowed coast guard cutter. Cowboy snapped the final leader on the line and spit a big wad of tobacco off the rear of the boat.

"Shit, Silk," he said. "They got us. We're goin' to be boarded sure as hell."

"You got any marijuana hidden, or any coho you haven't told me about?" asked Silk calmly.

"Nothin'," answered Cowboy. "We're as clean as a good whore. We just got interrupted, that's all. We're goin' to have to pull in all four lines."

"What a hassle!" said Silk. "Here they come."

As the sleek white boat closed on their stern, Silk could see the small cannon manned by a sailor, the American flag, and at least two men standing by. For a moment he forgot about Cowboy in his admiration for the ship. The next instant he realized Cowboy was standing next to him holding the big flopping salmon high above his

304

head. He shouted "here, you bastards," and hurled the coho directly in the path of the rapidly closing cutter.

"Now we've really done it," Silk said.

The pilot swerved to miss the fish and lost his boarding position. At a signal from the officer standing on the prow, the pilot pressed the throttle and the cutter slid past the troller so close that Silk felt he could almost reach out and touch her. The coast guardsmen were all smiles and flashed a thumbs-up signal that said everything was alright. Cowboy, at the same time braced against the fish box, and stuck both hands straight in front of him with the middle fingers extended in the international sign "Fuck you!"

A waist-high wake of water from the cutter swept over both men, but Cowboy held his position until the coast guard ship pulled next to the nearest fishing troller. They watched as the white clad sailors stepped aboard.

"What's the matter with you, Cowboy?" asked Silk smiling. "I've never seen you lose your temper like that before."

"I'm gettin' mad, Silk," said Cowboy. "This whole coast guard thing is turnin' into a bad joke. Here we've been workin' our asses off for two days. We've got two Chinook on ice. We never carry no dope. And them bastards want to stop us from workin' and search us. Well, it's beginnin' to get next to me. We're goin' down the tube, and they're out here helpin' us."

"Don't let it bother you," said Silk. "Anyway, you saved the day, Cowboy. They probably thought that big coho was a Chinook, and that anyone mad enough to throw a fish worth sixty dollars at them has got to be crazy. Even a coast guardsman with a machine gun doesn't want to mess with crazy people."

Cowboy bit off another large chaw of plug.

"They've boarded just about ever boat in the fleet now, and they haven't found one illegal fish, let alone any dope."

"I'm soaked," Silk said. "Let's pull in the lines and head in. What we need is a good shot of whiskey in a hot cup of coffee. There's not a chance in a hundred that we'll catch another keeper before dark."

Cowboy climbed down into the pit and shoved the gurdy forward. He was still grumbling about the coast guard.

"I've been doing some thinking, Cowboy, and a weird idea keeps running through my mind. I'll mull it over a little more before

I lay it out."

The two men had been sitting in the cabin of the CRACK-ANOON II since dusk. They had eaten a dinner of hot dogs with canned pork and beans, sourdough rolls, and two pots of black coffee, but neither man had cared about the food. Instead of the usual jocular banter that existed between two friends, tonight there was silence. When Cowboy had started to tell about the house on X Street, one of his favorite stories now that the experience was over, Silk had merely glared at him, and the western man with the long mustache had clammed up.

Silk suddenly took the Camel cigarette from the corner of his lips and extinguished it in a beanbag ashtray.

"Cowboy, tell me whether I'm on the right track or not."

"Shoot," said Cowboy, relieved that the silence was broken.

"As I see it," Silk said. "If we don't do something between now and next year, we'll never have another coho season. That'll be the end of fishing as we know it."

"That's right," agreed Cowboy. "You may as well kiss this here boat good-by."

"So we have to do something," said Silk.

"But what?" Cowboy asked. "You seen that coast guard cutter. They got us pegged. The PBFA didn't work; Tanner got your paperwork stole; this season's up; you ain't got a pot to piss in no more, Silk. What can you do by yourself?"

"It was you, Cowboy, who gave me an idea when you threw that coho at the coast guard this afternoon," continued Silk.

"Idea like what, Silk," asked Cowboy. "What's on your mind?"

"We'll have to break the law," said Silk. "All the Port Bane fishermen will have to break the law, but I think the thing will work."

"You've got my curiosity riled up now," said Cowboy. "I'm willin' to break the law, or even go to jail, for that matter, if it'll do some good. Let's have it."

"What if the whole fleet took one trip, and each fisherman filled his boat with coho. There are so many cohos each skipper couldn't help but catch at least a hundred."

"Simple," said Cowboy. "We'd all get fined for so much money we'd never be able to pay up."

"But where they would arrest one skipper who had a few coho

hidden on his boat, what would they do with three hundred boats full of coho?"

"Yeah," Cowboy grinned. "That would be a real problem, wouldn't it? What would they do?"

"Nothing," said Silk. "That's what they would do. Nothing. They would never arrest three hundred skippers. That would alert the public they were enforcing a corrupt law, a law so unjust that every fisherman was willing to put his life on the line."

"Keep goin'," said Cowboy, beginning to get excited.

"To insert another wedge, we won't keep the fish," said Silk. "We won't steal the fish. We won't sell the fish. We won't do anything more illegal than catching them. We'll bring them into port on a given day, and we'll give the salmon away to the public. We'll advertise that we're giving away free salmon. We'll notify the media, the governor, the state legislature, the old folks homes, and the schools. For one day, everyone in the state of Oregon can have a free salmon if he wants to come to Port Bane and pick it up.

If we can publicize this one overt act properly; explaining to the media why we are breaking the law, we will draw attention nation wide and bring pressure on Dumas and Washington D.C."

Cowboy pondered the proposition for some moments before he said,

"Count me in, Silk, let's do her."

By the time the CRACKANOON II slipped into her berth at Port Dock 3, the morning sun was already sparkling off the calm of the bay. Silk and Cowboy got in Silk's van and drove straight to the Albatross for a fisherman's breakfast. Cowboy was halfway through his meal when he noticed that Silk was only toying with his food and sipping on his third cup of coffee.

"You're acting kind of strange these days, Silk," he said smiling through a mouthful of egg.

"Strange, Cowboy?" asked Silk. "How, strange?"

"I'm not criticizing you, Silk. You know that. But you got to admit your behavior isn't the same as it used be when we was fishin' together. You're different."

"What do you mean different?" Silk asked looking directly into the eyes of his friend.

"Well, kinda like your mind's somewhere else. And your habits is different."

"I'm just the same as I always was," said Silk

307

"No you're not," said Cowboy. "Take comin' in. I steered the boat most of the way, and you was down in the fo'c'sle changin' clothes, washin' yourself, and putting that there lotion all over you again. Now you can't tell me you used to do that. In the old days you never took a bath, same as me, and you woulda' never used that sweet smellin' stuff. You come in to see Silvia, didn't you?"

"Don't be ridiculous, Cowboy," said Silk "We have business to take care of."

"Yes sir," Cowboy grinned. "That's the first woman in all these years I've seen you really like. She's been here about three months, and you been tryin' to get her attention ever since she got here. I'll bet you haven't even got to first base yet."

"You don't understand, Cowboy," said Silk, "I have to check with her after every trip, in order to keep her abreast of the situation and take her some of the money. She's just a partner and friend. I can't see anything wrong with having a woman as a friend and partner."

"Nothin's wrong with that," said Cowboy. "But I can tell you're crazy about that girl, even as far as bein' in love with her. You'd spend ever' minute in port with her if she'd let you."

"I don't think the lady is interested in fishermen," Silk said.

"Then why do you think she hangs around with a guy like you?" asked Cowboy.

"Like I said," said Silk. "It's purely friendship and business."

"What I'm thinkin' is that she's feelin' the same way 'bout you that you're feelin' 'bout her," said Cowboy.

"I doubt that," said Silk. "I know she has other plans, but you're right about one thing, I do have to see her this morning before I check on Semyonov. Could you unload the boat, pick up some ice, and buy whatever groceries we need? I should be back here by dark. I'd like to get the giveaway thing organized and leave by day after tomorrow."

"Sure," Cowboy said. "You go see the little woman and get this idea of yours organized. I'll have her shipshape when you get back."

When Silk arrived at Jodie's house, Jodie said Silvia had left for the beach at dawn, so there was nothing he could do but try to find her there. He parked in the tourist lot and looked north toward the lighthouse. Far down the beach in front of the motels he could see several couples, but past them, almost at the limit of his vision, there was a solitary speck. That speck had to be Silvia. Hell, it must be almost two miles. He slammed the door of the van and started running.

As he ran, the image of Silvia took shape. Her long slender legs, her waist-length sweet scented hair, the misty green eyes, her soft lips. He couldn't wait to touch her. By the time the tiny speck had turned into a full-blown figure, he was out of breath. Without thinking, he had been sprinting.

Just before he reached her, he stopped suddenly, turned his back, and put his head down to catch his breath. He couldn't let her see that he was anxious.

He was still in that position when he heard her low voice and felt her hand on his arm.

"Why, John," she said. "I'm surprised to see you. I thought you and Cowboy were fishing."

"I had to come in for some business," he panted, "and I just thought I'd stop by to see you. Jodie said you were walking on the beach, so I came on down."

"I feel honored," she said smiling.

When Silk straightened up and looked deep into her green eyes, he realized she knew everything there was to know about him.

Suddenly his emotions overwhelmed him and he swooped her up in his arms and kissed her passionately on the mouth. For an instant she melted against him and was still, but then her light laughter brought him back to reality.

Not knowing what else to do, Silk set her firmly back down on the sand and laughed himself.

"John," she said. "You are certainly impulsive this morning."

"I don't care," he said. "I wanted to touch you."

Silvia looked long and questionably at Silk as she took his big calloused hand as they strolled up the beach.

'Here I am thirty-five years old and never really been in love before,' Silk thought. 'Suddenly, it's hit me hard, and there's nothing I can do about it.'

Now that he had violated his own principles, the best he could do was to feign lightheartedness, to pretend that her holding his hand meant nothing to him.

"You look devastatingly lovely this morning." he said.

"Must be the morning dew," Silvia said. "And you as usual, John, look like a ragtag, but you smell divine. Are you certain that you take any food with you on your fishing trips? You look like you haven't touched a bite in five days. I worry about that."

"You can stop your worrying about food, Silvia," Silk said. "I eat like a horse."

"I was surprised to see you," she said.

"We came in early," Silk said. "I have an idea I want to talk over with the fishermen. I'll see what they think about it, then if they feel the plan is sound, I'll tell you the whole story. Can I give you a ride back to Jodie's?"

"Please," Silvia said. "Although it's beautiful, I have no more interest in walking on the beach."

Silk dropped Silvia off at Jodie's front door, checked his watch, and, having time, drove directly to the Albatross.

Of course the fishing community was aroused by Silk's idea. Here was a new project in which they could help control their destiny. The date was set for Labor Day because that vacation was the last big tourist inundation of Port Bane before the school season started and the winter rains began in earnest. An advertising committee was appointed with Hattie as chairman, and a 'save the fishermen campaign' was soon in effect. Jodie, who knew every woman in Port Bane, enlisted the aid of the women's pool teams, the women's church organizations, and the girl friends and wives of the fishermen. They divided these women into working groups, and in five days they had designed and printed leaflets and posters advertising free salmon. They contacted the organizations across the state who cared for the poor: Loaves and Fishes, the Salvation Army, the Elks, the Eagles, the Moose, and the Rotaries. They sent news bulletins to every newspaper and television station in the state; and last, they telephoned the governor, individual members of the state legislature, and Oregon's senators and representatives in Washington D.C.

By the time Labor Day rolled around, virtually every person in the state knew of the salmon giveaway in Port Bane. Unsure of exactly what this giveaway meant, NMFS, the Department of Fish and Wildlife, the canneries, the local police force and the United States Coast Guard were notified to watch out for any illegal activities. People in the know thought the Port Bane fishermen were out of their minds. Why would they give away free salmon? Those fish were their only source of income?

As luck would have it, the coast guard did not board a single troller the last two days before the giveaway, and three hundred boats came into port with an average of two-hundred illegal coho per boat. At midnight the unloading of the boats began. By dawn there were boxes, sacks, and tubs of the illegal fish stacked along the dock to be given away. The beauty of the plan was that only the bay front people and the fishermen knew what was in those containers.

At six A.M. the first flatlanders arrived to receive their free salmon, and it was noon before a game warden from the Fish and Wildlife Department admitted to his superiors that the fishermen were really giving away illegal coho.

By that time it was too late. Approximately 60,000 fish, give or take a few, had been given away, and of course the police department was not going to arrest everyone. The NMFS was called. The governor, and the state legislature were called. And it was decided by the bureaucrats to take a lesson from the Port Bane fishermen, to let silence do its work. They knew human nature. Out of fear everyone who received a free illegal salmon would keep his mouth shut. The fishermen would not try such a tactic twice. They had to make a living.

Instead of the earth-shaking act the fishermen wanted, statements were issued to the media explaining why the fishermen gave away their salmon. The statement read as follows: "The Port Bane Fish Feed was sponsored by the Port Bane Chamber of Commerce to attract more tourists across the state to visit the fishing community. The fishermen, desiring to enhance their market, have been willing to participate in this feed in order to advertise their product." The fact that the salmon given away were illegal was never mentioned.

The whole project was declared a 'fiasco' because only sixty thousand fish had been given away, and there were at least two-hundred thousand people in Port Bane over the Labor Day weekend. Another let down for the fishermen was the kind of people who received the fish. They went, for the most part, to people who could afford them, the upper and middle classes who had money enough to buy salmon anyway, and still afford the travel to Port Bane for a week-end vacation. The fishermen had never dreamed the poor people would be left out. Neither could they believe that not one state sena-tor, the governor, nor the head of any important organization attended the give-away.

The fishing season for the Port Bane fishermen dwindled to an innocuous end. Not a skipper or a deck hand in the fleet was mak-ing enough money to live. If the big Chinook were out in the ocean, they were not being caught by the Port Bane salmon trollers. The shortened coho season had dashed their hopes. Now the fishermen had a long cold winter to face without enough money to support their families or to maintain and repair their boats for the next year. The gloomy atmosphere pervaded the bay front, and each man reacted in his own way.

When the beer source for the Locals ran out, something had to be done. Rat and Bent applied for minimum wage jobs at the shrimp plant and got them. Standing in a line with women picking shrimp for $3.35 an hour was better than going without beer. Graveyard, however, could not lower his pride to such an extent. He put his last money in the nickel machine one night, then he walked up to the Rip Tide to pick a fight. He selected to insult a big tourist who looked like a heavy-weight wrestler. The two men got it on out in the street. Graveyard knew he could have whupped the hell out of the big bas-tard, but instead he let himself get beat into unconsciousness. Now that he looked and hurt so bad, no one could expect him to work.

Esteban was working at the port dock salvage yard disman-tling diesel engines from confiscated trollers. Working on boats was kind of like repairing old airplanes but not as difficult. As yet, he didn't know what a bad fishing season meant, but he did know that the people he knew were unhappy.

"Sheet," he said to an engine one day. "Sheet on you, you cabron. All my compadres ees sad, so I make for to be sad too. I quit heem."

Doc Ox quit pounding copper pennies into nails and began drinking heavily. Semyonov started having trouble with his writing. Without the tourists Claude's business seemed especially slow. One day he jumped off the dock and sat on the bottom of the bay until the air in his tanks ran out. When someone asked him why he wasn't paying attention to the time, his answer was simple.

"I was thinking."

On the day before the season's end relief came. Silk and Cowboy had been out for five days and had caught only five Chinook salmon. They were sitting disgusted in the cabin of the CRACKANOON II when a voice came over the radio.

"To everyone in the fishing fleet," the voice said. "You are all invited to a Costume party at Jodie's."

"Sounds good to me," said Cowboy sitting on the hatch with his head between his hands. "I'm sick of this here fishin'."

"We'll go in," said Silk. "It's the last day of the season anyway. We may as well get a jump on that party."

PART IV

FINIS

CHAPTER

29

COSTUME PARTY

The Costume party was big news on the bay front, and virtually everybody planned to attend. Like their work these people went all out for every occasion whether it was a birthday, a wedding anniversary, a death, Thanksgiving, Christmas, or New Years. If there were a party, they came, and a party at the end of this fishing season was a welcome relief to the fishermen in Port Bane.

Jodie's house was one of the few structures besides the Albatross that was large enough for such an occasion. And, before it was dark on the party eve, the guests began to arrive in a costume that he or she thought was appropriate for his or her character. Virtually no one had money to rent or buy expensive clothes, but they were dressed all the same. Rummage sales, second-hand stores, and free church clothes, imaginably selected over the years, were the order of the evening.

Jodie stood at the front door as the perfect Parisian 'femme fatal' in a 1930's full-length, black, sequined gown complete with a boa and a monstrous black hat covered with ostrich feathers.

The Locals, of course, arrived first with gallon jugs of red Cribari wine. Rat was dressed in an old tux too short for him so that he looked like Fagin from OLIVER TWIST. The costume was smelly and full of moth holes with the shirt sleeves two inches too long and his grandfather, high-topped shoes two sizes too small.

Bent came as a nuclear explosion. He was carrying an old World War II bombshell he had dug up somewhere, and he wore a string of firecrackers around his neck. A balloon in the shape of a mushroom cloud was fixed to the back of his collar with a clothes hanger.

Graveyard came as Santa Claus. He wore fishing boots over long red-flannel underwear, and his beard was dyed bright orange. When someone said that Santa had a white beard, Graveyard asserted that Santa had never before tried Purex bleach on his beard.

And so they came, Semyonov as a northwoods logger in logging pants, flannel shirt, boots and axe; Sue Ellen as a lady of the evening with heavy makeup and a sheath dress too tight to cover her bulk; Lillian as Marilyn Monroe in a blond wig; Silk as Red Beard the pirate with a patch over one eye; Silvia on his arm, dressed as an Indian maiden in buckskin dress and leggings with an eagle feather protruding from the leather band about her braided hair; Doc Ox as Frankenstein with artificial scars and metal plugs glued on to his neck; Claude, in an ancient deep sea diving outfit, with a screw on helmet and weights so heavy he could scarcely walk; Esteban as a Mexican Don Juan in beige female tights and matching buckskin shirt; and Cowboy dressed in a general's World War I helmet, black uniform, jack boots, and long gray trench coat.

The party was several hours old, and everyone was in pretty fair shape, when there was a knock at the door, and the final guest arrived. Cowboy opened the door and squinted for a moment at a big handsome man dressed in Gucci loafers, Harris tweed sports coat, matching tan slacks, and coat and tie. In his inebriated state he didn't recognize Buck Tanner, and of course, Tanner didn't recognize him in the flickering light of the fire.

"Madre di Dios," said Esteban peering around Cowboy's back. "Es el narco y...we're busted."

"I'm not a narc," said Tanner mildly. "I'm Buck Tanner, and I'm looking for John Silk. Is he here? Tell him Buck Tanner wants to speak with him."

"Well I'll be damned!" roared Cowboy. "If it ain't Tanner. Come in! Come in! Hell, I thought you was gone for good. Hey everbody! Guess who's here? It's Tanner."

Tanner became aware that once he was recognized the people in the room had ceased trying to hide the trays of marijuana and dried mushrooms under the tables and in the dark corners. Silk came up followed by Silvia and they warmly shook his hand.

"What are you doing back here, Tanner?" Silk asked. "We thought we'd never see you again."

"Once the crocodiles get you, you stay got," said Tanner. "And believe me I got got. Just like you told me, Silk. After your material was stolen, I kind of ended up on everybody's list of undesirables in Washington. My wife divorced me, and I was thrown out of the Justice Department. I didn't want to go back to Texas so I came on to Port Bane. You people don't need a good lawyer do you?"

Silk smiled and threw his arm around Tanner's shoulders.

"Hell yes, we need a good lawyer," Silk said. "We'll likely need several lawyers before this thing is over. Come on in. Let's get you a drink."

The party this year went like most parties in Port Bane. There were three incidents which made the occasion memorable. At about midnight Graveyard, who was drinking wine from a coffee cup, suddenly stiffened at the top of the stairs and tumbled like a bouncing ball to the bottom where he lay in a heap. Doc Ox felt his pulse, announced him still breathing, and supervised the two fishermen who carried him to a dark corner so he could sleep it off.

At 2:00 A.M. Cowboy began to take off his clothes; first the helmet and topcoat, then the uniform, then the long flannel underwear. At last he was standing alone in a circle of costumed people clothed only in his boxer shorts.

"Go ahead, Cowboy!" shouted Sue Ellen. "Take it off."

Cowboy stood entranced in his glory for only a moment, then he jerked down his shorts. Over his crotch was a black cardboard swastika.

"This fuckin' party is gettin' too fuckin' indecent for my kids," Wilks was heard to say. "We're goin' home."

After the laughter had subsided, the guests with children and families went home, and the hard-core individuals of the community gathered about the fireplace in a semi-circle.

About 3:00 A.M Claude staggered back in with a double barrel shotgun and fired both rounds into the ceiling. He didn't know why. He just felt like it. No one got very excited.

Someone turned off the lights, and Silk decided to split more logs in the flickering light like a Neanderthal man.

After the log splitting, Silk sat on the floor between Tanner and Silvia while they all listened to the silence. Except for the popping and crackling of the fire trying to get a grip on the damp wood, no one said a word. Finally Tanner, the newcomer, broke the silence.

"Silk," Tanner said. "I'm sorry I failed you in Washington D.C., but I guess this isn't the first time you folks have watched the bureaucrats win. How did the rest of the fishing season go?"

"Not too well," said Silk. "I doubt there will be two-hundred boats left to start next year's salmon season, If there's a season at all."

"I'm sorry as hell," said Tanner.

"It's not your fault, Tanner," said Silk. "It's SEABOOM, AQUAPROTEIN, and JOINT VENTURE. If those three culprits would simply disappear, we could have some control over our own destiny again."

"Talk, talk, talk," said Cowboy. "But we never do nothin' about it."

"Well, what else can we do, Cowboy?" asked Silk. "They can buy all the lobbyists they want to insure that the government passes laws to protect them. Who's going to listen to anything a commercial fisherman has to say?"

"Who said anythin' about talkin' anymore?" Cowboy retorted. "We've already wasted enough words. Let's just get some dynamite and blow them salmon pens to hell and be done with it."

"Cowboy, you're crazy," said Claude. "I work with explosives all the time, and I can tell you that blowing up a pen with those nets hanging down is more trouble than it's worth. You'd just get caught and go to jail, and even if you did manage to set off an explosion, I doubt that it would destroy more than one pen."

"I know how to get rid of them there pens," said a voice from the dark corner, and Graveyard came crawling toward the fireplace on all fours. "And you don't need no dynamite to do it neither."

"Oh yeah, Graveyard," laughed Bent. "Just how do you think you could kill off sixty-thousand coho?"

"Easy," said the big man, his orange beard glinting in the firelight. "I'd make them coho sick so they'd die."

"Sick?" asked Silk. "You mean poison them."

"Nope," said Graveyard. "When I was livin' up there on that Tauqulin Indian Reservation in Washington State, I got to know them Indians real good. They had a fish hatchery up to the head of the Tauqulin River where they raised sockeye salmon. While I was up there, all them salmon got a kidney disease, and they all died like flies. It's called BKD or IHN, or some other kind of initials, but they can't hardly cure it. Anyways, it killed all them sockeyes in the hatchery, and that hatchery ain't never opened since."

"But what does that have to do with the pens here?" asked

Silk, curious now.

"Ever'thin'," said Graveyard. "Them sockeye salmon in that there river's still got that disease, all of 'em. And if a sockeye salmon gets with a coho, then them cohos gets the disease and dies. Now, what's in them pens out there if it ain't cohos?

If I was goin' to do somethin' like killin' coho, I'd just go up on that there reservation, find me a Indian, and buy me some sockeye salmon. I'd bring 'em back, cut out the guts, the liver, and the heart, then go throw them guts, and liver, and heart in them pens. Them pen-reared coho's'll eat them guts and catch the disease. Every damned one of them'll die, and there ain't nothin' nobody can do about it. All you got to do is find you a Indian with a sockeye for sale."

Everyone burst out laughing at Graveyard's story, then Silk said seriously,

"That idea would never work, Graveyard. The owners of the pen would simply buy new eggs, hatch them, and restock the pens. You'd only hold them up for a time."

"You're wrong there, Silk," said Graveyard. "That disease them sockeyes has got is one bad-assed disease. Once them cohos get sick, they'd never be able to use them pens again. The germs would be all over everthin'."

"That's quite an idea, Graveyard," said Silk. "But that would still leave AQUAPROTEIN and their ocean ranch."

"That one would be easy, Silk," said Cowboy, getting into the imaginative game now. "I could fix that ocean ranch with no problem. What does it take to keep that thing operatin'?"

"A lot of things," said someone. "Money, people, machinery...."

"You're part right," said Cowboy. "But the main ingredient they got to have is electricity. It takes electricity to run them conveyor belts, them heat pumps, them incubators, them refrigerators, and whatever else they got over there. Cut off their electricity for twenty-four hours, and all them fish are dead."

"And how do you propose to cut off the electricity?" asked Silk.

"Easy as pie," said Cowboy. "Blow the main transformers. With no electricity juice them people are dead in the water."

"I don't know about using dynamite on that transformer sta-

318

tion, Cowboy," said Silk. "You know those transformers are sitting in a base of PCB to cool them off. That PCB oil is very highly toxic and exceedingly cancerous to those around it. If we blow those transformers, we're going to scatter that oil two hundred yards in every direction. The stuff's as dangerous as radiation, and those chemicals might hurt someone. They'd have to take up a foot or two of soil or concrete wherever a drop of PCB lands before they could even start to repair the transformer. I just feel the stuff is too toxic to take a chance on."

"You're the expert, Silk," said Cowboy. "It was just a idea, that's all."

"Hey, wait a minute," Bent said. "I used to work for a power company, and you don't have to use dynamite to blow out a transformer station. There's a lot better way."

"What are you saying, Bent," said Silk.

"All a transformer really is," said Bent, "is a device that transfers electric energy from one alternating current to one or more other circuits. In a step down transformer, like the one you're talking about, the device lowers the voltage that comes in so that power can be shot to other lines without burning them out. Now what is the metal that conducts electricity the best? Why, aluminum! All you have to do is to make some spears out of aluminum conduit and throw them onto the transformer lines. I tell you there'll be some fireworks. That aluminum will weld those wires and transformers together and short them out so that they can't ever be fixed. The beauty of the scheme is that aluminum is cheap and easy to come by, and aluminum won't bother that PCB at all."

"You mean like an aluminum javelin?" asked Tanner.

"Exactly," said Bent. "You just throw an aluminum javelin over the fence into the wires of those transformers, and the electricity will do the rest."

"But I know for a fact that they have several big back-up diesel generators for their auxiliary power," Silk said.

"I feex thees diesel engine for you," said Esteban. Like many Latin Americans, Esteban talked with his hands. He was fully animated now that the discussion had finally come around to a subject he understood. "I make for to pour water into thees engines. They try to start. Theen she's too late. The engines she ees gone."

"Esteban's right," said Graveyard. "Water'll destroy a diesel

319

engine. They'd have to buy new engines or rebuild ever' one of them. Yes sir, water'd put them out of business for a year."

Tanner was amazed. These people knew about mechanical processes he had never imagined.

"Well that's two out of three, Silk," said Cowboy jokingly. "But how're you goin' to stop foreign fishin'? There's a whole fleet of them Russian ships out there, and a lot more joint venture foreigners." Silk thought for a moment before he answered.

"I don't really see how we could stop whole fleets of ships without a navy," he said. "What we're talking about here is making a gesture to voice our dissatisfaction with the fishing conditions, of taking action that will get someone's attention. What we need to do is to call attention to the MURMANSK because we know she ran down Cap and Trevor. I'd like to get that nine-hundred foot monster off the ocean."

"Blow her up!" said Rat who had been biting his mustache in preparation for saying something.

"That's easier said than done," said Silk. "Besides, I liked the Russian fishermen. They're just out there trying to make a living like we are. And I'm sure Semyonov, here, has friends aboard. No, I don't think blowing her up is the answer. Right, Semyonov!" and he clapped the pensive man on the shoulder.

"No, taking life is not the answer," said Semyonov solemnly.

"Well, if you want to blow her up, I can do it," said Claude with enthusiasm. "I was an underwater demolitionist in the navy, and I can sink anything that floats. I've never thought of a project like this before, but I have access to a submarine, and she could do the job. She's a three man craft, a lock-out submersible, so she can hover under the water while a diver scoots out the hatch in the bottom and sets the charges. With a sub like that, blowing her would be easy, but I wouldn't like to kill anybody, and I don't speak Russian, so I couldn't tell them to abandon ship."

"I do not know whether or not you people are serious," Semyonov glanced at Silk. "But if you decide to sink the MUR-MANSK, I can help. If I'm given a good radio and enough time, I can get every man and woman off the ship. If I speak with them, they will obey my commands without question.

I find it both ironic and fitting that I should help sink my own ship as a protest for her transgressions." For a moment after the captain had finished speaking, there was a long silence in the room. Bodies shuffled uneasily. What had started out as a game with everyone using his imagination, had suddenly become serious. The fact that Semyonov would actually help blow up his own ship had a sobering effect. The ugly finger of reality had buried itself in the minds of the people present. Each person in the room was now thinking about how three such drastic steps could actually be pulled off.

"What do you think, Tanner?" asked Silk to the now unemployed Justice Department investigator. "Can you think of anything we could do that we haven't already thought of?"

"No, I can't," said Tanner. "But I can tell you this, if you actually destroy any one of these installations, you're getting yourselves in big trouble. Each action would be a federal offence, and you would have not only the local police after you, but the FBI and the CIA as well. If you were caught, you would spend a long time in prison."

"I don't give a damn about goin' to no prison," said Graveyard through an alcoholic haze. "Nothin's goin' to make no difference anyways. As she stands now, I don't have a ghost's chance in hell. If no one else is willin' to do somethin', well, I will. I'll go up on the Indian reservation and get me some sockeye guts and fix that SEABOOM once and for all."

"And Bent and I will go with you," said Rat. "We need a trip anyway."

"I maybe go feex thees compressors by myself, then I steal a plane and fly away," Esteban said.

"Wait a minute," said Silk. "Let's don't go off half-cocked. If you men are serious about this kind of involvement, I think you are going about the problem all wrong. I don't think two or three people taking action on their own is the way to go. We're kind of all in this mess together."

"I got a idea, Silk," said Cowboy. "Why don't we take out SEABOOM, AQUAPROTEIN, and that there Russian ship MURMANSK all at the same time. We'd really make a impact that way. I already sold my boat and I'm single and got nothin' to lose. I'm willin' to fight this thing anyway I can."

"You're wild, Cowboy," said Silk. "But you are making some sense."

"This whole fishing scam has destroyed my family and my future," said Tanner. "I don't believe in breaking the law, but if break-

321

ing the law is the only way left to call attention to this problem, count me in."

"I shall help," said Semyonov.

"I'm going to lose the CRACKANOON II anyway, if I try to sit out another fishing season," said Silk. "So, I'm free amd clear."

"Are we really planning to do this thing?" asked Claude.

"Sure looks like it," said Silk. "But we have to plan carefully."

"Damn right," said Graveyard. "Me and Rat and Bent has got the pens."

"And I got thees compressor," chimed in Esteban.

"And I have the supplies to take care of the transformers, the submarine, the dynamite and the place to work," said Claude. "I'll just hang a 'closed' sign on the door, and nobody will bother us."

"What about the PBFA?" asked Cowboy.

"No," said Silk. "The fewer people who know about this operation the better. We want the organization there to keep up the struggle after we're gone. We're just going to make a statement, and we're going to have to be damned careful. It's too late tonight to get anything accomplished, so everyone who wants to be involved meet at Claude's place the first thing in the morning, and for all our sakes, don't talk to anyone."

"There ain't no need for me, Bent and Rat to go to Claude's," said Graveyard. "What do you say, Silk, if the three of us head on up to Tauqulin so's we can make sure of gettin' some sockeye?"

"How are you going to get there?" asked Silk. "That's quite a distance."

"You let us worry 'bout that, Silk," said Graveyard. "We'll make it."

"O.K.," said Silk. "I'd let you have my van, but I know we're going to need the damned thing. Here's two-hundred dollars. You're going to need some money."

The party broke up then, and about a dozen very serious people went to their beds with a new responsibility lying on their minds. They were no longer merely fishermen or bay front residents. They were about to conspire against the democratic powers that were destroying a way of life.

Silk asked Silvia if she would drive down to the Port Dock 3 with him for a few minutes. Silvia nodded "yes", and they drove the

322

entire distance in complete silence.

When Silk parked the van overlooking the docks and Bane Bay, Silvia sat studying Silk for a long moment without speaking. She took the lighted cigarette he handed her and puffed at it.

"You're really going to do this thing, aren't you?" she said.

"I don't see there is a choice, Silvia," Silk said. "I can't turn my back and forget about everything that's happened, about losing Cap and Treavor, about what's happened to fishing, about the future of their lives.

What would Cowboy think, or Semyonov for that matter, if I suddenly quit and left them still fighting for this cause? You see, Silvia, they need me."

"You are right," said Silvia. "Without you, they wouldn't do a thing, and there this nonsense would end."

"But I don't feel taking action is nonsense," said Silk. "You've been here long enough to see what's happening. If the fishermen are ever going to make a statement, now is the time. We have to take steps now, or commercial fishing is a profession in the past. Can't you understand, Silvia?"

"Yes, I understand, John," said Silvia. "You are planning to sacrifice your future, and maybe your life, for some absurd gesture nobody in his right mind is going to fathom anyhow. I've never heard of such things as killing salmon, throwing spears at transformers, and blowing up ships. I've only read about them. Those are the things fiction is made of. Do you know who blows up ships, John? Terrorists blow up ships. Won't you be branded a terrorist?"

Silk forced himself to stay calm.

"I want to catch salmon, to sell them, and to make a living. I want to live the simple life of a commercial fisherman."

"That's what I fail to understand," said Silvia. "Why the commercial fishermen and the pen-rearing and ocean-ranching owners can't fish side by side. Why does it have to be one way or the other? Aren't you all catching the same kind of fish?"

"Yes, that's exactly what we're doing," said Silk. "The problem is that there are two sets of laws for the same business. Number one, the commerical fishermen cannot keep the salmon they catch unless they measure a certain length. That length is determined by the Fish and Wildlife Department at any given time. Any length shorter, the commercial fisherman must throw back into the ocean, dead or alive.

Number two, the commercial fishermen can only fish six months each year, and at times, only keep coho during designated weeks. This rule is usually general: if a fisherman catches two Chinook salmon he can keep one coho. On the other hand, the pen-rearing and ocean-ranching corporations can legally sell salmon of any length at any time of the year. Now, what is fair about two sets of rules?

Why do you find pan-sized coho salmon for sale in the markets out of season? No matter how the package is marked by the grocery stores, these fish are always coho raised in pens. They aren't caught by commercial fishermen because there is no season for them at this time of year. Some of these coho are no bigger than trout. Now, why can these salmon companies legally sell baby coho and we cannot? Tell me that! Where is the justice in a system that makes people in the same business play by different rules?

In reality, this struggle is about the future. Your future, my future, everyone's future. Even Tanner and the Port Bane people aren't aware of what they're fighting against, but I know and Semyonov knows. The people in power don't want grocery shoppers to hear about where these salmon come from. Why do you think these guys are so bent on getting the commercial fishermen off the ocean as quietly as possible? We're the only group who might stand in the way of their plans."

"What in the world are you talking about, John?" asked Silvia. "What don't I see and can't understand?"

"You don't understand the real struggle," said Silk. "You don't understand what's happening here on the west coast, and all over the world. Behind this little fracas here in Port Bane, there is a monstrous plan to eventually control the world.

For lack of a better term, we'll just call this plot by government and moneyed sources PROTEIN CONTROL. If you think that the United States sponsoring arms sales and drug trafficking is bad, imagine what it would be like if one government controlled all the food in the world? That's what these people like SEABOOM and AQUAPROTEIN have in mind. They aren't stupid. In fact, they're some of the brightest people around. They know the world's resources are rapidly diminishing, and they know that the ultimate struggle will be over food to keep the world's populations alive. This is all down the road, but you can bet your bottom dollar they've figured out the world's needs on their computers.

They know that eventually, he who controls the world's protein, will rule the world. Where is the greatest source of protein located? In the ocean. Dope doesn't make the world go around. Food does. What I'm trying to explain, Silvia...these people are already in on the ground floor. They're striving every day to control the protein this country produces as well as the protein which every other country produces. No, this endeavor is not about a way of life. This struggle is over food. Over this issue someone stole Tanner's material and ran him out of Washington. These people will do anything to make sure their plans and their investments aren't thwarted.

We have proved that we in the fleet can't even make a statement important enough for anyone in power to pay attention to. Therefore, we must take action now. Someone must make the public aware of what's really going on. Maybe, then, the democratic processes in this country can start working."

"But this issue is not what you talked about this evening," said Silvia. "You talked about fishing seasons."

"I tell you, Silvia," said Silk. "I'm probably one of the few people in the world who is on to this ultimate scam. We're not going to destroy ourselves in a nuclear holocast, and I doubt the holes in the ozone layer will kill us before we starve to death, or that AIDs will become so severe an epidemic that everyone dies. No, our resources are going to go first, and we're going to starve to death. We're going to find ourselves down on our knees begging the protein controllers for enough food to keep ourselves alive. Now you know why I never talk out loud about this subject. People would think I was crazy and put me in an asylum. Most people, including the fishermen, are simple. They don't want to think about such things so they rationalize the truth away. Now I've had my say, Silvia. Do you still think I'm crazy for wanting to take action against these bastards while there's still time? What I'm fighting for is the future of the individual,everytime. Who is going to be alive when this Armaggedon takes place."

"I don't know," admitted Silvia trying to understand the depth of the problem. "There's no way that I can encompass such an idea all at once. I just know you're getting ready to throw your life away for a principle that only you understand. If you carry through with

325

what you discussed tonight, you're going to be a fugitive for life."

"I know that," said Silk. "But if anyone out there listens, the sacrifice will be worth it. I never wanted to do anything like this, but now I have no choice."

There was silence in the van while both Silk and Silvia thought about the questions Silk was struggling with. Silvia had thought that Silk was simply a die-hard fisherman who loved the ocean, but the depth of him had erased that image. At heart he was a romantic and an idealist. At this moment she was simply overwhelmed.

"I really don't know what I feel anymore," Silvia said.

"What bothers me about the whole thing," Silk said, "is you and me. At first I thought I could win you with time, with giving you ample opportunity to get over your losses. Now this terrible thing has come up, and there is no time for us. There is no future together."

"Did you ever think life would come about like this?" asked Silvia.

"All I know about life right now is that I love you more than anything in this world," Silk said. "The wildest thoughts keep whirling through my head. I can even imagine us together throughout eternity. Something like Lillian talks about sometimes. She calls it soul mates, or destiny. Hell, I don't know what terms she uses. It's as if we've been together before, and I've known you for a long long time." Silk kept looking straight ahead as he talked, and his words were now a mere whisper.

"I already know how it would be to love you. How you would feel in my arms. And it would be that way, I know it would. Sometimes, I start shaking, and I think I'll die if I can't love you.

You'll go back to your life in San Francisco, and I'll go to wherever I end up. I regret, darling, that this world and I are screwed up like we are. In another age, in another time, we might have made a life together. Now there's no way out. Do you realize, Silvia, that this is probably the last time we're going to see one another? Why do you think I drove you here? I knew I had to tell you goodby."

"But how shall you disappear? Where shall you go?" asked a perplexed Silvia.

"We've not decided on a destination, but we will be needing the CRACKANOON II for the escape," Silk said. "Cowboy told me tonight he would furnish the money for her purchase, at whatever price you named."

"The CRACKANOON II is not for sale, John," Silvia said. "She's yours. She would not be at home with me, and I just as well take some part in all this nonsense too. I don't want the CRACK-ANOON II or the money. I don't want to be haunted by whatever may happen to you."

Silk put his arms around the woman he loved and felt her shivering. At last he released her and started the van.

"I guess this is it, darling," he said.

CHAPTER

30

SOCKEYE

While Silk and Silvia were parked at Port Dock 3, wrestling with the fishing problems and the emotional trauma of saying goodby, Graveyard, Bent and Rat were sitting beneath the docks, tipping up some Cribari wine and discussing how they could con Hattie out of her 1948 International pick-up for their trip to the Indian reservation. The discussion was intense because Hattie was such an unpredictable woman. But according to Graveyard, she could be had.

"The way I see it," Graveyard said wiping the wine off his mouth with one big paw, "There's just one way we got of handling this thing. We got to spread Silk's name. Hattie'd do anything for Silk. We got to show her if she borrows us the pick-up, she'll be helpin' Silk out. She don't need to know nothin' else 'til we're through."

"I know one thing," said Bent. "We better ask her before she gets busy. She looked a hole through me the other day when I tried to ask her for some salt to shake in my beer."

"We better start walkin' then," said Rat. "By the time we get there the Albatross ought to be open."

"Mull things over in your minds while we're walkin'," said Graveyard. "That way it won't seem new to you when we say somethin'. If you get any other bright ideas, you can spit it out on the way."

Hattie had just unlocked the Albatross door when Graveyard, Bent, and Rat walked in. They ordered three drafts and handed her a fifty dollar bill. Of course, Hattie became suspicious immediately. She knew that none of the three men had worked steadily during the fishing season, and they hadn't spent a cent of their own money in her tavern for three months. Graveyard, who was as sensitive as a cat, picked up on Hattie's thoughts. He decided to take the bull by the horns.

"Hattie," he said confidentially. "Silk sent us up here to ask a favor from you. He has this special project that needs workin' on, and he hired us to do the job. The problem is, we need somethin' to haul in. Silk's goin' to be usin' his van so we thought that maybe you'd loan us your old International pick-up truck you got parked out there in them blackberry bushes alongside your house."

Silk's name was magic to Hattie, but she still didn't trust Graveyard. When he wanted something, he usually went about it the long way around. Then, when he finally ended up with the object of his desires, the unlucky person who got involved with him came out on the short end of the stick. In his own way Graveyard was a con-artist, and you don't make money by making deals with such people.

"That old pick-up is completely out-of-whack and you know it, Graveyard," said Hattie hotly. "I've just been waiting for Pat to have it hauled off to the wrecking yard. The tires are flat, and the engine won't start. We decided not to bother fixing it up."

"But, Hattie, you can't lose on this here deal," he continued enthusiastically. "If you'll loan us the pick-up, we'll fix her so's she'll run, and we'll get some tires, too. You could get a hell of a price out of her if she's fixed up a little. Shit, I'll bet as she stands now, that old junkyard'd only give you twenty-five bucks for the thing. See, Silk give us the money to pay for the parts." Graveyard took out another fifty dollar bill and slapped it on the counter. He knew that, with Hattie, money talked.

"What is this special project that Silk has you men working on?" Hattie asked. Silk did a lot of business in her place, and the last thing she wanted to do was lose his money.

"For a couple of days we'll be hauling some stuff for 'em," Graveyard said. "Can't say no more than that."

"Well, alright," Hattie said at last. "I'll let you use the pick-up provided you have it running when you return it, and Silk will take full responsibility for you."

"We'll take full responsibility for ourselfs," Graveyard pointed his thumbs toward his chest. "Take it or leave it."

Early the next morning Graveyard, Bent, and Rat stopped at the Albatross for a hot cup of coffee and the key to the pick-up truck. Then they walked over to the McDougal house where the old relic

329

was parked. Rat carried Silk's toolbox on his right shoulder and a lug wrench in his left hand. Graveyard, who was in charge of the expedition, carried a gallon of wine under each arm. This was to give them sustenance and inspiration. Bent packed a gunny sack for the sockeye they would bring back, and a paper bag filled with a coffee pot, three tin cups, MJB coffee, a five pound loaf of welfare cheese, some paper plates and a skillet they could cook in if they stumbled across anything to cook. They planned to live off the fat of the land. To the three Locals the old International pick-up truck was a stroke of luck they hadn't counted on. Most people would have shuddered at the thought of driving the piece of junk around a block, but to them, having a vehicle to ride in for a four hundred mile round trip to an Indian reservation was tandem to riding in a chauffer-driven limousine.

The engine fired on the second turn, and a stream of black smoke enveloped the truck.

"Looks like she burns a little oil," said Graveyard, "but that don't matter none. We'll just buy five gallons of that cheap stuff at the truck stop and feed her a little oil along the way. I want to be at the reservation by dark."

By the time the old truck reached the small village of Tauqulin, it was well after midnight. To the Locals the trip had been uneventful. At a top speed of forty miles per hour they had covered the two hundred miles north on Highway 101 in a little less than thirteen hours. They had changed two tires, wired the muffler on again, and fed the engine three and a half gallons of oil. They had finished off the two jugs of Gallo wine at the three quarters mark, but when they stopped in Astoria to get more gas, they purchased a third gallon as well as two loaves of day old bread, a roll of bologna, a jar of generic peanut butter, and two plugs of Red Dog chewing tobacco.

"Now, we don't want to eat too heavy," Graveyard had warned. "We ain't here to party. We got business to take care of. 'Sides, I'd like to give Silk back part of that two-hundred dollars he give us."

"What about those sockeyes?" Bent asked. "We're not eating any part of those sockeyes are we?"

"I know for sure I ain't eatin' any sick salmon," said Rat.

"Stop your jawin'," Graveyard said. "No. We ain't eatin' none. What we're goin' to do is take the whole fish back. They'll keep better that way. 'Sides, I don't think we'd be too popular if any of them Indians seen us take the guts and liver out and throw the rest of the sockeye away. We got to remember, them Indians' got eyes, and they can track you down without you ever knowin' it."

His mind at ease, Rat went to sleep leaning against the door which was tied on with fishing line. Bent fell over against Rat's shoulder, and for the last hundred miles Graveyard glued his eyes to the black highway illuminated by the International's one dim light. He thought about the meaning of killing all those penned salmon, and the controversy going on in his head kept him awake until they reached what looked like the end of Highway 101.

Graveyard parked the old truck beside the road on the left bank of the Tauqulin River and nudged the two men awake.

"Looks like we're here!" he said. "Anyways it's as far's the highway's goin'. We got to find a place to stretch out some 'til mornin', then we can tell where we're at. Now come on, I'm beat to hell." And they walked a short distance to a soft mossy spot under a towering Sitka Spruce and fell asleep.

By dawn Graveyard was up and surveying the area for Indians. As far as he could tell, there were no Indians inhabiting this section of the river. He turned back toward the big spruce tree where Rat and Bent were still sleeping, and in short order had started a small fire.

For a moment he looked at Rat and Bent, still sleeping while he was working his head off, and he felt like cracking their teeth together. Instead, he picked up two tin cups and banged them together as hard as he could.

"Rise and shine, boys," he roared, beating the tin cups together again. "I've been scoutin' the area for some Indians, and there ain't none. We got to eat this here stuff and get goin'."

"Shit, Graveyard," Rat whined. "What're we goin' to do if we can't find no Indian?"

"We'll find one alright," said Graveyard reassuringly. "We ain't goin' back to Port Bane 'til we do what we set out to do. Now, shut up and eat. We're goin' to drive into Tauqulin and find us a tavern, and you can bet your bottom dollar there'll be a Indian sittin' in there somewheres. You can always find a Indian in a tavern."

There was a tavern in Tauqulin alright, but it was not like the Albatross in Port Bane. From the outside the small building looked like an unpainted cedar house with a sign that read 'beer' painted in white on one of the front windows. There were also old refrigerators, stoves, bottomless chairs, papers, cans, wood, toilets, tires, stripped-down cars, and other cast-off items scattered from the front porch to the back of the house. Graveyard, Bent and Rat opened the front door and eased inside.

An ancient Indian woman, who looked to be about seventy years old beneath her long stringy white hair, was the bartender. She weighed, Graveyard estimated, in the neighborhood of three hundred pounds. Rolls of fat seemed to sag halfway down her cheeks, and they could just barely see the pupils in her black eyes when she looked at them.

"This is the damndest place I've ever been in," Bent whispered to Graveyard as the three men stood uncomfortably near the front door. "Are you sure we're in the right place?"

"We're in the only place," growled Graveyard out of the corner of his mouth. "Take off your hat and sit down." He gestured to the chairs around the table.

As soon as the men were seated, the woman sidled over to the table and looked at them suspiciously, trying to focus through the folds of fat.

"You no Indian," she said in a squeeky voice. "Reservation no want white man."

"I been here before," said Graveyard unabashed. "We ain't Indians, but we got business with Indians."

"What business you got?" asked the old woman.

"We got to find us a Indian that'll sell us a sockeye salmon," said Graveyard.

"He no fish no more," she nodded toward a leather-faced Indian, sound asleep, next to a smouldering fireplace. "Too old."

Then the old woman smiled with her three yellow teeth: one tooth on top and two at either side on the bottom.

"You want drink?"

"Nope," said Graveyard. "But we'll take a gallon of that there Thunderbird to go."

The fat woman smiled again when Graveyard handed her a five dollar bill on the way out. As soon as they were settled in the old truck, Bent started asking questions.

"Where's all the Indians that's supposed to be here? Where's the bucks and squaws and teepees? And what about bows and arrows and buckskins, and feathers?"

"Beats me," said Graveyard. "I guess them old 'uns died out, and them young 'uns must of left. Maybe fishin's kind of gone to hell for them too."

The dirt road ran alongside the Tauqulin River about twenty miles before it ended in two deep dried ruts. The old truck bounced along the ruts and over rocks in the ruts until the right rear spring gave way and broke. Graveyard decided to stop there. To drive further might not be in their best interest. As they stood next to the truck taking turns with the jug of wine, a sense of failure lay heavy on their shoulders.

"I seem to remember them Indians used to fish up this way," said Graveyard. "But we ain't seen one damn Indian, let alone a Indian a fishin'. I'm scared to death to take Hattie's truck any further. Looks like we've struck out boys."

"I sure hate to let Silk and them down," agreed Bent. "But I don't think this truck can make it any further. We'll likely be lucky to get her back home."

"I see some smoke," Rat said pointing a skinny finger upstream. "Look. It's comin' from somewhere around that bend."

Bent squinted his eyes and stared into the distance.

"Rat's right. Looks like it's comin' from down there on the river, about half a mile, I'd say."

"This old truck ought to make it that far," said Graveyard. "Maybe we're goin' a get lucky after all. Hop in boys."

The truck clattered another half mile up the road and they parked it where the river made a wide bend and seemed to run beneath a cliff. There below them was a sandy beach, a small campfire, and a Winnebago Superchief. The occupants were no where in sight.

"That looks like some damned tourist to me," Rat said logically.

"Looks like we struck out again," Bent said. "We just as well go back."

"Nope," said Graveyard. "We just need one sick sockeye, and maybe that tourist's caught one. I know for sure any sockeye in that river is sick. It won't hurt to ask."

Graveyard jumped out of the truck and went charging down the hill. Rat and Bent followed because Graveyard had the jug of wine in his hand.

After a hazardous journey and a few scrapes and bruises from sliding, stumbling, and tumbling down the steep incline, the three men arrived at the Winnebago Superchief. Only sure-footed Bent had arrived unscathed.

"Looks like nobody's around," said Bent looking at the empty campsite.

"There's got to be," said Rat. "See, over there. There's two fishin' poles propped up on holders welded out of a black steel rod."

"Yeah, I see 'em," said Graveyard. "But what I'd like to know is how that tourist got that big motor bastard way down here."

Graveyard was interrupted by a voice behind a rock where the two poles were propped up.

"If you're fishing, this hole is occupied."

Graveyard walked around the rock outcrop and saw a man with a soft felt western hat pulled over his eyes. He was sitting comfortably on a portable canvas tourist chair.

"We ain't fishin', mister," Graveyard said. "We're a lookin'. We don't mean to bother you none, but we're tryin' to find a Indian with some sockeye salmon for sale. We just come down here to ask you if you'd seen any."

"In that case," said the man without looking up, "You've come to the right place. I'm an Indian, and I'll sell you some salmon if the price is right."

"Bullshit!" said Graveyard. "You ain't no Indian. You're a fuckin' tourist up here poachin' on Indian land."

"You're quite wrong, sir," answered the man suddenly pulling up his Stetsun and standing. "My name is Billy Whitefeather, and I belong on this reservation. My father's name was William Graywolf, and my mother was White Eaglewing. I'm what you call an educated Indian. I manage the I.B.M. district headquarters in Seattle."

Graveyard did a doubletake and sized the man up. He was about six foot two, dressed in a Levi's jacket, Levi's, and cowboy boots. He had black hair and black eyes, but his nose was straight and his hair was cut short. He looked like a white man except for a swarthy complexion. Graveyard decided it was best to play along.

"Well, I'll be damned," he said. "I guess you are a Indian at that. This here's Bent, and Rat, and I'm Graveyard. We're up here on a mission to buy a sockeye salmon from a Indian. Looks like the tribe ain't fishin' no more."

"You're correct about the fishing," said Billy Whitefeather. "Since the hatchery closed, the Tauqulin Indians have more or less gone out of the fishing business. I come here because I need a break from my business once in a while."

"Well," said Graveyard. "We won't bother you no more, Mr. Whitefeather. We'll just go on back home. Probably a bad idea anyways. We was just tryin' to help out a friend of ours."

Suddenly, Billy Whitefeather didn't want the men to leave. He had been fishing for three days and hadn't seen a solitary sole. Besides, there was something about the straightforwardness of these three men that was refreshing and intriguing to him.

"If you men have time," he said, carefully measuring his words. "I'd like to offer you a drink before you go. I have a couple of bottles of scotch in the camper that I've been saving for some proper occasion. I'll get it out if you think you can afford the time. Maybe we can talk about your problem. I might be able to help you."

Graveyard looked at Rat and Bent. He looked at the Indian and frowned as if he was considering a serious problem. He scratched his beard and rubbed the back of his neck and studied the beach sand for a long moment. There was one thing Graveyard had learned in his vast experience. When someone offers you a drink, never appear to be too eager to take it.

"We don't usually drank too much when we're a workin'," he said, "but seein' as how it's early, I guess we could have a little snort with you. And if you'll let us buy some of your salmon, why, we'd be ahead of the game right there. Sure, we'll have one with you. A good stiff scotch would help our spirits a lot, wouldn't it men?"

Billy Whitefeather opened the door to the camper and stepped inside. The men could hear him rustling around in the back part of the Winnebago. A moment later he came out with three more canvas chairs which he sat around the tiny fire and two full bottles of Johnny Walker Black Label. He handed each man a glass and poured a double shot, then he sat down in the fourth chair. Graveyard, Bent, and Rat waited politely until Billy Whitefeather had poured his own drink before they saluted each other and downed it together.

For a moment the men smacked their lips savoring the taste. Tears came to Graveyard's eyes.

"That's real sippin' whiskey," he said appreciably in a husky voice. "It's been a long time."

"Then you might share another one with me," said Billy Whitefeather.

"Maybe just one more," said Graveyard. "We still have to drive." Rat and Bent nodded their heads in agreement, and another round was poured.

"Where are you boys from?" asked Billy Whitefeather politely.

"We're from down in Oregon. Port Bane to be exact," answered Rat anxious to be included in the conversation.

"You mean you drove all the way from Port Bane to buy a salmon?" asked the Indian increduously. "I understood that Port Bane was the salmon center for west coast fishing. Why would a salmon from the Tauqulin Indian Reservation be better than one from Port Bane? That doesn't make much sense to me."

Graveyard knew he was trapped, but he remembered his old adage. 'Always tell at least part of the truth'.

"It's kind of hard to explain," he said taking a dainty sip from his whisky glass. "But we're up here on a kind of scientific expedition, kind of helpin' out a friend of ours who's workin' on a project he's got in mind. It's not just any old salmon that he needs; it's one of them sockeyes.

I used to live up here on the reservation, before the Indians kicked all of us white men off, and I was here when all them salmon from the hatchery took sick and died from that there liver and kidney disease called IHN or IPN or some kind of initials like that. My friend says if he could get ahold of one of them sockeye salmon with that disease, he might figure out how to use them in his project. He said it might help him to take care of some other fish he wants to take care of real bad. And he needs the whole salmon, guts and all."

Billy Whitefeather smiled and said,

"I just happen to have a few of those sockeyes on ice. They're freshly caught, and I haven't had time to clean them yet. I guess that's what you're looking for. I'll be glad to donate them to your project, but first I'd like to invite you boys to have lunch with me."

Graveyard had become very serious during his explanation. It took a lot of concentration to tell the truth and omit the truth at the same time.

"Both them offers is appreciated," he said. "We'll take the salmon, but we got to be headin' back to Port Bane for now. Maybe some other time."

It was mid-morning when the old pick-up rolled past the city limits of Port Bane. Something else had happened to the engine about twenty miles out, and now a sound like somebody was banging on a washing machine was coming from beneath the hood. They had just turned the corner on sixth avenue when there was a particularly loud crack, and the engine stopped. Graveyard got out and lifted the hood. He came back a moment later shaking his head.

"There's oil all over hell and a hole in that block I could put my fist in," he said dejectedly. "She must a throwed a rod. We'll have to push her the rest of the way."

The three men pushed the International truck the last half block, carefully guiding it to the same patch of blackberry bushes alongside the McDougal house. They transferred the salmon from the soggy box into the gunny sack Rat had brought and trudged wearily down the road toward the port docks. Rat said,

"Hattie's goin' to be madder then hell at us for tearin' a hole in that engine."

"She'll get over it," said Graveyard. "'Sides, she's got a new battery and a couple a extra tires she can sell, don't she? She's worth more now then when we got her. What matters is, we got the salmon! Silk'll be happier than shit, and that's what counts."

The three men carried their sockeye salmon through Port Bane and on down to Port Dock 3 where the CRACKANOON II rocked gently in slip eleven. Graveyard gave a shout, but there was no answer. They asked around the dock, but no one had seen Silk or Cowboy for a couple of days. They didn't dare freeze the salmon for fear of killing the infectious bacteria, and they didn't want to leave them sitting in the sun either. After some discussion, they decided something had to be done about the sockeye which were already unfit for a human to eat.

"Silk give us a job to do, so we just as well finish her up," Graveyard said. "We'll chop them sockeyes into pieces, guts, livers, and all, then we'll go out to them pens ourselves and see old Gus."

"We can chop them up alright," agreed Rat, "but how're we goin' to get way out there from here? Hell, them pens is halfway across the bay."

"That's the easy part," said Graveyard. "We'll borrow the life raft from off the CRACKANOON II here."

While Bent searched the boat for a five gallon plastic pail to put the salmon pieces in, Graveyard and Rat skillfully cut up the sockeye on the butcher's table at the end of the dock. Then the men unfastened the rubber raft and threw it over the side. When they finally shoved off, they looked like long-suffering sailors who needed to be rescued from the sea. Graveyard sat in the bow with the jug of wine, much like an officer in command of a lifeboat. Rat and the pail of salmon rested astern. And Bent manned the oars located in the center.

A short distance from the Port Dock, a sharp evening wind sprang up, and soon white-capped waves were slapping over the round rubber sides of the raft. At the halfway mark Rat began bailing water with his fishing hat, and by the time the raft nudged alongside the outer pens at SEABOOM, the men were soaked to the skin. Undaunted, they tied the raft to the bracket on the floating dock and climbed onto the heavy plank boardwalk carrying the bucket which held their deadly cargo.

"Alright, listen up," Graveyard whispered as they stood huddled in the darkness. "I'll walk up to the old shack at the end of the dock, and keep Gus busy 'til you're done. I've knowed Gus for years, and I know how to keep him talkin' about old times.

Now, you boys wait till you see me go inside, then you bust ass up and down them walkways. Make damned sure you get some of them salmon guts and liver in ever pen. Soon's you're through, give me a hollar. It's too damn bad I can't tell Gus what we're up to, but he might feel like he ought to try to stop us since he's got a job to hold down. 'Sides, he'll probably get the blame when them cohos start dyin', so it's best he don't know nothin'."

Graveyard knocked loudly on the metal door of the watchman shack, and it was opened immediately.

"Well, I'll be damned," said old Gus, his bald head glistening in the light of the bare electric bulb. "If it ain't Graveyard. What're you doin' way out here this time of night, and soakin' wet too? What'd you do swim out?"

"Nope," said Graveyard, motioning with his right hand behind his back for the boys to start the feeding. "A couple of the boys

338

rowed me out here. We was sittin' in town drinkin', and I thought 'hell, old Gus is out there in the middle of nowhere, tendin' them pens in that bay, without a drop of nothin' to warm his belly. I ought to go out there and give him a little snort. So here I am."

"Well, shit," said Gus. "That's awful nice of you, Graveyard. I sure could do with a good belt. Why don't you tell your friends to come in too, and we'll have a little party. It's late enough that none of them big boys'll come around, at least not 'til mornin'."

"Naw," Graveyard said stepping inside and closing the door. "Them boys have had enough for a while. 'Sides, they're doin' just fine, watchin' them fish. Boy, you sure do have a bunch of 'em suckers and all sizes too."

"'Bout seventy-thousand at last count," said Gus setting out a couple of dirty glasses from a cluttered shelf. Graveyard poured each glass full to the brim. "But they ain't real fish as far as I'm concerned. They're flabby as hell. Them bastards can't get no exercise in them pens, and all they eat is chicken pellets. They ain't like them big coho we used to catch."

"With all them fish out there to take care of, if I was you, I'd ask for a raise," said Graveyard.

"Can't do that," said Gus. "They'd just can me and put on somebody else who'd work cheaper. Now, they pay me five dollars an hour, and that ain't too bad since I can't fish no more with this here hook for a arm. Nope, I can't quit, and I can't ask for no raise. I got to keep a roof over my head."

"You can always come down to the dock and join us Locals," said Graveyard. "We live pretty good."

"Rheumatism's too bad," said Gus. "It's too damp under them docks. Now that I'm gettin' old I got to have my comfort."

A voice sounded through the open window interrupting the conversation.

"Hey, Graveyard. We're through."

"Through what?" asked Gus.

"Lookin' at them fish, I guess," said Graveyard quickly. "Tell you what I'm goin' to do, Gus. I'm goin' to leave the rest of this here jug with you. But don't get too drunk."

"It's been damn nice seein' you again," said old Gus opening the door. "I don't get much company out here. You come back anytime."

339

"Sure will," said Graveyard. "And when you're down at the bay front, stop in at the Albatross sometime. Most of the boys is still down there."

On the way back to Dock 3 and the CRACKANOON II, the men were in high spirits. They didn't mind the waves washing into the raft. They had completed their task, and they were anxious to find Silk and tell him all about it.

"We got some guts and liver in ever pen," said Rat happily. "Them cohos'll start dyin' off tonight, that's for damn sure."

"What did old Gus have to say?" asked Bent.

"Nothin' much," said Graveyard thoughtfully. "I told him I'd sure as hell quit that job if I was him. He don't know it, but he's goin' to have one hell of a mess when them fish starts dyin'. Do you know he's got to lift ever one of them dead fish out of the water with a pole net? There's seventy-thousand of them babies. Looks like he's goin' to need some help. Hell, I might even get me a job out there myself."

CHAPTER

31

ACEY! DEUCEY!

With a background like Claude's, the job of masterminding and synchronizing three impossible projects fell right into his bailiwick. He had the location; he had the equipment; and he had the expertise. By exercising his creativity he could blow out the transformer station on South Beach, sabatage the diesel back-up generators at AQUAPROTEIN, and make explosives powerful enough to blow up a 900 foot Russian ship. That he was an accessory to illegal activities never occured to him. He just loved brain teasers.

Except for Rat, Bent and Graveyard, who were off on their mission to the Indian reservation, the conspirators, Silk, Cowboy, Tanner, Semyonov and Esteban had gathered at Claude's workshop by 9:00 A.M., on the morning following the costume party. Claude had never gone to bed. He had played BEETHOVEN'S FIFTH SYMPHONY all night while he was thinking. He had already drawn a schematic of the operation, and he was explaining the step-by-step chart to those sitting around the pot-bellied, antique stove in his main barracks.

"At first, I couldn't see the complete picture," Claude said. "But when I drew it out on paper this morning, the whole thing became rather simple. I divided the job into four sections, and the Locals are already working on the first. If they can take out the cohos over at SEABOOM, all the rest of us have to worry about is AQUAPROTEIN and the MURMANSK. We can do both of those in one night. I've divided you men into areas of responsibility, so feel free to ask any questions or make any suggestions as I run through the plan.

First on our list are the compressors at AQUAPROTEIN. The men who are going to sabatage the compressors will be the ones who know the most about diesel engines: Cowboy, Esteban, and Rat. I included Rat, Bent and Graveyard in this scheme because I figure they'll be back from the Indian reservation in plenty of time to help us out here.

Cowboy, you will be in charge of this phase of the operation. Your job will be to enter the compound at a specified time, cut the fence, find the building with the diesel back up generators in it, syphon half the diesel from the tanks on the compressors, and fill them with water. Your crew will carry flashlights, wire cutters, and whatever else you think you'll need. You should pack your water in five gallon cans since there's no reason to count on having an easy access to a hose inside the compound.

The second project is the transformers. I know how to make aluminum spears, and I know they'll work if they can be hurled over the steel fence at a decent height so they will hit the power lines and fuse them together. I have plenty of half inch aluminum conduit to make the spears. We'll cut them into eight foot lengths and fill each end with three, 2 oz. lead fishing weights. That way we can insure the spears will land head first. After we place the weights, we'll have to hammer and grind those ends into points. The thing that's got me stumped is the trajectory. We need something on the other end to help balance and guide them in a straight line, kind of like feathers on an Indian arrow. I don't think feathers glued to the other end is the answer. Does anyone have a suggestion for a guide which we can connect to a metal pole?"

"I have some aluminum eagles I won at the state fair last year," Cowboy said. "They're s'posed to go on flag poles, but I didn't have no flags so I just put 'em in a cigar box. What would happen if you put them on the opposite end of the pole? That'd make them eagles be flyin' backwards, wouldn't it? You think that'd work?"

"If they were fitted in backwards?" Claude thought it through for a moment, "Yes, I think their wings would still catch the air. Those eagles just might give us the stabilizers we need. We'll try them."

"The other steps are simple," Claude continued. "Graveyard, Tanner and Bent will be assigned to this project. You men will have to go to the transformer station, measure the height of the fence, and mark out your positions. When you have these measurments, you'll know the proper height to throw your spears onto the transformer lines. The alumium spears will then fuse the lines together and short out the transformers. The repair crews will need at least a week to get her back into operation."

342

"Shall we meet here, Claude?" asked Silk.

"Just think of my place as the place of business," said Claude. "The location is perfect. We're close to the transformers, and we're close to AQUAPROTEIN. We can dock both the submarine and the CRACKANOON II here, and it's just a beeline through the jaws and out to the open ocean.

Silk, Tanner, Cowboy and Semyonov can sleep on the CRACKANOON II, and the rest of you can bunk down here at my place.

The big project is blowing that Russian ship," Claude continued. "I already have the DUNGENESS docked here. She's a three man submarine, and she can get the job done. I have enough C-4 explosive, magnets, timers, and detonators to blow up a forty story building. All we have to do is to build the package we'll attach to her hull.

Semyonov will navagate the submarine while Silk and I set the charges. As soon as we're back aboard the sub, Semyonov can radio the MURMANSK to tell the Russians to abandon ship. I'm not going to blow her unless every person is off that vessel.

Does everyone agree with this plan? Alright then, we'll work like hell for the next few days. I figure blast off time will be November 10th. Today is the 2nd, so that'll give us eight days. Everyone do his job, and we'll take out those bastards in one fell swoop. Any questions?"

"I have a question," said Silk. "Say, your plan works, Claude. What are you going to tell the police and the FBI when they come around?"

"I guess I'll tell them I blew everything to hell and back," Claude laughed.

"That's what I thought you'd say," grinned Silk. "But I disagree with your logic, Claude. You're right that someone will have to take the full responsibility for our actions, but that someone should not be you. You have a business here and a future.

As far as responsibility for the destruction goes, there is really no need for all of us to go to jail. Cowboy, Semyonov, Tanner, and I have been talking it over, and we have decided to be the culprits.

I'll pay off my dock fees and move the CRACKANOON II over here for some repairs. We'll load her down with fuel and supplies for a long trip, and after the transformer and the compressors are destroyed, Cowboy and Tanner will take the CRACKANOON II out to sea and wait for us at a rendevous point.

When the three of us finish blowing up the Russian ship, Semyonov and I will transfer from your submarine to the CRACK-ANOON II. As soon as we're on board, we'll take off for parts unknown. You can bring the submarine back in beneath the water, and later, if you're questioned, you can say you've been working on your treasure ship project.

We'll go out first and set the charges on the sunken liner so it'll look like you've been working there. Knowing our bureaucracy, the police forces will take at least a day to figure out the sabotage was done deliberately. By that time the four of us aboard the CRACKANOON II will be beyond the 200-mile limit and into international waters outside U.S. jurisdiction. We'll give no destination so the people here will be innocent if they're questioned about where we went."

"Wheech way weel you go?" asked Esteban. "Maybe I go weeth you."

"That's not possible, Esteban," said Silk. "I'd like to take you all, but there simply isn't room on that boat for everyone. You must continue your lives here like nothing ever happened. The four of us have nothing to lose."

"Do you have a destination in mind?" asked Claude.

"I don't know where we'll go, Claude," Silk said, "but you'll have to admit we have one hell of a good plan."

"We'll leave it at that then," Claude said, taking charge again before there could be a disagreement about self-sacrifice. "Let's get to work, men. We'll make the spears first. Tomorrow Semyonov, Silk, and I will work out on the DUNGENESS. Tanner, you, Cowboy and Esteban had best run out to the transformers and AQUA-PROTEIN. Make a schematic of the locations and a list of everything you'll need. The next day Silk, Semyonov, and I will take the CRACKANOON II out to the Russian fleet so we can study the probable coordinates of the MURMANSK. After that, we'll take the DUNGENESS out to the sunken liner where I've been working. As Silk said, that job will give me an alibi. Graveyard, Bent and Rat will be back with the sockeye by then, and as soon as they finish infecting the pens, they can work with us here. I'll tow the barge out on the 5th and anchor it over the sunken liner. Since November 10th will be blast off night, Silk, Semyonov, and I will have to leave on the DUNGENESS seven hours early in order to reach our destination and get the detonators set on the MURMANSK. At zero hour we will blow the ship, and rendezvous with the CRACKANOON II on time. The sooner we pull this thing off, the less chance of word getting around. Are there any questions?"

"I think you've covered it, Claude," said Silk. "We can fill in the gaps as we come to them. You did one hell of a job working this project out. The thing is going to work."

"Of course it's going to work," said Claude proudly. "I always know what I can do and what I can't do. And I never set out to do something that can't be done."

Talking about what must be done and doing it were two different things. The physical tasks were much more difficult than first met the eye. Claude found himself spread thin. He not only had to be the teacher, he had to do the work. He showed Tanner, Esteban and Francois how to cut the aluminum conduit on the band saw, insert the lead weights, grind the pole to a point, and indent it with a chisel so the weights wouldn't slide back to the other end. As soon as the spear was completed, he stuck the aluminum American eagle in the opposite end. Contact cement held the piece snug, and the weight of the projectile was perfectly balanced. Now that they understood the manufacturing steps, all the men needed to do was to build nineteen more spears like Claude's prototype.

Tanner and Cowboy, meanwhile, took Silk's van out to the transformer station to survey the size and convenience of the installation. On their return trip they drove back by the AQUAPROTEIN complex, surveyed the fence Cowboy would have to cut through, and located the building where the compressors were kept. They decided the breaking and entering had best be left to the night of the sabotage.

While these two projects were going on, Claude showed Semyonov and Silk their functions in the submarine. Three days wasn't all the time in the world to memorize everything there was to know about a new craft. Somehow, the two men had to learn enough for Semyonov to take charge of the vessel while Silk and Claude went out through the bottom hatch, set the explosives on the hull of the Russian ship, and returned to the submarine.

The DUNGENESS was one of a new breed of submarines called a lock-out submersible. The after portion of the pressure hull was a sphere in which the pressure could be brought to ambient and a hatch in the bottom opened to allow a diver to enter and exit while the craft was submerged. As Silk stood staring at the odd-shaped sub, he thought that with a little imagination and her two mechanical claws—the claws were folded now on either side of her hull—the DUNGENESS did resemble her name. One could say she was an oversized mechanical crab because she was a creature who scavenged man-made leavings from the bottom of the ocean.

The bright yellow submarine rubbed gently against the dockside as she was pushed by the soft waves of the incoming tide. She was twenty-nine feet long, seven feet wide, and eight feet tall. Powered by three electric motors, her maximum speed was eight knots. Inside her steel hull was an instrumentation sphere which contained seats for the pilot and co-pilot. In the center was a viewport with eight windows, and aft was the diver lock-out chamber. The three cubicles had a central access hatch. To enter the submarine, one had to climb through the hatch on top of the conning tower, resembling a sodapop can attached to a crab.

She was equipped with nine fixed lights ranging in power from 500 to 2500 watts which could be individually controlled. She was also furnished with a sonar, a high powered VCR, a T.V. recorder, and an intercom system. She could carry enough food and oxygen to stay submerged for seventy-two hours. She was originally designed for checking dam faces and bridge abutments for cracks and erosion, or for checking underwater pipelines and telephone lines, or for videotaping sunken ships.

However, the Marine Salvage and Survey Company had purchased her and put her to work doing more serious jobs, such as demolishing sunken ships which blocked a channel, or retrieving valuable papers or merchandise from old wrecks which lay in the more shallow waters along the continental shelf. In the future the Marine Salvage and Survey Company planned to use the submarine for mapping offshore well-drilling operations. They never, even by stretching their imaginations, figured that someone would use the underwater craft to blow up a Russian ship.

On November 5, Claude towed the barge out to the edge of the ocean shelf and anchored it over the old sunken wreck of the

VICTORIA PRIDE. The VICTORIA PRIDE had sailed from Liverpool, England toward Vancouver B.C. in 1934. A heavy nor'wester had struck, and she went down. Experts insisted that the ship carried one million pounds of British currency, but treasure hunters had searched for years and never found the safe.

Undaunted by the failure of so many divers, Claude studied chart after chart of the sunken vessel. After several years he announced he had discovered where inside the ship the safe was located.

He told the Marine Salvage and Survey Company he had the barge and the necessary equipment for diving. All he needed was the use of their submarine. The Marine Salvage Company was unconvinced that Claude knew the whereabouts of the money, but they leased him their underwater craft anyway. If he found the booty, their ten percent would make the gamble worth the effort. They had no idea Claude's salvage operation would provide him with a perfect alibi for the time he helped sink the MURMANSK.

While Claude was towing the barge and anchoring it over the VICTORIA PRIDE, Silk and Semyonov in the CRACKANOON II located the rusty Russian mothership some thirty miles due west of Port Bane. From her present coordinates, Semyonov figured the likely position where she would be located on blast off night. Now the definite rendezvous point where Claude, Silk and Semyonov would meet with the CRACKANOON II could be set. From this location the crew would navigate north and west for parts unknown.

Because Claude was a diver and often involved in underwater demolition work, he had no trouble purchasing the C-4 plastic explosives to blow up the giant MURMANSK. The plan was to place four one pound bricks directly under the engine room, and ten one pound bricks directly under the large fish hold in the center of the ship. The explosives were to be mounted on two magnetic plates with waterproof glue, capped in a series, then detonated by pushing a button on an LED timing system.

On November 5, the Locals reported in great detail the success of their mission. They had thrown the chopped-up sockeye pieces, especially the heart, livers, and guts, into the pens on the 4th, and no one had caught them. Those "stinkin'" coho should definitely start dying within two or three days.

Bent made it a point to speak with Silk about his expertise with shooting sling shots and throwing the javelin. Silk realized immediately he was dealing with a mountain man. He assigned Bent to teach Tanner and Graveyard how to run and throw a spear so it would hit the target.

On November 6 Graveyard borrowed Silk's van to pick up enough food, clothing, blankets, cigarettes, and liquor to supply the four men on the CRACKANOON II for two weeks. As soon as he had loaded the supplies in the van, he stopped by the Albatross for a quick beer. He was sitting at the bar silently sipping from his pitcher when he heard a fisherman two stools down say thousands of coho were mysteriously dying at the pen-rearing operation out in Bane Bay. Graveyard grinned with pride. He tipped up his pitcher and chugged the rest of his beer. This was good news to Graveyard, and he could hardly wait to tell the story. The Locals had proved he knew what he was talking about.

When Silk, Semyonov, and Claude heard that the coho were already dying, they looked at each other in shock. They had worked day and night to get this plan ready, and now there was already an unforeseen circumstance. They had made the aluminum spears for the transformers. They had the bolt-cutters to cut the chain-link fence. They had a spool of soft wire and wire-cutters to repair the fence once it was cut. They had flashlights and the five gallon containers for water. And last, they had acquired a set of lock picks for Esteban who insisted he needed them to open the door of the compressor building. But, now, they had to move their time schedule forward. Questions about the dying coho would be asked immediately. The transformers and compressors must be destroyed before anyone of their crew came under close scrutiny.

After much discussion among the men, zero hour was moved up three days to November seventh. The destruction of the transformer station and the dismantling of the compressors at AQUA-PROTEIN would occur at exactly 11:00 P.M. Claude figured thirty minutes was plenty of time for Tanner and Cowboy to return to the dock from the transformer and AQUAPROTEIN. At 11:30 P.M the CRACKANOON II with Cowboy and Tanner on board would leave Claude's dock for its ocean rendezvous with the submarine. The detonation of the MURMANSK was moved to 4:00 A.M. on November 8. By that ungodly hour the three men on the submarine would have reached their position, planted their explosives on the huge hull, and radioed for the Russian crew and workers to abandon ship.

After the MURMANSK was blown up, the submarine would rendezvous with the waiting CRACKANOON II precisely thirty minutes later. Semyonov and Silk would quickly board the troller, and Claude would turn the DUNGENESS toward the salvage barge. By dawn Claude would be hard at work on his salvage effort, and the CRACKANOON II with the four fugitives on board would be to hell and gone out to sea.

Graveyard blinked his eyes and scratched his beard while the schedule was being changed. After Claude had finished his explanation, the big man cautiously raised his hand.

"What you say is good with me," he said, "but I'm damned mixed up with all this different kind of activity and time change stuff. Maybe you better write my part out on a piece of paper so's I don't make no mistakes."

"Me too," said Bent and Rat in unison.

"Sheet!" said Esteban. "I make for to follow you, Cowboy."

At 9:00 P.M. on the evening of November 7, Silk stood on the dock studying the DUNGENESS heaving at her lines like she was anxious to get under way. Curious how life suddenly took a twist. In short order he would be leaving Port Bane forever, leaving Silvia and everything he had loved behind. He was about to commit an unpardonable act against his own government. He, Claude, and Semonyov would soon sink a great Russian ship.

He looked at Graveyard, Bent, Rat, and Esteban. He might never see these old friends again. Why did life have to come to this? Where was the reason a man should be placed in his position? Why should a man have to surrender the people he loved? Life for him had become so desperate, so corrupt, so extreme. Truth, for him, seemed completely foreign. Somehow there was no way out. He had to take this final step. He looked at Claude and Semyonov smoking one last cigarette before they got underway. Claude was walking up and down the length of the submarine rechecking the equipment for the last time, and Semyonov was standing at the end of the dock looking calmly at the bay. There seemed nothing else to say so Silk threw his cigarette into the water and stepped on board.

"She's as ready as she's ever going to be," he announced. "We have two thermos' of coffee, and plenty of sandwiches to eat while we're traveling. Let's cast off."

"Too bad we have no vodka," said Semyonov. "We will need to make a toast for our accomplishment. At last I am to take an action against the Russian government."

"We'll drink vodka when we meet with the CRACKANOON II," said Silk grimly. He felt a lump rise in his throat. Silvia had not even shown up to say goodby.

"Hey, Silk," called Graveyard as DUNGENESS slowly eased from the dock. "What do you want us to do with your van while you're gone?"

"Here, catch the keys,"said Silk. "It belongs to you and the Locals now."

Claude grinned as the men settled themselves into position. Claude and Semyonov were sitting at the control panel, and Silk was standing with his head and shoulders sticking out of the hatch.

"I've had a sense of hate for one of those Russian ships for years," Claude said to Semyonov. "And now I've finally got the chance to blow one of the fuckers out of the water. I can't wait."

"Yes, but the Russian fishermen aren't bad," said Semyonov. "They are generally good men, like you and me. I only regret we are taking their jobs away from them."

"I want them off that ship when she blows. I don't want to be responsible for killing anyone," said Claude.

"They'll be off," promised Semyonov.

As they pulled away from the dock, Cowboy and Tanner gave the thumbs up signal.

"See you on the CRACKANOON II," said Tanner.

"Acey! Deucey!" yelled Cowboy.

At the same time the DUNGENESS pulled away from the dock, Jodie and Silvia were sitting in Jodie's front room drinking a cup of tea. The women were silent. Jodie crossed her legs and uncrossed them. She got up and paced the length of the room several times. Suddenly, she stopped in front of Silvia, tapped her foot, put her hands on her hips, and blurted out a statement she had intended to keep to herself.

"You know, Silvia, I just can't understand you," she said. "Silk loves you, and I know damn well you love him. Why didn't you two ever get together? God knows, I've given you a dozen opportunities. Now he's gone. And you'll never see him again. Never."

Jodie watched the unchanged robotic stare on Silvia's face.

"Too bad about you. Too damned bad," Jodie said. "I really wonder what you've got in that stubborn head of yours. For a pretty woman you don't have brain one."

"I don't want to hear this, Jodie," Silvia said severely. "Can't you see I don't want to talk about John Silk?"

"You know what Lillian told me, don't you?" Jodie said. "She said you and Silk were in a karmic relationship. She said you were soulmates!"

Silvia stood up, walked to the hall closet, and put on her raincoat.

"Where are you going?" asked Jodie.

"I'm going for a walk," said Silvia.

"Do you want me to come along?" Jodie asked jumping up. "I'll grab my coat."

"No, no," said Silvia. "I'll be fine, Jodie. Don't worry."

CHAPTER

32

ROYAL FLUSH

As soon as the DUNGENESS had disappeared in the dark of the bay with Silk, Semyonov and Claude aboard, the remaining seven men walked back up to Claude's workshop to wait until time to leave for the transformer station and AQUAPROTEIN. The aluminum spears, the five-gallon cans of water, and the wirecutters were already loaded in the van.

Graveyard jingled the keys to the truck in his pocket. He had come into this adventure with nothing, and now he owned Silk's van. He could hardly believe his good fortune. That Bent and Rat were part owners never entered his mind. Only images of him driving the Locals and other friends to the Albatross every morning swam before his eyes. He might even go down and try to get one of them driver's licenses. The men sat around the glowing pot-bellied stove watching the hands of the antique clock above Claude's workbench tick off the minutes. They had blackened their faces with charcoal from Claude's stove, and in their heavy coats, accentuated with dark stocking caps, they looked like a marine battalion waiting for its first combat. Each man was lost in his own thoughts. Only Tanner was aware of the months and years in prison he would spend if they were caught. At exactly 10:15 P.M. the men trooped out to the van in single file, and the over-loaded vehicle pulled onto the road which led past the ocean ranch to the Bane Bay transformer station. Graveyard was driving with Cowboy and Tanner sitting in the front seat beside him. On the floor in the rear of the van, huddled amidst the tools, were Bent, Rat and Esteban. The van rolled past the thirty acre ocean ranch until it came to an intersection. A narrow gravel road wound on up a steep hill to a dead end where the transformer station stood. This man-made edifice supplied AQUAPROTEIN and South Port Bane with its electric power. The van made a U-turn and parked beside the gravel road halfway between AQUAPROTEIN and the transformer station. From this central location the men would have to walk.

Inside the van, there was total silence. Nobody moved. Then a flashlight lit up the interior, and one by one the men scrambled out and began unloading the equipment. Overhead, the heavy overcast parted, and for a brief second, the new moon accentuated the fact that the night was as black as carbon. The men worked silently and swiftly in the darkness. In short order three human figures, laden with water cans, struck out toward the AQUAPROTEIN ocean ranch. The three remaining shadows, each with a bundle of aluminum spears on his back, started climbing the steep hill toward the transformer station.

Cowboy, Rat, and Esteban walked rapidly along the outskirts of the highway until they came to the south fence of AQUAPROTEIN. The terrain was rugged. Cowboy carried one five gallon can of water plus the bolt cutters while Esteban and Rat shared the weight of the other can between them. Every five hundred yards or so either Rat or Esteban had to rest. At the fence corner they took a north course across the heavily grassed sand dunes until they reached a northwest corner pole. From here they moved due west until they arrived at their destination, a spot in the fence north of the large, metal building where the back-up diesel compressors were stored. Cowboy, with his uncanny sense of direction, never missed a turn.

As soon as they were in position, Cowboy went to work with the bolt cutters. Esteban and Rat had regained their breath by the time he had cut a rectangular hole in the heavy chain-link fence large enough for a man to walk through.

"I hope they don't have one of them pit bulls patrolling this fence at night, or shepherds," said Rat grunting as he dragged the heavy five gallon can through the fence.

"Sssh," whispered Cowboy. "We already checked that out twice, Rat. There's just one night watchman, and he's sittin' in the office out front."

"If uno pero grande mak por mi, I keel heem weeth my knife," said Esteban determinedly as he slid through the fence.

"There won't be any dogs," said Cowboy. "And if there is, they won't be the kind you'd like to fight. O.K. You guys pick up that can and let's get a move on. By the time we finish our job, all hell's goin' to break loose on that hill. I'd like to be in a position to

353

watch them fireworks."

They reached the rear door of the big corrugated iron building, and Esteban took a set of lock picks out of his pocket. While Cowboy held a flashlight, he inserted a long, thin blade and twisted. The lock clicked, and the door opened. Esteban gave a short hoot of triumph,

"Eeet work!"

"Shut up, for hell's sake," whispered Rat looking for dogs or a shadow that moved. "Somebody'll hear you."

"Lo cinento," said the Mexican. "Paro mi open door eezy, no? I tell Silk I feex locks in Mexico City, but he no hear mi."

Cowboy was already inside flashing his light on the two large compressors. He screwed off one diesel cap, and stuck a finger inside.

"The tanks are full," he said. "We're going to have to siphon at least five gallons out of each tank. Rat, look around and see if you can find a bucket over by those drums." He unwound a coil of rubber hose from his belt, inserted it in the tank, and began to suck.

Rat came back with a plastic grease bucket and kept watch while Cowboy and Esteban took turns on the hose. They filled the bucket three times and poured the diesel into a half empty fifty gallon barrel so there would be no spilled fuel as evidence. When they figured they had drained at least ten gallons of diesel, they emptied a five gallon can of water into each tank, replaced the tank caps, and returned the bucket to its workbench.

"Now let's get to hell out of here," said Cowboy.

They picked up their empty cans, relocked the door, and ran across the field to the hole Cowboy had cut in the fence. While Esteban and Rat squatted low on the sand behind him, Cowboy unwound a coil of wire he had carried inside his coat pocket. With sure, deft movements, he quickly wired the piece of chain-link fence back into place.

"Got her," he said at last. "Nobody's ever goin' to see that patch unless they go over her foot by foot. Let's get back to the van."

By the time Tanner, Graveyard, and Bent had climbed the steep hill to the transformer station, they were out of breath. They lay the bundled spears on the ground, and Tanner shined a penlight on his wrist watch. The climb had only taken a quarter of an hour.

354

"We're ahead of schedule," he said to the men. "We have fifteen minutes to wait until 11:00 P.M. That will give us plenty of time to get into position."

"I think we ought to throw them now," said Graveyard, "get the thing done with and get the hell out of here. We don't even know if this idea'll work or not. I never heard of anybody blowin' a transformer with aluminum spears before."

"I suppose nobody's ever done it before," said Tanner. "But Claude said it would work. All we can do is try it and wait to see what happens."

"It'll work," said Bent. "I never heard of Claude ever being wrong before. He's a genius. Whatever he says works, works."

Tanner shown his flashlight through the fence. The beam of light illuminated the solid bulk of the transformers and the heavy electrical wires draping from glass insulators affixed to tall wooden poles. The beam stopped on a red sign that read "DANGER: HIGH VOLTAGE, KEEP AWAY."

"This ain't no kid's game," said Graveyard. "What'll happen if one of them sonsabitches really explodes? A big enough explosion might blow us all to hell with it. That big fence ain't up there for nothin'."

"I don't know much about transformers," said Tanner. "I do know when lightning strikes them there's an outage, but I've never heard of anyone being injured."

"Just the same," said Graveyard. "I'm goin' to keep my distance from that there fence. I'm goin' to throw my spears from a place further back."

"These spears'll carry plenty far," said Bent. "Just take a run and throw them as high as you can so they'll arch just right and come down top of the wires."

"We still have a few minutes to wait," said Tanner checking his watch, "Let's run through the plans once more, Bent."

"I'll go around behind on the lower side of the hill," said Bent. "At Tanner's signal, I'll run up the hill and throw my spears from that angle. Graveyard, you take the right side and Tanner, you take the left. Tanner, when it's zero hour, you signal three times with your flashlight, and we'll all take a run and throw our spears at the same time. Aim for the wires which lead to the first transformer. As soon as the first one goes out, we'll take the next one down the line."

"Time to split up and get in our positions," said Tanner. "Now, wait for my signal. We don't want to be too early or the repair trucks will spot Cowboy and his crew coming up from AQUAPROTEIN. How long do you figure it'll take a repair truck to get here once the damage is done?"

"I figure about thirty minutes," said Bent, "so we'll have plenty of time to get back to the van."

"I hope Cowboy and his crew aren't late," said Tanner.

"Don't worry none about Cowboy," said Graveyard. "He's probably already through. Not much to his job."

Each man picked up his bundle of missiles and walked swiftly to his position. Bent, who was the most agile, reached his location first.

Tanner untied the strings that held his spears and lay them out in a neat row with the points facing the transformer fence. He placed one spear on his right shoulder and wrapped his arm around it. His watch showed a minute to go, and he stared mesmerized as the second hand moved around toward the twelve.

For a moment he wondered what he was doing in Port Bane with these crazy people, preparing to destroy a transformer station. With his luck they'd probably get caught, and he'd go to jail again. He could see Vivian's face when she read the Washington Post: "Ex-Justice Department field investigator indicted for destroying transformer in Port Bane." He missed Vivian and the boys. He felt the emptiness in the pit of his stomach, and he longed to see them.

He forced himself to think about Silk and Semyonov out on the submarine with Claude. These men were preparing to sink a giant Russian ship. They had taken the dangerous job and left the easy one for him. He couldn't even imagine such a project, and he sure as hell wouldn't want to be in their shoes.

The second hand moved to the twelve. He flashed his light toward the heavens, on and off three times, then he pocketed the flashlight, grasped the spear in what he thought was a javelin grip, and ran toward the fence.

Three silver missiles shot simultaneously high into the air above the fenced enclosure. They reached a maximum height before they arched downward toward the wires leading to the transformers. There was a sizzling swoosh, and sparks shot a hundred feet into the air. For a moment the large transformers, the conductors, and the long legged steel power poles stood out in relief like STAR WARS' mechanical horses.

The wires hissed. Silver, blue and red slivers shot into the air. Larger flashes crackled fire above and around every insulator. Sparks leaped from one transformer and quickly raced to the next. The enclosure became an electric battlefield.

The lights in south Port Bane went out. The sparks were no longer flying by the time the men hurled their last spears. Everything seemed dead except the rising smoke from burning insulation. The men threw the last four spears anyway, just for good measure.

Across the bay the lights in north Port Bane were still shining. Somewhere a siren began to wail. The four men had gathered together outside the fence. They exchanged handshakes and began running down the road.

By the time they reached the van, Cowboy and his crew were sitting inside waiting for them. Graveyard jumped in the driver's seat and cranked the engine to life.

The van coasted silently down the hill until it reached the paved road which would take them back to Claude's turnoff. At this juncture Graveyard pressed heavily on the gas pedal, and the van picked up speed. To turn on the lights would give any observer a clue that there had been a vehicle on the road shortly after the transformer station was blown. During these hair-raising minutes the men were silent as if their talking might give the whole scam away.

Graveyard, taking advantage of the two lanes, grasped the steering wheel with both hands, and took the turn-off into Claude's circular drive on two wheels. They coasted past Claude's house to the parking area above the dock. Down below, the CRACKANOON II was idling, her sharp prow pointed outward toward the bay.

There were sighs of relief and congratulations before the men realized they were not alone. On the dock, standing next to the white troller, was the silent, huddled figure of an intruder.

"Who's that person next to the boat, Cowboy?" gasped Rat staring through the windshield of the van.

"Damned if I know," said Cowboy. "But nobody's s'posed to be there."

"We're caught already," Rat mourned. "It's the police or the FBI."

"I hardly think so," said Tanner logically. "No one has had

357

time yet to find out what we've been doing."

"Well, what're we goin' to do?" asked Graveyard. "We can't just sit here all night. You guys got to get down to that boat so's you can take off. Silk'll be waitin' for you. And we got to get the hell across that there bridge. Them repair trucks'll be comin' pronto, and we don't want them to see this van. I say throw the fucker in the bay and be done with it."

"I'll get out and see what's going on," said Tanner. "If that person is not supposed to be here, we'll ask him to leave. If he refuses, I'll give you a signal and try to hold him. Otherwise, I'll wave you on down, Cowboy."

The men held their breath as they watched Tanner talking to the figure. Shortly, he waved with both arms for Cowboy to come on down. Cowboy was already out of the door and running toward the boat.

"I don't understand what in the world is goin' on," said Graveyard. "Why ain't Tanner kicking the hell out of that guy?"

"I guess he's friendly," said Bent.

The men watched as Cowboy walked up to Tanner and the smaller figure. Then he did a most unexpected thing. He hugged the unknown interloper before he ran toward the CRACKANOON II. The men gaped as Cowboy quickly untied both bow and stern lines, climbed aboard, and disappeared into the cabin. A moment later he gave Tanner a hand up, and the troller began to slide away from the dock.

As the men watched, the lone figure suddenly began running alongside the CRACKANOON II with his arms outstretched. At the last instant a broad shadow on board extended his hand and pulled the fellow over the rail.

"That's the damnedest shenanigan I ever seen," said Graveyard starting the engine. "Now what in the world do you suppose that fella was doin?"

"We'll never know," said Bent. "But it looks to me like somebody's hitched himself a ride with Cowboy and Tanner. Nothin' we can do, Graveyard. Let's get out of here before the cops come."

Until she was thirty miles out at sea, the DUNGENESS submarine ran as smoothly on the surface as any fishing troller. In the distance the men could see the moving lights of the Russian fleet. In the center of these circling trawlers they knew the refrigerator/factory ship, MURMANSK, would be steaming a steady course. She would be easy to locate. The question in Silk's and Semyonov's mind was whether or not two heavy-duty bombs could sink such a colossus. Only Claude seemed sure.

"We're close enough," said Semyonov. "We do not want to be seen on a ship's radar."

Claude cut the electrical motors, and Silk could feel the forward progress stop as the submarine began to drift.

He checked his watch.

"We have a little less than two hours to wait," he announced to Semyonov. "Care for a game of chess?"

"I think I'll take a nap," said Claude. "I haven't had much sleep lately, and I have to work later tonight."

Time dragged. Claude sat crumpled in the center of the viewport snoring while Silk and Semyonov sat side by side in the pilot's sphere. Each man stared silently at the chessboard though neither man concentrated on the game. At last the hands of the tiny clock on the instrument panel indicated the time was 2:30 A.M.

Wordlessly, Silk folded up the board and replaced the little figures in their box. In another few minutes he might have beat Semyonov for the first time since they had met. However, he knew the captain had not been paying attention to his moves.

Semyonov engaged the electric motors, and the DUNGE- NESS moved in a slow circle until her snout pointed due west toward the center of the Russian fleet. Claude, feeling the movement of the submarine, awoke and stretched.

"Time to get dressed," he said to Silk. "We don't want to be late for the party."

"Everything is ready here," said Semyonov in a calm voice. "We will be close enough to submerge in thirty minutes."

Silk stared into the blue eyes of the captain. He could see determination mirrored in their intensity. He was amazed at the authority and the ability of the man, his coolness in handling a new kind of boat. He never doubted for a moment that Semyonov could command any ship in the world.

While Silk and Claude donned their wet suits in the cramped space of the viewport, Semyonov filled the ballast tanks, and the DUNGENESS slipped silently beneath the black surface of the ocean. The rest of their journey would be made underwater.

"Diving," he said over the intercom. "I shall level at a hundred and twenty feet. We are still a mile from the target."

Silk and Claude crawled through the narrow hatchway into the diver lock-out chamber. Here they strapped on their scuba tanks, fastened their weight belts that would keep them at zero buoyancy while they swam at sixty feet, and slipped into their frog fins. The waiting now became tense.

"Let us know when we're in position, and we'll fill the chamber," said Silk over the intercom.

"I will let you know," came Semyonov's calming voice.

Again there was interminable waiting, then...

"We are now in position," said Semyonov.

Claude pulled the lever, and the two men stood silently while ocean water quickly filled the chamber.

"Let's go," said Claude.

They slipped on their diving masks, cleared their vision, then Claude quickly turned the heavy metal wheel which opened the bottom hatch. One at a time the men slid into the stygian depths, each carrying a heavy magnetic package slung in a pouch over his shoulder. For a moment they treaded water while they switched on their diving spotlights. Claude checked his compass, pointed his finger, and they swam west through clear water lit by the expanding light beams.

Suddenly there was a huge metal barrier surging through the water before them. The hull of the giant ship seemed to stretch forever into the darkness. The divers stopped for a moment before they parted company. Silk swam for the spot astern beneath the engine room, and Claude in the opposite direction toward the center of the ship beneath the main fish hold. In less than five minutes there were two metal plunks as the magnetic plates attached to the explosives adhered to the rusty steel surface. The two divers came together beneath the ship, and again Claude checked his underwater compass and pointed.

The divers located the hovering yellow bulk of the submarine and eased themselves through the bottom aperture. As soon as the hatch was secured and the sea water forced out of the diver's lock-out chamber, Silk and Claude took off their masks, grinned at each other, and shook hands.

Silk switched on the intercom and spoke. "They're in place,

Semyonov. Take her out."

The DUNGENESS moved underwater until Semyonov felt he was beyond the range of Russian radars, then he emptied the ballasts, and the yellow submarine broke through to the surface. The men dressed and slid into the pilot's sphere where Semyonov periodically checked his watch. Minutes went by before Silk said,

"O.K. Semyonov. Get on the horn and tell them they have exactly thirty minutes to get off that ship."

The captain began to speak in Russian, the tenseness of the situation sounded in the tone of his voice. He waited several minutes, then he repeated the message. An excited voice came over the speaker. Shortly thereafter other Russian voices joined the diatribe.

Semyonov waited until there was a break in the conversation before he began methodically to repeat his message at five minute intervals. A particularly loud command sounded over the speaker, and Semyonov heaved a sigh of relief. He smiled at Silk.

"They have listened to reason at last," he said quietly. "They are abandoning the ship."

"I can't understand why they took so long," said Silk, cursing beneath his breath for not learning Russian.

"You have to understand the Russian mind," said Semyonov. "At first they didn't believe there was a bomb. I assured them there was. Then they wanted to take their chances because the lives of the fishermen do not matter to the government. I explained to them there was no chance to save the ship. That got their attention. Since they couldn't save the ship and the cargo, they decided to save their lives. I assured them making this decision was the only politic thing to do. Still they tarried, so I just demanded they get the hell off the ship."

Thirty minutes went by, and again Semyonov spoke into the radio.

"For hells sake," said Claude. "We've passed our deadline. We have to blow the bastard and go."

"No," said Semyonov. "We cannot. We will not destroy lives. We must wait a few moments more."

Silk could imagine the confusion aboard the giant Russian ship.

"We'll wait," he told Claude.

More time passed. Semyonov took a deep breath and gave the signal.

"Now!" he said thrusting his thumb up.

Claude pushed the red button on the LED remote, and the men stared through the viewport. The entire horizon became white light as the two charges exploded simultaneously.

"Holy shit," muttered Claude in a subdued voice. "Did we do that? That's beautiful. We must have blown the fuel too. I never did see such an explosion."

Semyonov blew the ballast tanks, and the DUNGENESS submerged. They would not surface again until they rendezvoused with the CRACKANOON II.

The autopilot was set at a depth of a hundred feet, and the men relaxed as they lit cigarettes.

"What happened out there, Semyonov?" Silk asked. "What caused the delay?"

"Just another example of Russian inefficiency," said Semyonov, rather taciturn Silk thought. "Some of the old life boats on the upper decks were untrustworthy, so several trips had to be made in the ones that would float. At the last moment the captain did not want to leave his ship. Like me, he knew the MURMANSK would be the last ship he would ever command. I had to wait until he had finally been removed, forcibly, I might add. Simply stated, when I was sure everyone was off the ship, I gave the signal."

While they were talking, long volleys of excited Russian dialogue kept coming over the radio. After a particularly long explanation, Semyonov smiled.

"Tell us," said Silk.

"They say," said the captain, "that the MURMANSK has been split in half, and she is sinking. There is no loss of life, but the economic ramifications are deplorable. We have succeeded."

Claude grunted his approval. "I knew I made those charges big enough," he said. "When I plant a charge, I plant a charge."

Semyonov became taciturn once more. "The sinking of such a large ship is a serious matter to those people," he said. "None of them will ever fish again. We are the happy ones because we have succeeded in our mission. In the Soviet Union no failure is allowed; only success is tolerated. I feel sad for the people aboard my ship, but I also feel happy that such a destructive machine as the MURMANSK has disappeared, and with no loss of life. I am vindicated."

Silk, feeling the compassion in Semyonov, changed the subject.

"We are fifteen minutes late," he said. "How long until we rendezvous with the CRACKANOON II? We want to be out of here before we're caught in a fifteen hundred mile manhunt."

"We are almost there," said Semyonov. "Do not worry my friend. Your boat will be there, and we shall escape. For me, escaping does not matter. I should have died with the MURMANSK."

"Staying alive matters to me," said Claude. "I want to be working on that liner by 5:30 A.M. The charges are already set. I'll be in that safe by noon and a rich man by evening."

"I doubt that," said Silk. "Someone has already taken the two million dollars if the money was on the ship. There are just too many treasure hunters to let such a haul lie dormant for all these years."

As if to accentuate his words, Semyonov blew the ballast tanks, and the submarine broke onto the surface. Directly in front of them lay the waiting CRACKANOON II.

Quickly, Silk opened the hatch cover. He and Semyonov carried the thick box to the deck of the submarine. Silk broke the seal, and they threw the inflatable raft overboard. As she filled with air, the three men hugged one another. Claude held the lines while Silk and Semyonov slipped overboard into the pitching craft. Each man grasped a set of oars, and they paddled toward the darkened troller which was rolling easily between the swooping swells.

The moment they were aboard the troller, Silk told Cowboy to give Claude a holler on the radio and tell him they had arrived safely. The two men stood on the deck watching the distant yellow submarine. At first she appeared to be floating, then she slid beneath the sea. Claude had received the message.

When Semyonov and Silk entered the cabin, Cowboy was standing at the wheel and Tanner sitting at the table smiling broadly. Cowboy put the CRACKANOON II on automatic pilot and turned to face his guests. Silk wasn't sure, but the drooping mustache seemed to be twitching like Cowboy was about to laugh. The four men relaxed while Cowboy popped the cork on a bottle of dry champagne. He filled five glasses, and the men toasted their success.

"Well, how did it go?" Silk asked his friend. Cowboy just stood there with a funny grin on his face.

"Fine," Tanner said speaking for the Texan. "Just fine. Our

part went according to plan. Those aluminum spears fused the transformers, and the diesels have five gallons of water in each tank. As far as I know, everyone got back to town safely. What about the MURMANSK? We heard the explosion and saw the flash. The Russians have been talking steadily on the radio ever since."

"She's on the bottom now," said Silk, "and no one was injured."

The men sipped at their champagne, and Cowboy pushed the throttle until the troller was cruising steadily at nine knots. He checked the charts and deftly twirled the wheel until the compass was pointed due west. Only then did he pour himself another glass of champagne. He ignored the fifth glass still sitting on the table.

Silk felt the remorse a man feels when a project is over, and he doesn't know what lies ahead. He took a sip of champagne and glanced up at Cowboy and Tanner. They seemed to be waiting to tell him something.

"Alright," he asked at last. "What's wrong? Is there a problem with the boat? You guys act like you have something to tell me. What went wrong?"

"Nothing went wrong," said Tanner looking down at his big hands.

"Yep," said Cowboy, "nothin' went wrong."

"Well, what is it for God's sake?" asked Silk suddenly becoming alarmed at his friends' mysterious behavior.

"Ahem. We've picked up another passenger," said Tanner.

"That's right," said Cowboy. "Another passenger we wasn't expectin'."

"You guys didn't let that crazy Mexican slip on board, did you?" Silk tried to hold his temper.

"Nope," said Cowboy, a mischievous smile lighting up his face. "It ain't Esteban."

"Hello, John," said a low, musical voice.

Silk whirled around and stared into a pair of olive green eyes. "Silvia!........How in God's name..!"

"Ahem," said Tanner. "She's going with us."

"Yeah, Silk," grinned Cowboy. "She's goin' too."

364

OTHER BOOKS BY THE AUTHORS:

¿QUIEN ES?, The True Story of Billy The Kid.

CROCKETT, The True Story of a Cowboy.

SADIE, The True Story of a Western Lady.